Technological Advancements and Artificial Intelligence Innovations

Editor

Daniel Asante Otchere
Research Computing Software Engineer (AI/ML)
Institute of Computational and Data Science
The Pennsylvania State University, PA, USA

CRC Press
Taylor & Francis Group
Boca Raton London New York

CRC Press is an imprint of the
Taylor & Francis Group, an **informa** business
A SCIENCE PUBLISHERS BOOK

First edition published 2027
by CRC Press
2385 NW Executive Center Drive, Suite 320, Boca Raton FL 33431

and by CRC Press
4 Park Square, Milton Park, Abingdon, Oxon, OX14 4RN

CRC Press is an imprint of Taylor & Francis Group, LLC

Library of Congress Cataloging-in-Publication Data (applied for)

ISBN: 978-1-032-81909-9 (hbk)
ISBN: 978-1-032-82811-4 (pbk)
ISBN: 978-1-003-50634-8 (ebk)

DOI: 10.1201/9781003506348

Typeset in Times New Roman
by Prime Publishing Services

Dedication

With a heart full of gratitude, I dedicate this work:

To God, my eternal rock and guiding light, whose presence has steered my journey.

To my father, a retired assistant commissioner of Ghana Police, Daniel Asante Otchere, for instilling in me an unwavering work ethic and an unyielding dedication to excellence. You have been a perpetual source of inspiration, and I am profoundly grateful for your steadfast support.

To my mother, Jane Duah Otchere, whose boundless love and steadfast encouragement have been my foundation. Thank you for being my pillar and for your boundless faith in me, even in moments of self-doubt.

To my beloved, Vicentia Nana Aba Insaidoo, my partner, my confidante, my best friend, and my most ardent cheerleader. Your unwavering love and support have been instrumental to my success; you are, truly, a gift from heaven.

To my siblings—Nicholas, Elliot, Yvonne, and Thelma—thank you for your unwavering belief, your enduring support, and the motivation you provide. I am endlessly grateful for the joy and laughter you bring to my world.

This book stands as a testament to the love, guidance, and support of these remarkable individuals. My gratitude to you all, for sharing my path and being my eternal inspiration.

Preface

The twenty-first century is witnessing a profound transformation in the ways knowledge is created, organized, and applied. At its core, this transformation is driven by the unprecedented convergence of computational technologies, artificial intelligence (AI), and data-intensive methodologies. This confluence of disciplines is reshaping scientific inquiry, industrial practice, and societal engagement, challenging conventional boundaries and redefining what is possible across a remarkably broad spectrum of domains. It is against this backdrop of accelerated innovation and complexity that this volume, *Technological Advancements and Artificial Intelligence Innovations,* has been conceived.

The genesis of this volume arises from a clear and pressing need: modern scientific and industrial systems generate data at volumes, velocities, and varieties that far surpass human cognitive capacity for analysis. Simultaneously, the systems themselves, from energy networks and geological formations to molecular and biological systems, exhibit nonlinearity, interdependence, and stochastic behavior, defying traditional modeling and analytical techniques. In this context, AI has emerged as a transformative force. No longer confined to niche applications, it has become an essential enabler of predictive, adaptive, and efficient solutions for complex problems. Yet, the rapid proliferation of AI has frequently outpaced a holistic understanding of its principles, architectures, and ethical dimensions. The resulting knowledge gap necessitates a synthesis of theory, practice, and insight, a role that this volume seeks to fulfill.

Central to the editorial vision of this work is the recognition that knowledge in contemporary research ecosystems is inherently interdisciplinary. The modern landscape is no longer organized around rigid disciplinary silos. Instead, it thrives on the interaction and cross-pollination of methods, models, and perspectives. Techniques developed in one domain often catalyze breakthroughs in another: advances in computational geoscience inform molecular modeling; AI frameworks for industrial optimization enhance predictive models in public health; and sensor technologies originally developed for nanoscale analysis find application in energy and environmental systems. By foregrounding this interconnectedness, the volume underscores the need for a scholarly approach that embraces complexity, celebrates cross-disciplinary insight, and balances theoretical innovation with practical impact.

The power of AI and computational technologies lies in their capacity to address problems that traditional methods struggle to resolve. Conventional modeling approaches, whether statistical, mechanistic, or heuristic, are constrained by simplifying assumptions, extensive parameterization, and labor-intensive data processing. Machine learning and AI, in contrast, can detect patterns, relationships, and predictive signals directly from complex datasets, providing insights that would otherwise remain inaccessible. This capability is particularly critical in domains characterized by high dimensionality, uncertainty, and variability. Yet, while computational power offers remarkable potential, the interpretability, transparency, and ethical deployment of these systems remain essential. This volume emphasizes research that integrates methodological rigor with practical relevance, ensuring that AI and computational tools serve as extensions of human expertise rather than opaque substitutes.

The scope of this volume reflects both the breadth and depth of contemporary technological innovation. It spans traditional engineering disciplines, computational and data sciences, the physical sciences, and life sciences. In petroleum engineering, for instance, AI-driven predictive models are redefining reservoir management by reducing reliance on large-scale simulations and improving accuracy in forecasting production performance. In geoscience, advanced computational methods are transforming digital outcrop modeling, enabling automated interpretation of complex stratigraphic and structural features. In the chemical and materials sciences, machine learning accelerates the discovery and optimization of compounds with desired properties, from energy storage solutions to carbon capture technologies. In biomedical and nanoscale research, AI integrates with sophisticated sensing technologies to enable high-throughput, precise, and autonomous molecular analysis. Across all these domains, the unifying theme is the deployment of computational intelligence as a tool for amplifying understanding, accelerating discovery, and driving impactful applications.

A critical dimension of modern research emphasized in this volume is the integration of intelligent workflows and multi-agent systems. Early AI applications were often limited to single-step, reactive interactions. Today, the frontier of AI involves dynamic, goal-oriented systems capable of perceiving complex objectives, planning sequences of actions, executing operations, and iteratively refining strategies in response to feedback. Multi-agent systems further extend this paradigm by enabling collaborative problem-solving among specialized AI agents, effectively simulating a distributed team of experts. Such systems exemplify the fusion of computational intelligence with practical problem-solving, creating platforms capable of addressing challenges that exceed the capacity of any single agent or traditional workflow. Understanding these architectures is increasingly vital for both researchers and practitioners seeking to harness AI for complex, real-world problems.

Data infrastructure and visualization also play a foundational role in contemporary research. Complex datasets, if not effectively managed and interpreted, yield limited insight. By developing interactive dashboards, visualization frameworks, and data pipelines, researchers can uncover patterns, detect anomalies, and translate raw information into actionable knowledge. This is critical across domains, whether for flow assurance in subsea pipelines, optimization of energy systems, or molecular analysis at the nanoscale. Such infrastructure transforms datasets into dynamic assets, enabling real-time decision-making, informed experimentation, and accelerated operational outcomes.

Equally important is the focus on ethical deployment, transparency, and interpretability. As AI becomes more pervasive, questions of trust, fairness, and accountability grow in urgency. This volume reflects a commitment to research that is not only innovative but also responsible. By emphasizing explainable models, rigorous validation, and contextual sensitivity, the contributions collectively promote a culture of responsible technological stewardship. Readers are invited to engage critically with both the promises and limitations of computational tools, developing the judgment necessary to apply these technologies in socially, economically, and environmentally sensitive contexts.

The interdisciplinary nature of the research in this volume further reinforces its timeliness and relevance. Today, the most pressing scientific, industrial, and societal challenges, from climate adaptation and energy sustainability to precision medicine and infrastructure resilience, demand integrated approaches. AI and computational methods provide the connective tissue linking disciplines, enabling insights to traverse traditional boundaries, and catalyzing innovations that are both theoretically robust and practically impactful. By emphasizing these interconnections, the volume encourages readers to cultivate a mindset that values cross-domain thinking, methodological adaptability, and the translation of abstract models into concrete solutions.

Another defining characteristic of this work is its attention to real-world applicability. Research is not presented as a purely theoretical exercise; rather, it is situated within contexts that highlight operational relevance. Examples range from predictive modeling in reservoir management to high-throughput molecular sensing, from automated safety monitoring in industrial operations to AI-enhanced agricultural optimization. By illustrating how advanced computational and AI methodologies directly improve efficiency, reduce risk, and inform decision-making, the volume bridges the often-perceived divide between academic investigation and applied practice. It demonstrates that scientific rigor and practical utility are complementary rather than mutually exclusive.

The volume also serves as a reflective lens on the evolution of contemporary research ecosystems. The integration of computation, AI, and interdisciplinary collaboration signals a paradigmatic shift in the conduct of science and engineering. Scholars and practitioners are increasingly expected to operate within multidimensional spaces of expertise, drawing from diverse methodologies, leveraging complex datasets, and collaborating across disciplinary, institutional, and geographic boundaries. This volume embodies that evolution, offering a curated vision of how technology and intelligence converge to redefine what is possible in research and application.

For readers, the aspiration of this volume is twofold. First, it seeks to provide a comprehensive perspective on the current landscape of AI-enabled technological innovation, highlighting methodologies, frameworks, and insights that define the state-of-the-art. Second, it aims to cultivate critical awareness of the broader implications of these innovations, from interpretability and scalability to ethical responsibility and societal impact. By balancing conceptual depth with practical illustration, the volume equips readers with the knowledge, perspective, and discernment required to navigate a rapidly evolving technological world.

In conclusion, *Technological Advancements and Artificial Intelligence Innovations* represents more than a compilation of research; it is a testament to the transformative power of interdisciplinary inquiry, computational intelligence, and thoughtful application. The volume captures the dynamic interplay between theory and practice, innovation and responsibility, and discovery and impact. It celebrates the capacity of AI and computational technologies to extend human understanding, improve operational outcomes, and address challenges of global significance. At the same time, it recognizes the limitations, uncertainties, and ethical considerations inherent in the deployment of these powerful tools.

By integrating advanced methodologies, cross-domain perspectives, and applied relevance, this volume offers readers a road map for engaging with the future of science, engineering, and technology. It empowers scholars, engineers, and industry leaders to not merely observe the ongoing transformation but to participate actively, responsibly, and knowledgeably in shaping its trajectory. The chapters collectively illustrate that the intersection of intelligence, computation, and technological innovation is not a passive frontier; it is an arena of opportunity, challenge, and profound societal significance. In embracing this vision, the volume aspires to inspire curiosity, inform practice, and foster the next generation of research that is both intellectually rigorous and materially consequential.

By positioning this collection at the confluence of technological advancement and applied innovation, the book aims to illuminate a path forward for both the researcher and the practitioner. It underscores the necessity of methodological rigor, interdisciplinary collaboration, and ethical responsibility, while celebrating the creative, transformative, and problem-solving potential inherent in AI and computational technologies. Scholars, practitioners, and forward-thinking leaders who engage with these pages will find not only a compendium of cutting-edge research but also a narrative of inquiry, discovery, and impact, a testament to the enduring capacity of science and technology to reshape our understanding of the world and to provide solutions to its most pressing challenges.

Daniel Asante Otchere

Contents

CHAPTER

1

Introduction

This book examines the accelerating convergence of advanced computation, data-intensive engineering, and artificial intelligence as a foundational driver of scientific discovery and industrial transformation. It surveys contemporary AI paradigms; supervised, unsupervised, and self-supervised learning; deep neural architectures for perception and representation; foundation models; and emerging agnatic and tool-augmented systems, and situates them alongside enabling technologies such as ubiquitous sensing, edge-to-cloud computing, digital twins, high-throughput experimentation, autonomous laboratories, and nanoscale characterization. Across multidisciplinary case studies spanning energy systems, geoscience, materials and chemical engineering, agriculture, and public health, the volume highlights how modern AI pipelines are architected end-to-end: problem formulation, data acquisition and governance, feature/representation learning, model training, evaluation, and deployment into decision workflows. Emphasis is placed on validation in complex, real-world environments, including uncertainty quantification, robustness to distribution shift, interpretability and explainability, and human-in-the-loop design for high-stakes settings. The book also addresses practical implementation constraints that frequently determine success in applied AI: data quality and provenance, class imbalance, label scarcity, privacy and security considerations, scalability across heterogeneous compute and instrument platforms, interoperability with legacy engineering tools, and lifecycle maintenance through monitoring and continuous improvement. Ethical and responsible adoption is treated as a technical requirement, with attention to transparency, auditability, reproducibility, and risk-aware governance. By integrating methodological foundations with implementation patterns and domain exemplars, the text provides readers with a cohesive technical roadmap to current innovations, such as physics-informed learning, multimodal fusion, and digital-twin-driven optimization, and to emerging research directions shaping the next generation of intelligent systems.

DOI: 10.1201/9781003506348-1

Navigating the Confluence of Technology and Intelligence

This part of the twenty-first century is witnessing an unprecedented convergence of technologies, sciences, and industries that is reshaping the very foundations of how humanity understands and interacts with the world. At the heart of this revolution lies the powerful and synergistic convergence of advanced computational technologies and artificial intelligence (AI). This confluence is not merely augmenting human capability; it is actively reshaping the very fabric of scientific inquiry, industrial operation, and economic competition. From artificial intelligence and agent-based models to nanoscale sensing and molecular engineering, the scope of current research is staggering in terms of diversity, multidisciplinary approach, inclusion, and depth. Yet, in this landscape of constant innovation, there exists a pressing need for works that not only capture the essence of these technologies but also reshape the ways industries operate, scholars conduct research, and societies engage with technology. *Technological Advancements and Artificial Intelligence Innovations* emerge as a timely and essential intervention of this dynamic landscape, offering a comprehensive examination of state-of-the-art developments, presenting a vital navigational chart for complex and rapidly evolving landscape, and highlighting their tangible impact across critical scientific and industrial domains, amid the urgent call for sustainability and resilience in practice. It is a comprehensive research volume designed to dissect, explain, and demonstrate how these disruptive forces are being harnessed to solve some of the most persistent and challenging problems across a diverse spectrum of critical domains, from the depths of the earth to the intricacies of the human body, and from the scale of single molecules to the global dynamics of pandemics and climate change.

The genesis of this book is rooted in a clear and pressing necessity. The digital transformation sweeping across global industries is generating data at a volume, velocity, and variety that far surpasses human cognitive capacity for analysis. Simultaneously, the systems we seek to understand and optimize, be they global energy networks, geological formations, or biological pathogens, have grown too complex for traditional modeling and analytical techniques. In this context, AI has undergone a critical metamorphosis. It has evolved from a promising auxiliary tool, often relegated to niche applications, into a core, indispensable technological pillar. It is the only viable engine capable of driving the next wave of innovation and efficiency, providing the adaptive, real-time, and predictive power required to navigate modern complexity. However, this breakneck adoption has often outpaced a holistic and nuanced understanding of its underlying principles, its practical implementation architectures, its limitations, and its profound ethical ramifications. A significant knowledge gap has emerged between the potential of AI and the widespread expertise required to deploy it responsibly and effectively. This book is a direct and deliberate response to that gap. At its core, the volume brings together research that exemplifies the fusion of theory and innovation, bound by a shared commitment to practical relevance. The scope ranges from petroleum engineering and computational modeling to bio-inspired sensing and epidemiological forecasting. This wide horizon reflects the reality of modern research ecosystems where knowledge is no longer siloed. Instead, it exists in vibrant interaction, where methods and insights drawn from one domain catalyze breakthroughs in another. What we are seeing is a gradual dissolution of disciplinary borders, replaced by a porous and dynamic web of interconnections that enable deeper understanding and transformative solutions, all while maintaining a critical eye on the implications of these powerful technologies.

Consider, for instance, the growing role of artificial intelligence in scientific practice. Traditional modeling often demanded an exhaustive set of assumptions, human intuition, and simplifications that, while necessary, introduced limits to predictive power. With the rise of machine learning and artificial intelligence, these barriers are being dismantled. In the domain of reservoir engineering, predictive modeling has historically relied on conventional statistical decline-curve analysis, material balance, and numerical reservoir simulation, all of which often demanded extensive datasets, strong prior assumptions about reservoir heterogeneity, and labor-intensive history matching. These approaches, while foundational, frequently yielded forecasts with significant uncertainty. Today, AI-driven models are mitigating these limitations by leveraging historical production trends, well log data, seismic attributes, and petrophysical properties to generate predictive insights with markedly improved accuracy. This shift has reduced the reliance on large-scale forward simulations, minimized uncertainty in decline-curve extrapolations, and lowered error margins in reservoir performance forecasts, fundamentally transforming predictive capabilities in reservoir engineering. A prime example is the development of an artificial neural network (ANN)-based correlation for predicting oil formation volume factor (FVF) using a white-box technique. The oil and gas industry has long relied on empirical correlations derived from limited datasets, often leading to significant predictive inaccuracies in reservoir engineering and production planning. Today, they can deploy data-driven models that learn complex, nonlinear relationships directly from the subsurface. While ANNs have been applied to this problem before, they have traditionally been treated as “black boxes,” highly accurate but inexplicable, which erodes trust and hinders field adoption. This chapter tackles this critical limitation head-on. This transformation improves accuracy while also redefining how uncertainty is managed in decision-making. The blending of computational intelligence with engineering expertise is emblematic of the broader theme of this book: the interweaving of emerging technologies with industry-critical challenges.

The theme of intelligent systems extends further into discussions of workflow design and optimization that defines the frontier of AI itself. In the era of rapid digitalization, one cannot underestimate the significance of agentic workflows and frameworks that empower autonomous systems to coordinate, adapt, and act within complex environments. The exploration of agentic workflows and multi-agent systems is pivotal to this foundation. This represents a quantum leap beyond the single-step, prompt-response interactions that characterize early generative AI applications. Agentic workflows embody a new philosophy of AI operation: multistep, iterative, and adaptive processes orchestrated by autonomous AI agents. These agents can perceive a complex goal, devise a plan, execute a sequence of actions using external tools and application programming interface (APIs), evaluate outcomes, and refine their approach in a recursive loop until the objective is met. This shift from static scripting to dynamic, goal-oriented autonomy dramatically improves AI's effectiveness on open-ended, real-world tasks. The discussion is elevated further by examining multi-agent systems (MAS), where a collaborative team of specialized AI agents, each an expert in a specific function (e.g., data retrieval, analytical processing, strategic planning), work in concert. This simulated "team-of-experts" approach, communicating in natural language, leverages a form of collective intelligence that can solve problems exceeding the capability of any single monolithic agent. Their relevance is not abstract; they resonate directly with the needs of industries where conditions shift unexpectedly, from drilling environments to large-scale data processing pipelines. By embedding agency into workflows, researchers are paving the way for systems that do not simply execute tasks but also interpret goals, prioritize constraints, and respond dynamically to evolving contexts. Understanding these platforms and the protocols that enable agent-to-agent communication is essential for anyone looking to build the next generation of automated, complex problem-solving systems, making this chapter a cornerstone of the book.

This theme of overcoming industry-wide limitations through technological innovation is further explored in Chapter 2 (A Review of the Prospect of Utilizing AI Techniques in Digital Outcrop Model Geological Mapping). Geological mapping is a foundational yet notoriously time-consuming and subjective process in geoscience. While the advent of photogrammetry and light detection and ranging (LiDAR) has created rich 3D digital outcrop models (DOMs), the interpretation of these models—identifying stratigraphic sequences, faults, and folds—remains a manual task reliant on expert geologists. In a world awash with digital twins, virtual models, and interconnected infrastructures, the importance of accurate digital mapping cannot be overstated. This chapter reviews the transformative potential of applying AI, specifically deep learning models such as deep neural networks (DNN) and point cloud segmentation techniques, to automate this process. By adapting methods used in 3D urban mapping to detect buildings and roads, the authors propose a framework where AI can learn to identify and classify geological features directly from point cloud data. The capacity to construct precise digital representations of complex systems, whether physical assets, molecular interactions, or industrial processes, enables simulation, optimization, and intervention at scales once impossible. The concept of DOM mapping emphasizes not only fidelity but also the interpretability and transferability of models across domains. For instance, the same principles that undergo digital mapping in geoscience can be translated into computational biology or materials science, underscoring the universality of the approach. Yet, the limitations here are equally important to acknowledge: issues of data integrity, model validation, and interoperability remain as persistent barriers. This book does not shy away from such discussions but instead situates them as essential hurdles that must be addressed if the promise of digital modeling is to be fully realized. This application promises to revolutionize field geology by enhancing the speed, efficiency, objectivity, and scale of geological mapping, ultimately leading to more accurate reservoir models and reduced exploration risk.

Not all innovations are born in silico, however. The transformative power of technology extends beyond resource extraction to the global challenge of environmental sustainability. The chemical and molecular dimensions of industry continue to demand equal attention, especially as the global community confronts the twin challenges of energy security and climate responsibility. The comparative analysis of machine learning techniques for estimating CO_2 absorption in ionic liquids directly addresses the urgent need for advanced carbon capture and storage (CCS) technologies. Carbon dioxide absorption remains at the heart of carbon capture and sequestration strategies, and novel approaches to understanding its mechanisms are essential. Traditional absorption models, while useful, often failed to capture the nuanced interplay between chemical solvents, phase behavior, and thermodynamic variables. Identifying ionic liquids with high CO_2 absorption capacity is key to CCS, but experimental methods are prohibitively resource-intensive and slow. By integrating computational insights with experimental methods, recent advances have deepened our grasp of CO_2 absorption dynamics, offering more efficient and sustainable pathways for large-scale deployment. Such work has immediate implications: industries under increasing pressure to decarbonize now have access to enhanced models that enable both economic and environmental optimization. It is precisely this blending of innovation and impact that anchors the chemical sciences within the broader narrative of technological transformation. This chapter exemplifies how AI can accelerate material discovery and optimization for environmental technologies, drastically reducing the time and cost required for laboratory experimentation.

The biological and biomedical sciences, too, make their mark here, reflecting the growing importance of cross-pollination between physical sciences and life sciences. The scope of this book extends to the very cutting edge of analytical science, exploring technologies that operate at the nanoscale. Nanopore sensing, for instance, exemplifies how tools once rooted in fundamental physics have become indispensable in biological diagnostics and molecular characterization. Chapter 9 (Nanopore Sensing for Single-Entity Analysis: Challenges, Enhancements, and Future with AI and Computational Methods) delves into a transformative technology that allows scientists to detect and characterize molecules, nanoparticles, and viruses at the single-particle level, which carries profound implications, from genomics to material science. Yet, challenges remain (issues of signal-to-noise ratios, reproducibility, and scalability continue to temper the field's transformative potential). This chapter goes beyond explaining the principles of solid-state and biological nanopores. It thoughtfully examines the technical challenges of technology and then illuminates how its integration with computational modeling and AI is poised to overcome them. AI algorithms are being deployed to automate the intricate interpretation of signal data, enhancing throughput, precision, and reliability. This fusion of nanoscale hardware and intelligent software is paving the way for autonomous experimentation and high-throughput diagnostics, representing the future of analytical chemistry and molecular biology. By presenting nanopore sensing within the larger landscape of technological innovation, the book underscores the necessity of continued research that not only pushes the boundaries of sensitivity but also ensures accessibility and reliability for real-world applications.

Perhaps nowhere is the imperative for accurate, explainable, and timely AI application more critical than in the domain of public health. The role of computational biology and epidemiological modeling is also given rightful attention, particularly in light of recent global events. The COVID-19 (Coronavirus Disease 2019) pandemic demonstrated with unforgiving clarity the limitations of existing forecasting models and the dire need for adaptive, accurate, and interpretable approaches. Chapter 7 (Explainable Deep Learning and Multi-Scale Temporal Analysis for COVID-19 Outbreak Prediction in Nigeria) presents a masterful framework for operating in resource-constrained environments. Predicting infectious disease outbreaks in settings with sparse data and limited surveillance infrastructure is a monumental challenge. On the one hand, models grounded in network theory, agent-based simulations, and machine learning enabled unprecedented insights into viral spread, informing policy and public health interventions. On the other hand, the uncertainties inherent in data collection, parameter estimation, and human behavior highlighted the fragile underpinnings of such predictions. The contributions in this area emphasize not only the technical achievements but also the epistemic humility required in modeling living systems, where complexity often resists neat abstraction. This chapter stands as a testament to the fact that in high-stakes applications, AI's value is inextricably linked to its interpretability and integration into human decision-making processes. What emerges is a balanced understanding: technology can empower, but it cannot replace the necessity of robust data, context-specific interpretation, and interdisciplinary collaboration.

The energy industry provides further illustration of technology's transformative power through innovations in safety systems. The application of AI to enhance safety and prevent catastrophic failure in high-risk operations is compellingly demonstrated in Chapter 8 (An Automated Kick Detection System Based on Optimized Data Analytics). Automated kick detection, a critical safeguard in drilling operations, represents a convergence of sensor technology, real-time analytics, and machine learning. Kicks, uncontrolled influxes of formation fluids into a wellbore, pose serious risks to both safety and economic performance. Traditionally, detection relied on operator vigilance and basic threshold-based systems, which were vulnerable to delays and misinterpretation. Automated systems, by contrast, leverage continuous streams of sensor data and sophisticated algorithms to detect anomalies in real time, reducing response time and enhancing safety. This chapter presents a sophisticated AI-driven solution that moves beyond reactive monitoring to predictive analytics. By leveraging data from a drilling simulator, including equivalent circulating density (ECD), well temperature, and pressure, and processing it with advanced mathematical techniques such as Akima interpolation for precise well path modeling, the system builds a highly accurate digital twin of the drilling process. These advances stand as tangible examples of how research translates directly into improved operational resilience, safeguarding both human lives and industrial assets. This chapter is a quintessential example of mission-critical AI: a system designed to operate in real time, providing early warning to drillers and thereby acting as a vital safeguard for human life, environmental protection, and asset integrity.

Beyond specific domains, the book also showcases quintessential innovations that challenge the very architecture of machine learning itself. Chapter 5 (CitricNet: Citrus Quality Grading Using Computer Vision) addresses a fundamental agricultural challenge: the inaccurate and subjective manual grading of fruit quality. The proposed solution, CitricNet, combines YOLOv5 for object detection with EfficientNetV2S for classification to automatically assess the quality of oranges based on smoothness and color. The emergence of models such as CitricNet reflects the constant drive to refine neural architectures for both performance and interpretability. While traditional networks have achieved remarkable feats, they often remain opaque, raising concerns about trust, explainability, and overfitting. Novel architecture offers pathways toward models that are not only powerful but also more aligned with the needs of human-centered decision-making. Whether applied to image classification, molecular property prediction, or industrial optimization, such advances in network design speak to the dynamism of the field and its willingness to interrogate and reinvent its foundational tools. Deployed in a real-world Indonesian plantation, this system demonstrates how accessible, off-the-shelf AI models can be strategically combined to create powerful tools that boost agricultural efficiency, reduce waste, and increase market value. It is a perfect illustration of how AI is democratizing technological advancement and creating tangible impact even in traditional sectors.

Taken together, the contributions in this book offer more than isolated innovations—they constitute a mosaic of technological progress and intellectual ambition. What unites them is not a shared methodology or a single discipline but rather a commitment to bridge the gap between abstract theory and practical relevance. At the same time, the limitations and challenges interwoven throughout these discussions are equally instructive. Data scarcity, model validation, interpretability, scalability, and interdisciplinary translation remain pressing hurdles. Yet, by acknowledging these limitations rather than obscuring them, the book cultivates a scholarly honesty that enhances its value. Readers are not presented with a utopian narrative of frictionless progress, but with a realistic appraisal of both achievements and the work yet to be done.

For scholars and practitioners alike, the timeliness of this collection cannot be overstated. The current era is one where technology simultaneously dazzles with its potential and unsettles with its pace. Industries face mounting pressures, from climate change and resource constraints to safety demands and geopolitical uncertainties. In this context, research that offers both innovative methodologies and practical pathways forward is indispensable. The contributions within this volume exemplify precisely that dual commitment, making the book a valuable reference not only for academic inquiry but also for industrial application and policy design.

In conclusion, "*Technological Advancements and Artificial Intelligence Innovations*" stands as a testament to the transformative power of interdisciplinary research and technological innovation. By weaving together themes as diverse as digital object modeling, agentic workflows, neural network architectures, nanopore sensing, carbon dioxide absorption, waxy crude restart pressures, outbreak modeling, and automated safety systems, it captures the dynamism of modern scientific inquiry. It has been meticulously crafted to be more than a passive repository of information, emerging as an active, dynamic resource designed to educate, inspire, and critically engage. More importantly, it demonstrates how these innovations, while grounded in theory, radiate outward into real-world impact, reshaping industries, informing policy, and enriching our collective understanding of complex systems. Scholars, engineers, researchers, policymakers, and forward-thinking industry leaders with a deep and nuanced understanding of the current technological landscape will find in these pages not only a compendium of cutting-edge work but also a narrative of science and engineering in motion, driven by curiosity, constrained by reality, yet constantly striving toward solutions that are both elegant and effective. By seamlessly integrating the theoretical underpinnings of next-generation AI architecture with groundbreaking, practical applications across a deliberately chosen range of fields, and by maintaining a steadfast commitment to the principles of transparency, explainability, and practical utility, this book equips its readers with the knowledge and foresight required to do more than just witness the ongoing revolution. It empowers them to actively, knowledgeably, and responsibly participate in shaping its future, ensuring that these powerful technologies are developed and deployed to the greatest benefit of science, industry, and society at large.

CHAPTER

2

A Review of the Prospect of Utilizing AI Techniques in Digital Outcrop Model Geological Mapping

Jude O. Ojero[1*] Daniel Otchere[2] and Sebastian Nnorom[1]

[1] Department of Earth and Environmental Sciences, University of Manchester, M13 9PL, United Kingdom

[2] Institute for Computational and Data Sciences, Pennsylvania State University, PA 16802, United States

Introduction

In recent times, field mapping methodologies have been advanced by the application of remote sensing technologies such as LiDAR and photogrammetry (Pringle et al., 2010; Lanen et al., 2010; Hodgetts, 2013). From these, digital outcrop models (DOMs) are constructed that provide high-resolution 3D digital representations of geological outcrops. The application and benefits of employing digital outcrop data in geological studies range from filling up the gap of our understanding of reservoir rock heterogeneity between small-scale structures seen in core and geophysical logs and large-scale structures observed in seismic to obtaining of geostatistical data for the generation of more realistic geocellular models. Also, accurate measurements of sedimentary structures from inaccessible locations can easily be performed (Lanen et al., 2010).

Other benefits include: (i) being able to view the data from many different angles, as well as being able to swap rapidly between different scales of view, aiding geometries and features not usually visible from a land-based viewpoint to be detected; (ii) increased sample size, that is, the area from which we can make measurements is greatly increased; (iii) rapid data collection; and (iv) generation

*Corresponding author: judeojero@gmail.com

DOI: 10.1201/9781003506348-2

of new attribute (data in digital form can enable new attributes to be generated to highlight subtle features, help with interpretation, and provide the basis for automated mapping approaches) (Hodgetts, 2013).

With the current AI revolution, geological mapping could be enhanced to further transform the field of geology by improving the speed, efficiency, and accuracy of the mapping process. Conventional AI methodologies, when applied to point cloud data, provide automatic feature extraction and identification of objects. Techniques such as DNN and point cloud segmentation have been used to detect objects in point cloud data (Unal et al., 2021; Lopez et al., 2023; Zhou et al., 2025). These methods employ self-learning algorithms that learn complex spatial features as point clouds are labelled. After the extraction of complex spatial features from the point cloud, the neural network applies these learned features to classify the whole point cloud or certain regions of it as distinct objects. These approaches are employed in 3D mapping and urban planning, facilitating the detection and categorization of roads, trees, buildings, and other objects (Peixoto and Centeno, 2023; Li and Lu, 2024).

Nonetheless, similar methods have been applied in one way or another in research projects that use DOMs for analysis. For example, rock discontinuities and structural data have been mapped using both automated and semi-automated methods from 3D LiDAR and photogrammetry models (Gigli and Casagli, 2011; García-Sellés et al., 2011; Lato and Vöge, 2012; Vöge et al., 2013; Riquelme et al., 2014, Chen et al., 2016; Zhou et al., 2024). These works have used techniques such as random sample consensus (RANSAC) in conjunction with fuzzy K-means clustering, density-based spatial clustering of applications with noise (DBSCAN) plus principal component analysis (PCA), and planar regression with successive filters to extract geometrical features (such as orientation, thickness, curvature, and spacing) from point cloud data. The findings of these investigations demonstrate the usefulness and some of the potential of AI technology in DOM-based geological mapping. They also identify key challenges and record areas that require further development.

Given the need for additional improvement and to address other subjects in geological mapping, it was essential to review the current state-of-the-art approaches for object detection in point cloud data. In this paper, we look at some common point-based and voxel-based approaches, as well as several other notable methods. The goal was to evaluate the prospects of applying these methods in DOM-based geological mapping.

Point-Cloud-Based Object Recognition Techniques

Object detection involves the process of locating and identifying items in a three-dimensional space such as image or point cloud data, and is aimed at facilitating intelligent perception and comprehension of the environment (Hao et al., 2024). It is a key element of computer vision, which is used to identify and categorize target objects in digital images (Salmane et al., 2023). 3D vision techniques are used in point-based 3D object detection to extract features and predict object spatial information in the point cloud by processing its individual points (Guo and Hu, 2024; Hao et al., 2024). According to Hao (2022), point-cloud-based 3D instance segmentation algorithms and object detection in point clouds have a close relationship.

The sparse and unstructured nature of point cloud data makes object recognition a difficult task (Gao et al., 2025). Point- and voxel-based techniques are the two main categories of traditional 3D detection techniques, and as the field develops, new image-based techniques are becoming increasingly well-known (Lyu et al., 2025, Zimmer et al., 2022). These two approaches are different in that the point-based method handles the raw point cloud directly, whereas in the voxel-based method, the point cloud data is first converted into a 3D voxel grid before feature extraction and processing for every voxel performed (Zhang et al., 2024).

Point-Based Methods

This technique significantly improves the speed and accuracy of point cloud classification and segmentation by directly learning point features from raw point clouds (Shi et al., 2019). The groundbreaking PointNet series—PointNet, PointNet++, and PointRCNN—has notable examples of point-based frameworks.

PointNet

Proposed by Qi et al. (2017a), this technique, a unified architecture (Figure 1), takes a point cloud as input, transforms it with the T-Net module, learns each point by sharing the complete connection, then combines its features into global features using the single symmetric max pooling function (Qi et al., 2017a; Zhang et al., 2023). In order to choose points from the point cloud that are interesting or instructive and to encode the rationale behind their selection, the network efficiently learns a set of optimization functions or criteria. As previously indicated, the network's final fully connected layers either forecast per-point labels (shape segmentation) or for the entire shape; the learning optimal values are aggregated into the global descriptor (shape classification) (Qi et al., 2017a). The PointNet module applies a symmetric function to changed elements in a point set in order to approximate a generic function defined on the set:

$$f(\{x_1,\ldots, x_n\}) \approx g(h(x_1),\ldots, h(x_n)) \tag{1}$$

where f: $2^{R^N} \rightarrow R$, h: $2^N \rightarrow R^K$ and g: $R^K \times \ldots \times R^K \rightarrow R$ is a symmetric function (Qi et al., 2017a).

A multi-layer perceptron network approximates h, while a composite of a max pooling function and a single variable function approximates g (Qi et al., 2017a). The output generates a global feature of the input set that a multi-layer perceptron classifier or SVM (Support Vector Machine) can be trained on for

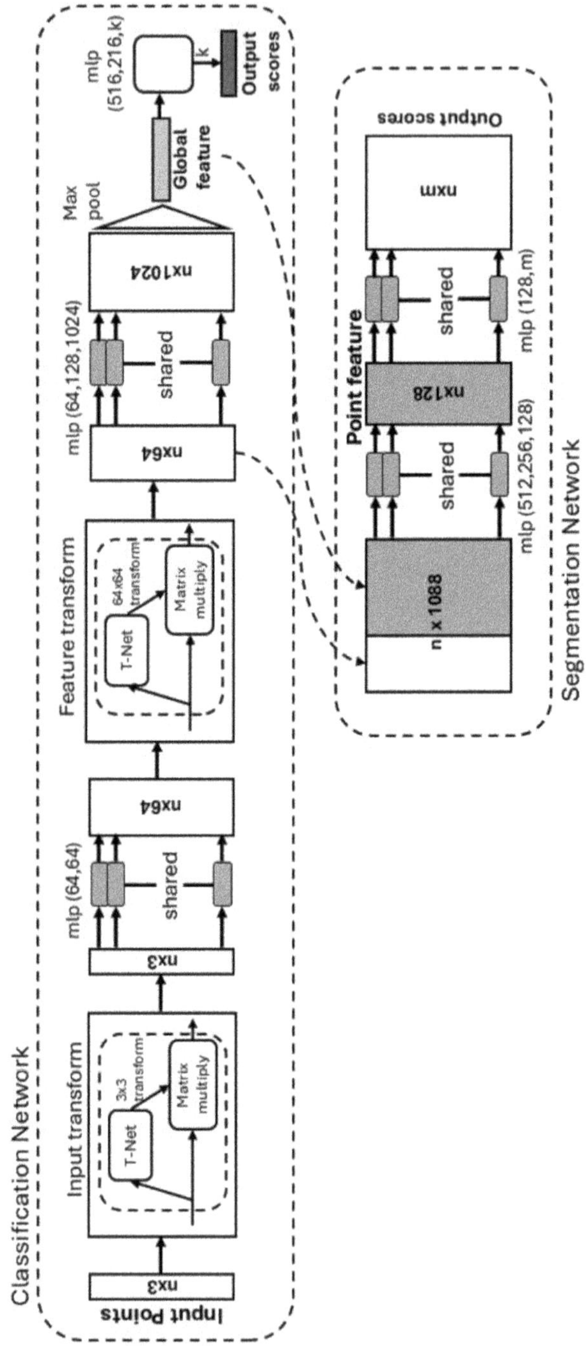

Figure 1: The PointNet architecture (adopted from Qi et al., 2017a).

classification. Following the computation of the global point cloud feature vector, each point feature is concatenated with the global feature to feed it back to the per point features. The combined point features are then used to extract new per-point features, thereby enabling the network to predict per-point quantities that depend on global semantics as well as local geometry (Qi et al., 2017a).

PointNet++

By integrating better local structures capturing and hierarchical feature learning, PointNet++ builds on PointNet (Yan et al., 2022; Zhang et al., 2024). In this method, a farthest point sampling (FPS) algorithm is used to sample points, which are then grouped using ball queries into local regions, and PointNet applied in those regions (Qi et al., 2017b). The network learns to adaptively weight patterns observed at various scales and incorporate multi-scale features based on the input data, with the use of random input dropout during training (Figure 2) (Qi et al., 2017b). With this approach, point clouds are processed in a hierarchical fashion, with sampling, grouping, and PointNet layers making up each layer (Zhang et al., 2023). This enhances performance in tasks such as segmentation, object detection, and classification by enabling the model to learn multi-scale characteristics (Karthik et al., 2024).

PointRCNN

PointRCNN, introduced by Shi et al. (2019), is a cutting-edge approach that processes raw point clouds directly in a two-stage framework (Figure 3), producing high-quality 3D proposals, which are then refined in canonical coordinate for accurate identification (Shi et al., 2019; Fukitani et al., 2022). The first stage is based on whole-point cloud segmentation, segmenting the raw point cloud by learning point-wise features using PointNet++ with multi-scale grouping in order to simultaneously create 3D box proposals from the segmented foreground points (Shi et al., 2019). This 3D proposal is constrained using a bin-based regression loss that regresses locations of a 3D bounding box (represented in the LiDAR coordinate system from the bird's-eye view as object center location, object size, and object orientation) from foreground points.

In order to learn more precise properties of each proposal, the second stage refines the 3D box locations and foreground object confidence by pooling points and their corresponding features from stage 1 based on each 3D proposal's location and transforming them to their equivalent 3D proposal canonical coordinate system by rotation and translation (Shi et al., 2019). Depth information loss of each object is however experienced due to the canonical transformation; as such for further box and confidence refinement, the transformed local spatial points are combined to their global semantic features.

Voxel-Based Methods

Voxel-based methods encode spatial information by employing volumetric

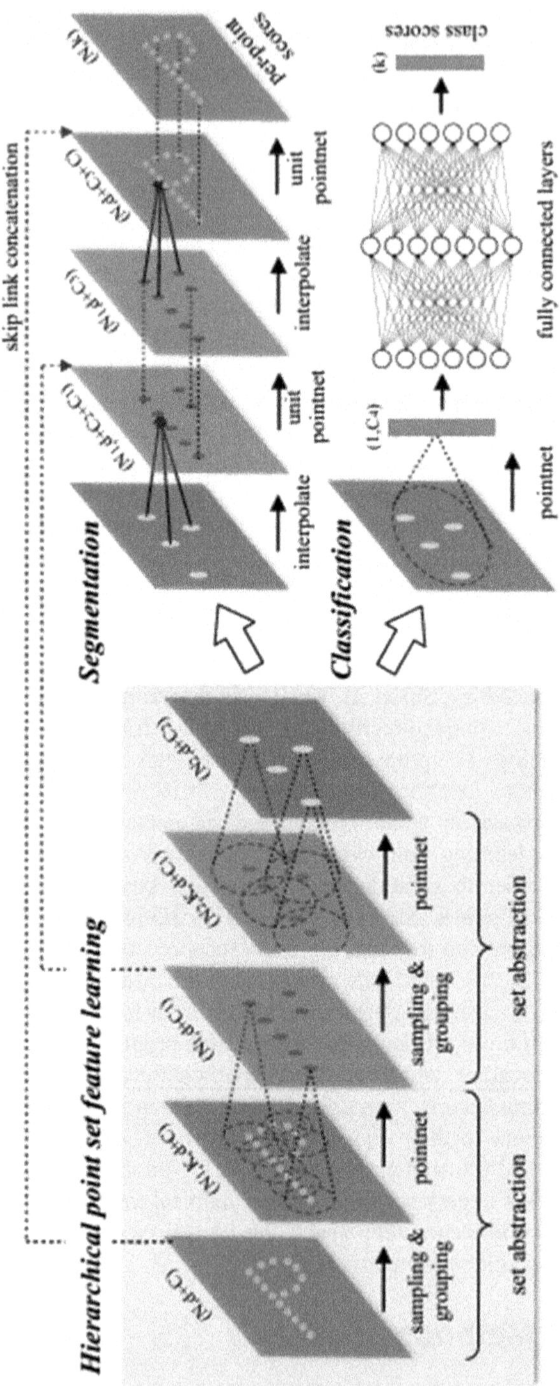

Figure 2: PointNet++ architecture based on hierarchical feature learning (adopted from Qi et al., 2017b).

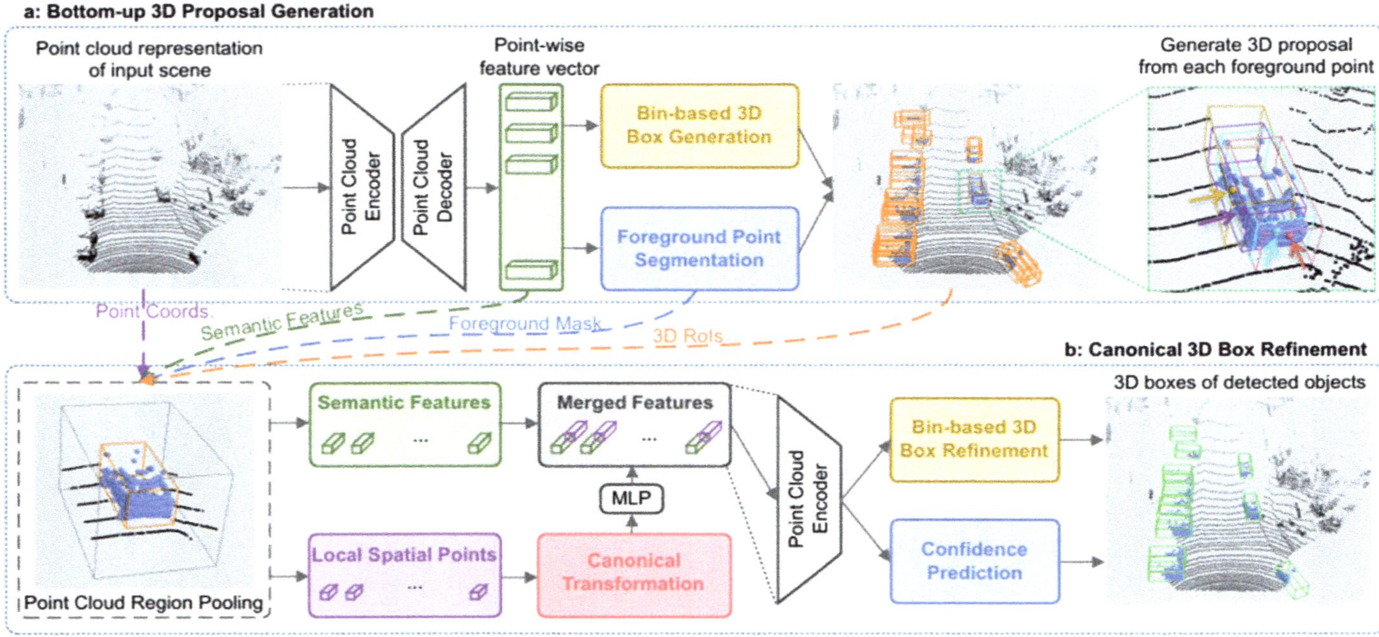

Figure 3: The bottom-up PointRCNN architecture with the first stage generating 3D proposals and its refining in canonical coordinate in the second stage (adopted from Shi et al., 2019).

representations (Karthik et al., 2024). With this method, a 3D point cloud model is converted into voxel form, which roughly represents an object's shape. Each voxel block has a collection of related points, and the voxels are then classified using 3D CNNs (Convolutional Neural Network) (Zhang et al., 2023). Popular examples of voxel-based approaches are VoxNet, VoxelNet, and SECOND.

VoxNet

VoxNet, developed by Maturana and Scherer (2015), is the earliest deep learning voxel-based framework created precisely for 3D object detection in volumetric data. The VoxNet system has two fundamental components (Figure 4): a spatial occupancy volumetric grid, composed of 3D lattice of random variables representing the environment state, which maintain a probabilistic estimate of occupancy based on incoming sensor data and past information, and a 3D CNN class label predictor that consists of an input layer, convolutional layers, pooling layers, and fully connected layers (Maturana and Scherer, 2015). This makes it possible to classify objects across a variety of datasets quickly and accurately.

VoxelNet

Introduced by Zhou and Tuzel (2017), this architecture uses 3D CNN for extracting and classifying features after voxelizing point cloud data into homogeneous 3D grids, fusing a feature learning network, convolutional middle layers, and a region proposal network (Figure 5) (Zhou and Tuzel, 2017; Zhang et al., 2024). In an end-to-end manner, it simultaneously learns a discriminative feature representation from point clouds and makes precise 3D bounding box predictions (Zhou and Tuzel, 2017). By splitting the point cloud into evenly spaced 3D voxels and encoding each one using stacked voxel feature encoding (VFE) layers, VoxelNet creates a high-dimensional volumetric representation of the point cloud. 3D convolution then further merges local voxel properties. The detection result is then generated by a region proposal network (RPN) after it has consumed the volumetric representation (Zhou and Tuzel, 2017). The VoxelNet technique, hence, enables efficient information capture of 3D shape and direct operation on sparse 3D points.

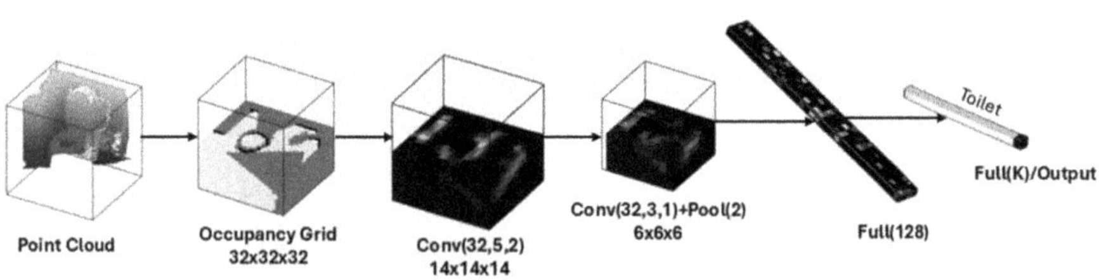

Figure 4: The VoxNet architecture showing key components and layers (modified after Maturana and Scherer, 2015).

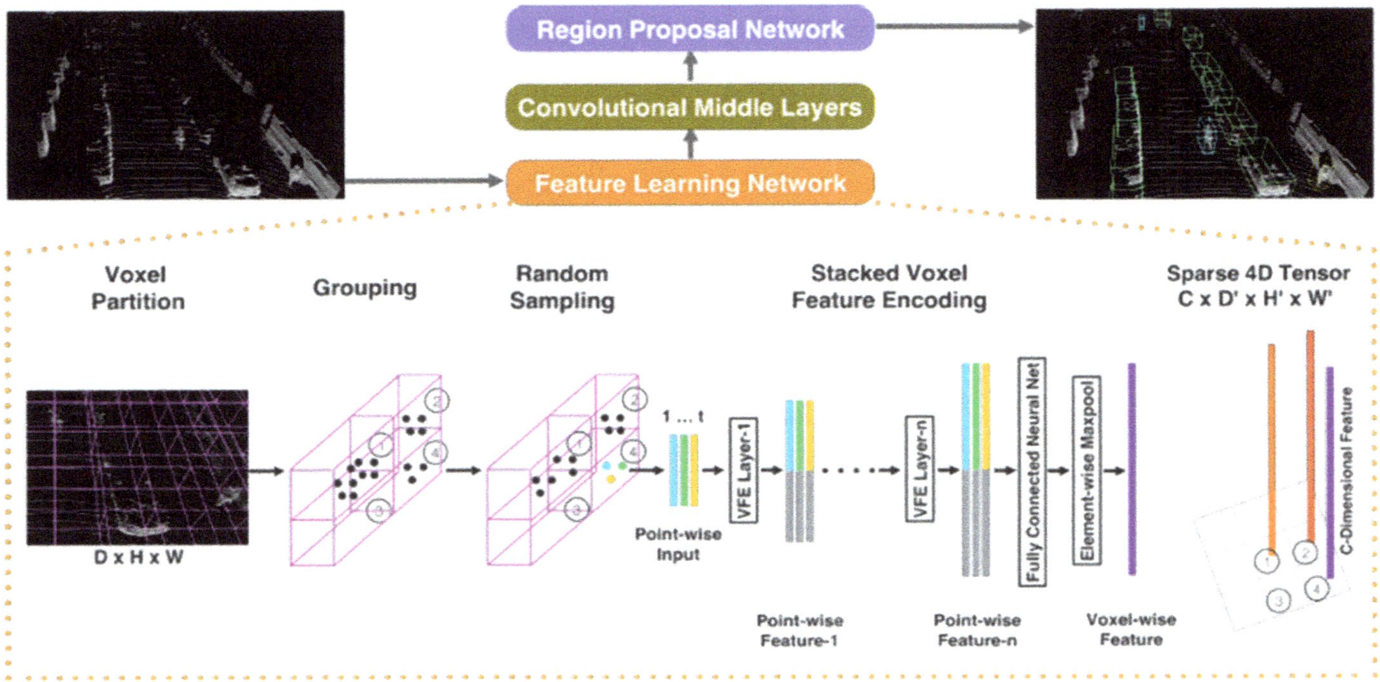

Figure 5: The VoxelNet architecture consisting of a feature learning network, convolutional middle layers, and a regional proposal network (adopted from Zhou and Tuzel, 2017).

Sparsely Embedded Convolutional Detection (SECOND)

Yan et al. (2018) presented SECOND, an enhanced voxel-based technique that uses sparse convolutional operations to efficiently decrease computational effort and increase processing speed (Zhang et al., 2024). The network architecture (Figure 6) consists of a voxel feature extractor, a middle sparse convolutional layer, and a region proposal network (Yan et al., 2018). A voxel representation of the point cloud data is first generated, then voxel-wise features are extracted using the voxel feature encoding layer. The sparse convolutional network takes this output and learns information about the z-axis, then transforms the sparse 3D input into an image-like 2D bird's-eye view data. Lastly, class, regression offsets, and direction are predicted with a single-shot multibox detector (SSD)-like RPN architecture (Yan et al., 2018).

Other Notable Methods

PointPillar

The introduction of PointPillars by Lang et al. (2019) marks a substantial breakthrough in effective 3D object detection for applications including autonomous driving, providing an end-to-end 3D object detection training network. This technique, consisting of three stages (Figure 7), uses PointNet to learn point cloud features arranged in vertical columns or pillars. The point cloud learnt features are then converted to a pseudo-image. After that, the pseudo-image is processed into high-level representation and 3D bounding boxes are predicted using a 2D object detection pipeline (Lang et al., 2019; Guo et al., 2020).

Point Voxel-RCNN

Point voxel (PV)-RCNN integrates the point-based and voxel-based methods, leveraging the voxel-based method's capacity to effectively produce high-quality 3D proposals, as well as the point-based method's ability to conserve accurate location information (Guo et al., 2020; Shi et al., 2021). Designed by Shi et al. (2021) to boost detection performance, the architecture (Figure 8) hosts a 3D voxel CNN and a two-stage framework consisting of a voxel-to-key point scene encoding and a point-to-grid RoI feature abstraction. Basically, to produce high-quality proposals, the input point clouds are fed into a 3D sparse convolutional network after initially being voxelized. Then, using a voxel set abstraction module, the learnt voxel-wise features are encoded into a small set of key points. Lastly, the RoI-grid points are used to aggregate the key point attributes to refine proposals and predict confidence (Guo et al., 2020; Shi et al., 2021).

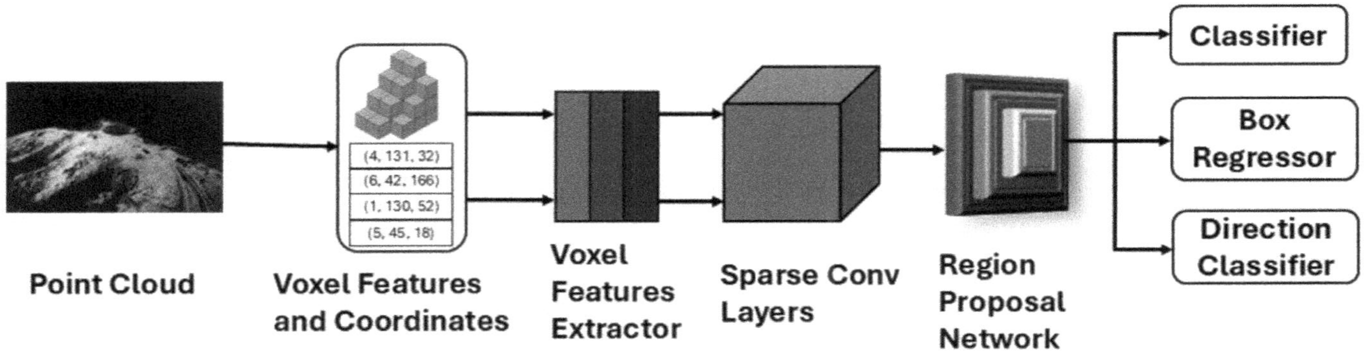

Figure 6: The SECOND network architecture framed using a voxel feature extractor, a middle sparse convolutional layer, and a region proposal network (modified after Yan et al., 2018).

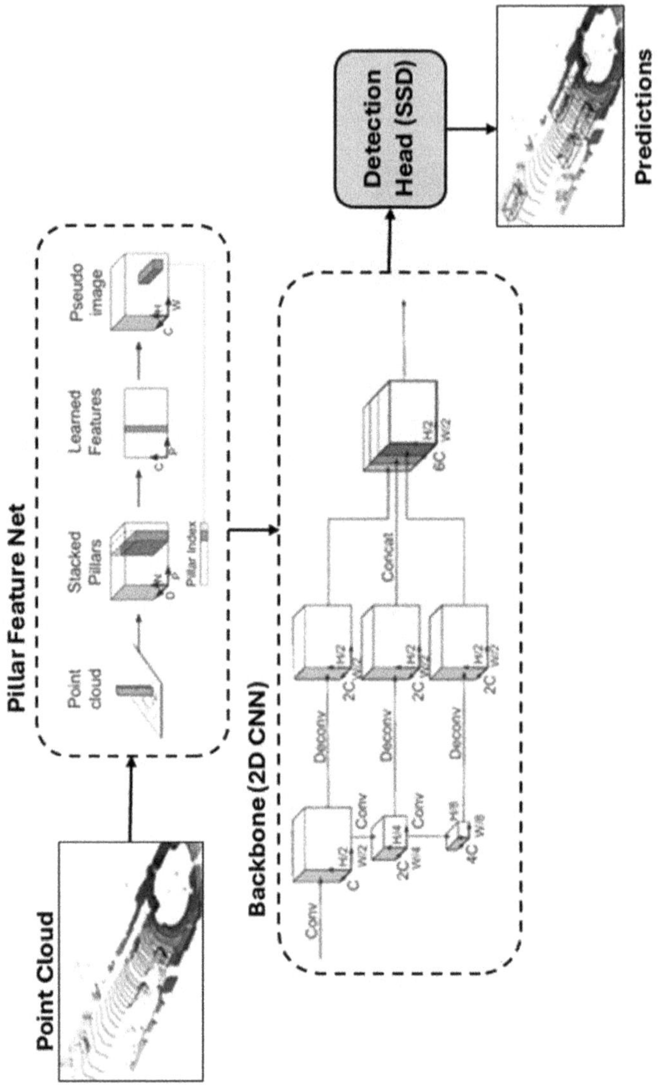

Figure 7: The main elements of the PointPillar architecture (adopted from Lang et al., 2019).

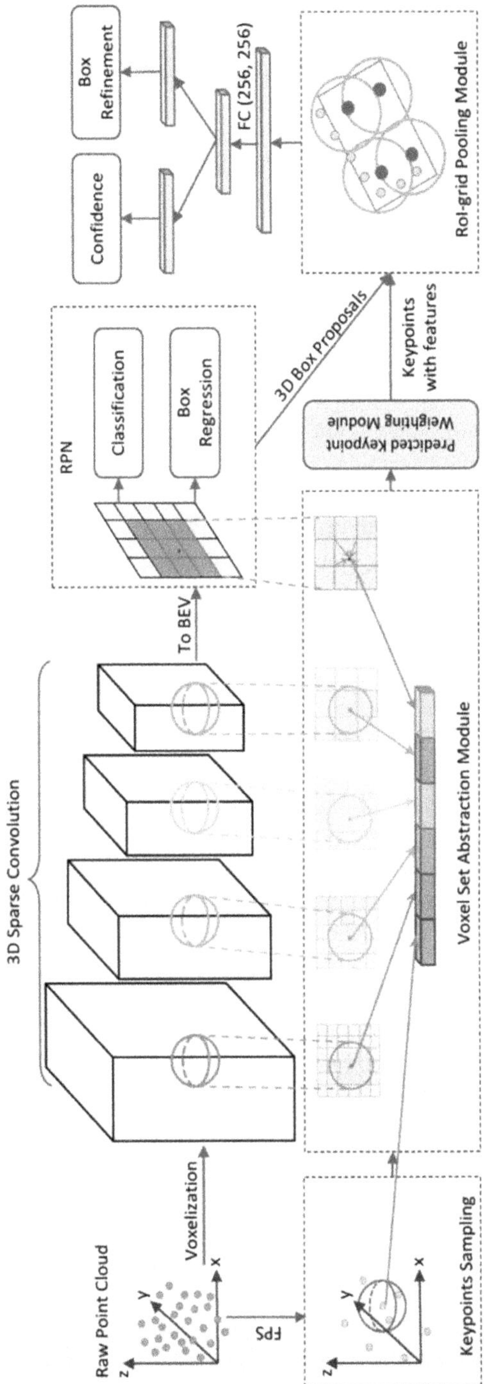

Figure 8: The framework of the PV-RCNN algorithm, integrating point-based and voxel-based networks (adopted from Shi et al., 2021).

Potential in Digital Outcrop Model Mapping

Digital outcrop models, or DOMs, have become increasingly common in geological research in recent years (Bellian et al., 2005; Pringle et al., 2010; Howell et al., 2014; Panara et al., 2022). These 3D surface representations of outcrops are produced through the use of laser scanning or photogrammetry techniques and contain millions of 3D point data (Bellian et al., 2005; Bistacchi et al., 2022). They offer several benefits, chief among them being the ability to measure geological formations in hard-to-reach places and gather more samples (Hodgetts, 2013). Their effectiveness has also been demonstrated in several works (Bellian et al., 2005; Xavier et al., 2009; Nesbit et al., 2018; Ojero et al., 2022; Tavani et al., 2024), highlighting the significance of DOMs in geological mapping.

With today's cutting-edge point cloud object detection algorithms, DOMs could become even more valuable by automating the mapping process. Automated tasks could include characterizing lithology and sedimentary structures, identifying structural features such as faults, fractures, intrusions, and bedding planes, and measuring structural geometry. For these activities, it is, therefore, crucial to determine the most effective approaches, weigh the advantages and disadvantages, and pinpoint areas for development.

Characterizing Lithology and Sedimentary Structures

Lithology classification using AI technologies has been carried out in several works (Thomas et al., 2011; Ran et al., 2019; De Lima et al., 2020; Dos Anjos et al., 2021; Zhenlong et al., 2023; Abdullah et al., 2024). These works have been entirely image-based, some of which are not fully automated. With DOM point cloud data, the automatic identification of lithology and sedimentary structures can take an exciting new direction. Considering the small-scale nature of these features, point-based methods, because of their capacity to preserve fine details, offer a profound solution. Relying on PointNet++ hierarchical point set feature learning, both fine-scale details such as lithology and larger-scale stratigraphic relationships can be detected. Because of the hierarchical architecture, sampling and grouping points at different resolutions are required. Thus, from complex local patterns to broad global structures, the network can discern features at various granularities thanks to this multi-stage sampling (Fatseas et al., 2024).

Another really promising approach for this task is PV-RCNN. PV-RCNN strikes a balance between large-scale efficiency and detail preservation by utilizing the PointNet-based networks' flexible receptive fields and the effective learning and superior proposals of 3D voxel CNN (Shi et al., 2021). With this multi-scale capacity, the voxel backbone allows for the capture of large-scale structures, while the point-based refining detects finer textures.

Identifying Large-Scale Geological Structures

Large-scale geological structures such as faults, fold, fracture, and intrusions are key elements for understanding the geology of an area, providing valuable information especially on tectonic history. Therefore, it is crucial to capture them in DOMs. Works such as Marques et al. (2021) that highlight the efficiency of using deep learning models to extract geological structures from images provide a basis for automating large-scale structure detection in DOM. Their work makes use of two CNN image segmentation models (SegNet and U-Net) to detect fractures from UAV-based (Unmanned Aerial Vehicle) digital photogrammetry outcrop images. They show that CNN models enable quality extraction of fracture features from images. The automatic identification of these large-scale geological characteristics can be advanced to a new level by applying PV-RCNN and the voxel-based technique to DOMs. As we previously pointed out, PV-RCNN has a multi-scale capacity, giving it ability to detect local and global features. Voxel-based methods, on the other hand, perform well with large-scale features due to their multi-scale processing, organized representation, and robustness to sparse data.

Measuring Structural Geometry

The task of automatically gathering geometrical measurements of geological structures from DOMs is challenging. Traditionally, RANSAC in conjunction with fuzzy K-means clustering has been used to extract geometrical features (such as orientation, thickness, curvature, and spacing) from point cloud data (Hammah and Curran, 2000; Kong et al., 2020; Zhou et al., 2024; Wang et al., 2024). However, by using image segmentation, feature learning, and geometrical fitting, contemporary point cloud object detection algorithms (such as PV-RCNN, PointNet++, or VoxelNet) can produce comparable or even better outcomes. For example, feature learning can draw distinction between faults, bedding planes, and other geological structures above and beyond mere geometrical properties from raw point cloud data. Moreover, clustering depends on local closeness, whereas algorithms such as PV-RCNN and PointNet++ can capture fine details and global features. Following the detection of points that correspond to a specific surface, such as traditional clustering workflows, plane fitting is carried out using either RANSAC plane fitting technique or PCA.

Challenges and Future Directions

Although modern point cloud object detection methods offer a remarkable way to automatically identify geological features from DOMs, there are a number of challenges that hinder these approaches. A good example of this is the resolution and quality of the DOM point cloud data. The resolution and quality of the data are critical factors in object detection from point cloud data. Therefore, to identify geological features in point cloud data, especially small-scale structures, the model needs to be high in resolution and devoid of data gaps caused by parts of the outcrop being inaccessible, occlusion, and shadow zones (Barazzetti, 2018; Duan et al., 2021). Moreover, further problems with data quality occur due to the complex nature of geological features. Complex processes create the majority of geological features, which can occasionally result in irregular and overlapping structures, varying lithology that looks similar, and partially exposed structures (Thornton et al., 2018; Wang et al., 2024). This can, therefore, significantly affect the automatic extraction of geological information from DOMs.

The scarcity of annotated training datasets is another good example of a potential challenge (Aribido et al., 2021; Lin et al., 2024). The approaches we discussed are primary reliant on supervised learning, thus annotated geological data is essential. By using these labelled datasets, the algorithms are able to identify features of interest in the point cloud data through learning structural geological patterns. However, the laborious and costly process of manually labelling geological features in point cloud datasets may limit the application of automated DOM-based geological mapping methods. Besides, high memory and computational costs are an additional challenge in automating DOM-based geological mapping (Wang et al., 2025). Point cloud datasets can contain millions or billions of points that facilitate the accurate representation of the reconstructed scenes. This generates a massive volume of data, which presents significant issues to storage and transmission systems (Gonçalves et al., 2019).

To address these challenges, it is essential for future works to explore the most effective ways to eliminate them. Techniques such as self-supervised learning framework which is employed in eradicating the time-consuming labelling of training datasets (Aribido et al., 2021; Li et al., 2025), improved data collection and processing techniques data used to minimize data gaps, remove noisy points, and enhance quality (Duan et al., 2021; Feng et al., 2024), and point cloud compression algorithms applied to reduce computational cost without losing resolution (Xu et al., 2018) should be examined.

Conclusions

The current state-of-the-art techniques used for object detection in point cloud data have enormous potentials for automating DOM geological mapping processes. The methods we reviewed have elements that make them capable of carrying out automated tasks such as characterizing lithology and sedimentary structures, identifying structural features such as faults, fractures, intrusions, and bedding planes, and measuring structural geometry. Specifically, components such as segmentation, feature learning, and 3D CNN enable these techniques to extract both complex local patterns and broad global structures. Therefore, even in situations where geometrical boundaries have insufficient sharpness or where geometrical features have similar signatures, it is possible to make distinctions from raw point cloud data that go above and beyond mere geometrical qualities.

Nevertheless, several challenges exist that limit the application of these methods in DOM-based geological mapping. Some of these challenges include the resolution of the DOM point cloud data, the complexity of geological features, scarce annotated training datasets, and high computational cost and memory. Addressing some of these limitations can be as easy as employing multi-scale feature extraction approaches or supplementing traditional plane fitting methods with feature learning to handle complex geological structures, while some like annotating training datasets can be extremely difficult. Overall, there is a high probability that experiments conducted using these techniques will produce exceptional outcomes in DOM-based geological mapping.

References

Abdullah, M. A. M., Mohammed, A. A. & Awad, S. R. (2024). RockDNet: Deep learning approach for lithology classification. *Applied Sciences*, 14(13).

Aribido, O. J., AlRegib, G. & Alaudah, Y. (2021). Self-Supervised Delineation of Geological Structures using Orthogonal Latent Space Projection. arXiv:2108.09605.

Barazzetti, L. (2018). Point cloud occlusion recovery with shallow feedforward neural networks. *Advanced Engineering Informatics*, 38, 605–619.

Bellian, J. A., Kerans, C. & Jennette, D. C. (2005). Digital outcrop models: Applications of terrestrial scanning lidar technology in stratigraphic modeling. *Journal of Sedimentary Research*, 75(2).

Bistacchi, A., Mittempergher, S. & Martinelli, M. (2022). Digital outcrop model reconstruction and interpretation. *In:* A. Bistacchi, M. Massironi & S. Viseur (eds). *3D Digital Geological Models: From Terrestrial Outcrops to Planetary Surfaces.* s.l. Wiley.

Chen, J., Zhu, H. & Li, X. (2016). Automatic extraction of discontinuity orientation from rock mass surface 3D point cloud. *Computers & Geosciences*, 95, 18–31.

De Lima, R. P., Duarte, D., Nicholson, C. & Slatt, R (2020). Petrographic microfacies classification with deep convolutional neural networks. *Computers & Geosciences*, 142(104481).

Dos Anjos, C. E., Avila, M. R., Vasconcelos, A. G., Neta, A. M., Medeiros, L. C. et al. (2021). Deep learning for lithological classification of carbonate rock micro-CT images. *Computational Geosciences*, 25, 971–983.

Duan, Y., Yang, C., Chen, H., Yan, W. & Li, H. (2021). Low-complexity point cloud denoising for LiDAR by PCA-based dimension reduction. *Optics Communications*, 482(126567).

Fatseas, K., Bekooij, M. J. G. & Sanberg, W. P. (2024). Optimizing PointNet++ and DBSCAN for Object Detection in Automotive Radar Point Clouds. *2024 21st European Radar Conference* (*EuRAD*). Paris, France.

Feng, Y., Xu, Y., Xia, Y., Brenner, C. & Sester, M. (2024). Gap completion in point cloud scene occluded by vehicles using SGC-Net. *ISPRS Journal of Photogrammetry and Remote Sensing*, 215, 331–350.

Fukitani, K., Shin, I., Lu, H., Yang, S., Kamiya, T., Nakatoh, Y. & Serikawa, S. (2022). 3D Object Detection Using Improved PointRCNN. *Cognitive Robotics*, 2, 242–254.

Gao, Y., Wang, P., Li, X., Sun, B., Sun, M., Li, L. & Di, R. (2025). PillarFocusNet for 3D object detection with perceptual diffusion and key feature understanding. *Scientific Reports*, 15(8776).

García-Sellés, D., Falivene, O., Arbués, P., Gratacos, O., Tavani, S. & Muñoz, J. A. (2011). Supervised identification and reconstruction of near-planar geological surfaces from terrestrial laser scanning. *Computers & Geosciences*, 37(10), 1584–1594.

Gigli, G. & Casagli, N. (2011). Semi-automatic extraction of rock mass structural data from high resolution LIDAR point clouds. *International Journal of Rock Mechanics and Mining Sciences*, 48(2), 187–198.

Gonçalves, M., Agostini, L., Palomino, D., Porto, M., & Correa, G. (2019). Encoding efficiency and computational cost assessment of state-of-the-art point cloud codecs. *IEEE International Conference on Image Processing* (*ICIP*). Taipei, Taiwan.

Guo, Y. & Hu, H. (2024). Multi-layer fusion 3D object detection via lidar point cloud and camera image. *Applied Sciences*, 14(4), 1348.

Guo, Y., Wang, H., Hu, Q., Liu, H., Li, L. & Bennamoun, M. (2020). Deep Learning for 3D Point Clouds: A Survey. arXiv:1912.12033.

Hammah, R. E. & Curran, J. H. (2000). Validity measures for the fuzzy cluster analysis of orientations. *IEEE Transactions on Pattern Analysis and Machine Intelligence*, 22(12), 1467–1472.

Hao, M., Zhang, Z., Li, L., Dong, K., Cheng, L., Tiwari, P. & Ning, X. (2024). Coarse to fine-based image–point cloud fusion network for 3D object detection. *Information Fusion*, 112(102551).

Hao, N. (2022). 3D object detection from point cloud based on deep learning. *Wireless Communications and Mobile Computing*, 2022(1).

Hodgetts, D. (2013). Laser scanning and digital outcrop geology in petroleum industry: A review. *Marine and Petroleum Geology*, 46, 335–354.

Howell, J. A., Martinius, A. W. & Good, T. R. (2014). The application of outcrop analogues in geological modelling: A review, present status and future outlook. *Geological Society*, London, Special Publications, 387, 1–25.

Karthik, R., Inamdar, R., Sundar, K. S., Cho, J. & Veerappampalayam, S. (2024). Point cloud-based 3D object classification with non-local attention and lightweight convolution neural networks. *Institute of Electrical and Electronics Engineering*, 12, 158530–158545.

Kong, D., Wu, F. & Saroglou, C. (2020). Automatic identification and characterization of discontinuities in rock masses from 3D point clouds. *Engineering Geology*, 265(105442).

Lanen, X. T., Hodgetts, D., Redfern, J. & Fabuel-Perez, I. (2010). Application of digital outcrop models: Two fluvial case studies from the Triassic Wolfville Fm., Canada and Oukaimeden Sandstone Fm., Morocco. *Geological Journal*, 44(6), 742–760.

Lang, A. H., Vora, S., Caesar, H., Zhou, L., Yang, J. & Beijbom, O. (2019). PointPillars: Fast Encoders for Object Detection from Point Clouds. arXiv:1812.05784v2.

Lato, M. J. & Vöge, M. (2012). Automated mapping of rock discontinuities in 3D lidar and photogrammetry models. *International Journal of Rock Mechanics and Mining Sciences*, 54, 150–158.

Li, H.-S. & Lu, Y.-L. (2024). 3D object detection based on point cloud in automatic driving scene. *Multimedia Tools and Applications*, 83(5), 13029–13044.

Li, M., Yan, X.-S. & Wu, Q.-H. (2025). A self-supervised deep learning framework for seismic facies segmentation. *Expert Systems with Applications*, 288(128290).

Lin, L., Zhong, Z., Li, C., Gorman, A., Wei, H., Kuang, Y. et al. (2024). Machine learning for subsurface geological feature identification from seismic data: Methods, datasets, challenges, and opportunities. *Earth-Science Reviews*, 257(104887).

Lopez, D., Haas, C. & Narasimhan, S. (2023). Specific object finding in point clouds based on semantic segmentation and iterative closest point. *Automation in Construction*, 156(105116).

Lyu, J., Wang, S., Qi, Y. & Chen, L. (2025). Data fusion-based joint 3D object detection using point clouds and images. *Electronics*, 14(12).

Marques, A., Racolte, G., De Souza, E. M., Domingos, H. V., Horota, R. K. & Motta, J. G. (2021). Deep learning application for fracture segmentation over outcrop images from UAV-based digital photogrammetry. *IEEE International Geoscience and Remote Sensing Symposium* (*IGARSS*). Brussels, Belgium.

Maturana, D. & Scherer, S. (2015). VoxNet: A 3D convolutional neural network for real-time object recognition. *IEEE/RSJ International Conference on Intelligent Robots and Systems* (*IROS*). Hamburg, Germany.

Nesbit, P. R., Durkin, P. R. & Hugenholtz, C. H. (2018). 3D stratigraphic mapping using a digital outcrop model derived from UAV images and structure-from-motion photogrammetry. *Geosphere*, 14(6), 2469–2486.

Ojero, J. O., Hodgetts, D. & Burnham, B. (2022). Employing a statistical framework to constrain orientation measurement inaccuracies from digital outcrop models: Insights from sandstone intrusions. *Journal of Structural Geology*, 157(104567).

Panara, Y., Menegoni, N., Carboni, F. & Inama, R. (2022). 3D digital outcrop model-based analysis of fracture network along the seismogenic Mt. Vettore Fault System (Central Italy): The importance of inherited fractures. *Journal of Structural Geology*, 161(104654).

Peixoto, E. B. D. A. & Centeno, J. A. D. S. (2023). Detection and classification of vertical objects in point cloud data for urban planning and management. *IOSR Journal of Applied Geology and Geophysics*, 11(6), 1–11.

Pringle, H. K., Brunt, R. L., Hodgson, D. M. & Flint, S. S. (2010). Capturing stratigraphic and sedimentological complexity from submarine channel outcrops to digital 3D models, Karoo Basin, South Africa. *Petroleum Geoscience*, 16, 307–330.

Qi, C. R., Su, H., Mo, K. & Guibas, L. J. (2017a). PointNet: Deep Learning on Point Sets for 3D Classification and Segmentation. arXiv:1612.00593.6

Qi, C. R., Yi, L., Su, H. & Guibas, L. J. (2017b). PointNet++: Deep Hierarchical Feature Learning on Point Sets in a Metric Space. arXiv:1706.02413v1.

Ran, X., Xue, L., Zhang, Y., Liu, Z., Sang, X. & He, J. (2019). Rock classification from field image patches analyzed using a deep convolutional neural network. *Mathematics*, 7(8), 755.

Riquelme, A. J., Abellán, A., Tomás, R. & Jaboyedoff, M. (2014). A new approach for semi-automatic rock mass joints recognition from 3D point clouds. *Computers & Geosciences*, 68, 38–52.

Salmane, P. H., Velázquez, J. M., Khoudour, L., Mai, N. A., Duthon, P. et al. (2023). 3D object detection for self-driving cars using video and LiDAR: An ablation study. *Sensors*, 23(6).

Shi, S., Guo, C. & Jiang, L. (2021). PV-RCNN: Point-Voxel Feature Set Abstraction for 3D Object Detection. arXiv:1912.13192v2.

Shi, S., Wang, X. & Li, H. (2019). PointRCNN: 3D Object Proposal Generation and Detection from Point Cloud. arXiv:1812.04244v2.

Tavani, S., Corradetti, A., Rizzo, R. E. & Seers, T. D. (2024). Best practices towards the digitization of 3D traces from virtual outcrop models. *Journal of Structural Geology*, 186(105222).

Thomas, A., Rider, M., Curtis, A. & MacArthur, A. (2011). A novel approach for automated lithology extraction from core photographs. *First Break*, 29, 103–109.

Thornton, J. M., Mariethoz, G. & Brunner, P. (2018). A 3D geological model of a structurally complex Alpine region as a basis for interdisciplinary research. *Scientific Data*, 5(180238).

Unal, O., Gool, L. & Dai, D. (2021). Improving point cloud semantic segmentation by learning 3D object detection. *IEEE Winter Conference on Applications of Computer Vision* (*WACV*), 2949–2958.

Vöge, M., Lato, M. J. & Diederichs, M. S. (2013). Automated rockmass discontinuity mapping from 3-dimensional surface data. *Engineering Geology*, 164, 155–162.

Wang, C., Luo, Y., Wang, K., Cao, Y., Tao, X., Geng, D. et al. (2025). Breaking barriers in 3D point cloud data processing: A unified system for efficient storage and high-throughput loading. *Expert Systems with Applications*, 227(126983).

Wang, Y., Zhou, J., Chen, Q., Chen, J., Zhu, C. & Li, H. (2024). Automatic interpretation of geometric information of discontinuities and its influence on the stability of highly-jointed rock slopes. *Journal of Rock Mechanics and Geotechnical Engineering*, 17(9), 5479–5492.

Xavier, M. T., Hodgetts, D., Redfern, J. & Fabuel-Perez, I. (2009. Applications of digital outcrop models: Two fluvial case studies from the Triassic Wolfville Fm., Canada and Oukaimeden Sandstone Fm., Morocco. *Geological Journal*, 44(6), 742–760.

Xu, Y., Ke, Z., Lanyi, H., Zhiqian, J. & Wenjie, Z. (2018). Introduction to point cloud compression. *ZTE Communications*, 16(3), 3–8.

Yan, Y., Mao, Y. & Li, B. (2018). SECOND: Sparsely embedded convolutional detection. *Sensors*, 18(10), 3337.

Yan, Z., Lei, P., Liangxi, L., Xiangyu, L., Fanzhi, Z. & Yuexia, Z. (2022). Research progress on three-dimensional object detection of laser point clouds. *Computer Science and Exploration*, 16(12), 2695–2717.

Zhang, H., Wang, C., Tian, S., Lu, B., Zhang, L. et al. (2023). Deep Learning-Based 3D Point Cloud Classification: A Systematic Survey and Outlook. arXiv:2311.02608v1.

Zhang, W., Dong, X., Cheng, J. & Wang, S. (2024). Advanced point cloud techniques for improved 3D object detection: A study on DBSCAN, attention, and downsampling. *World Electric Vehicle Journal*, 15(11), 527.

Zhenlong, H., Jikang, W., Jinrong, S., Xinwei, L. & Wentian, Z. (2023). Intelligent lithology identification methods for rock images based on object detection. *Natural Resources Research*, 32, 2965–2980.

Zhou, J.-W., Chen, J.-L. & Li, H.-B. (2024). An optimized fuzzy K-means clustering method for automated rock discontinuities extraction from point clouds. *International Journal of Rock Mechanics and Mining Sciences*, 173(105627).

Zhou, Zhou, W., Liu, K., Jin, W., Wang, Q., She, Y., Yu, Y. & Ma, C. (2025). Advancements in deep learning for point cloud classification and segmentation: A comprehensive review. *Computers & Graphics*, 130(104238).

Zhou, Y. & Tuzel, O. (2017). VoxelNet: End-to-End Learning for Point Cloud Based 3D Object Detection. arXiv:1711.06396.

Zimmer, W., Erçelik, E., Zhou, X., Ortiz, X. J. & Knoll, A. (2022). A Survey of Robust LiDAR-Based 3D Object Detection Methods for Autonomous Driving. arXiv:2204.00106.

CHAPTER
3

Agentic Workflows and Multi-Agent Systems: Platforms and Protocols in 2025

Hamzah Bawah[1*] and Daniel Asante Otchere[2*]

[1] Motorway, London, England
[2] Institute for Computational and Data Sciences, The Pennsylvania State University, State College, Pennsylvania 16802, USA

Introduction

Artificial intelligence has moved beyond single-step prompt-response interactions toward agentic workflows—multi-step, adaptive processes orchestrated by AI agents. An AI agent is an autonomous software entity that can perceive its environment, plan a sequence of actions, use tools, and learn from feedback to achieve a complex goal, thereby operating within a multi-step, iterative cycle. This shift from static scripts to dynamic orchestration dramatically improves AI's effectiveness on complex, real-world tasks by allowing continuous improvement and adjustment.

These agents, which function as the core components of agentic workflows, can be categorized by their operational paradigms and degree of autonomy. Retrieval agents are engineered with specific instructions governing their behavior and objectives, operating within a constrained set of predefined rules. Task agents are intrinsically linked to deterministic workflows, driven by knowledge bases and rules-based automation to execute repetitive processes. In contrast, autonomous agents exhibit a higher degree of independence, capable of dynamic planning, learning from environmental feedback, and adapting their approach without constant human intervention. It is this capacity for autonomous, iterative execution

*Corresponding authors: hamzahjbawah@gmail.com; ascotjnr@yahoo.com

DOI: 10.1201/9781003506348-3

that enables agentic workflows to break down complex problems, refine actions over time, and outperform single-shot models on intricate tasks. Early case studies have shown that even less advanced models can outperform more powerful ones if they employ an agentic, iterative approach; for example, GPT-3.5 solved coding problems with higher accuracy when following an agentic step-by-step workflow, surpassing GPT-4's zero-shot performance (Ng, 2024; Shinn et al., 2023).

The practical application of these agents is dictated by the complexity and variability of the business process. For instance, processing routine expense reports that follow a strict, unchaining path is a suitable case for a deterministic cloud flow. However, scenarios requiring dynamic analysis, such as managing complex support tickets that need real-time classification from multiple data sources, or reviewing loan applications against evolving regulations, are ideal for autonomous agents. In these latter cases, the agent's ability to orchestrate a multi-step workflow, adapt to new information, and make context-aware decisions is essential, underscoring the fundamental shift toward adaptive, agentic AI systems in modern enterprise automation (Völker & Oberhauser, 2024).

Central to agentic workflows are AI agents. An AI agent is an autonomous software entity (often powered by a large language model, or LLM) capable of perceiving instructions or environment input, deciding on actions, and executing those actions—sometimes calling external tools or services—toward achieving a goal (IBM, 2023). Agents maintain state (through memory of past steps or context) and can use reasoning strategies such as chain-of-thought planning and self-reflection to handle non-deterministic, open-ended tasks (Wang et al, 2023). In a typical agentic workflow, an AI agent iteratively: (1) interprets the user's objective, (2) devises or updates a plan, (3) calls upon tools or data sources as needed, (4) evaluates intermediate results, and (5) adjusts its approach until the goal is met or failure is determined. This contrasts with a static script or single prompt: the agent has a degree of *autonomy* within limits set by developers or users (for example, being allowed to call certain APIs, but not others, or to only operate within a sandbox). The result is a more adaptive and powerful workflow. Indeed, as IBM notes, by enabling generative AI to handle intricate, multi-step processes, organizations can achieve improved efficiency and scalability while reducing reliance on human oversight.

Multi-agent collaboration is an extension of this paradigm, wherein multiple specialized agents work in concert. Instead of a single monolithic agent trying to do everything, a team of agents can each contribute their expertise or capabilities. For instance, one agent might be skilled in data gathering, another in analysis, and a third in planning—together they can solve a complex problem more effectively than any alone. Such *multi-agent systems* (MAS) have long been studied in AI, but the emergence of LLM-powered agents has revitalized interest in collaborative agents that communicate in natural language (Tran et al., 2025). Multi-agent collaboration involves multiple agents or language models interacting to complete complex tasks, often simulating a group of experts in different roles working together. By distributing subtasks among agents, these systems can leverage diverse skills (e.g., an image-recognition agent working with a planning agent and a database-querying agent). Notably, research suggests that multi-agent systems can solve problems that exceed the capability of any single agent, exhibiting a form of *collective intelligence*. As we move toward more sophisticated AI deployments, multi-agent teams could become powerful solutions for complex, interdisciplinary challenges—collaborative AI rather than isolated AI.

This article focuses explicitly on two complementary layers of agentic AI:

(i) Platforms and frameworks that enable the construction, deployment, and orchestration of autonomous agents.
(ii) Inter-agent communication protocols that enable discovery, coordination, and secure integration with tools and services.

By separating platform concerns (runtime, tool integration, life cycle management) from protocol concerns (identity, message formats, discovery, and governance), we aim to clarify how these layers interact and where open technical gaps remain. We maintain the original review structure and extend it through 2025, but emphasize open research gaps that motivate our comparative analysis: (a) scalability—how agent populations and chained LLM invocations scale in latency and cost under production loads; (b) robustness and failure modes—detection and recovery strategies for unresponsive or Byzantine agents and cascading workflow failures; (c) security and trust—identity, least-privilege access, and defenses against prompt-tool injection; and (d) interoperability and governance—practical mechanisms and standards for mixing vendor agents across organizational boundaries. These gaps are revisited in each section to frame platform- and protocol-level trade-offs (Tran et al., 2025).

Platforms and Frameworks for Agentic AI

Dozens of new frameworks and platforms have appeared to help developers create AI agents and incorporate them into workflows. Broadly, these tools abstract the complexities of prompting, tool integration, and agent orchestration into higher-level libraries or services (Xi et al., 2025). Early examples in 2023, such as AutoGPT and BabyAGI, sparked public interest in autonomous AI agents that could recursively plan and execute tasks. AutoGPT in particular (an open-source project) demonstrated how an LLM-based agent (using GPT-4) could attempt to achieve a user-defined goal by generating sub-goals, using the internet and local files as tools, and self-correcting via feedback loops. While groundbreaking, these early projects often lacked robustness. In response, more structured frameworks emerged from both academia and industry.

We categorize entries in this section as: (i) production-grade platforms/frameworks—toolchains and runtimes intended for operational use, with features such as logging, life cycle management, monitoring, and connectors (examples: LangChain, AutoGen in its production-oriented releases, and CrewAI); and (ii) research prototypes—experimental systems designed for evaluation and concept demonstration (examples: HuggingGPT, early BabyAGI variants, and generative agents). This distinction clarifies evaluation criteria (robustness, monitoring, service level agreements (SLAs) for platforms versus algorithmic novelty and emergent behavior for prototypes).

One notable framework is Microsoft's AutoGen. AutoGen is an open-source library that allows developers to compose multiple agents that converse with each other to solve tasks (Wu et al., 2023). Each agent in AutoGen can be configured with a persona and specific tools or APIs it can utilize. AutoGen provides a flexible runtime for defining conversation patterns between agents (and optionally human users), enabling complex workflows such as an "Assistant" agent and a "Developer" agent co-creating code, or a chain of reasoning where one agent verifies or optimizes the work of another. Because AutoGen agents can use both natural language and code to communicate, the framework blurs the line between traditional programming and AI-orchestrated automation. Empirical studies have shown the effectiveness of AutoGen on tasks ranging from math problem-solving to coding and decision-making. This demonstrates the benefit of a structured multi-agent conversation: it injects modularity and specialization into the problem-solving process. Quantitative evidence is limited: community benchmarks report increased latency and cost with agent counts greater than 20; formal end-to-end throughput/accuracy numbers remain to be published in peer-reviewed venues. Although AutoGen simplifies composing conversational agent networks, it currently exposes several practical limitations for large-scale, production use: (1) failure recovery is limited—the runtime provides minimal automatic rollback or agent re-assignment semantics for long-running workflows; (2) latency and cost scale nonlinearly as the number of agents increases, because each subtask may invoke separate LLM calls; and (3) observability and SLAs are immature in the community releases, making enterprise adoption contingent on additional infrastructure for monitoring and governance. Empirical reporting on throughput and latency is limited in public literature and remains an important open area for benchmarking (Tran et al., 2025).

Another popular toolkit is LangChain, an open-source framework that provides building blocks for LLM applications, including agent implementations. LangChain is often described as a "LEGO set" for AI, letting developers chain together LLM calls, tools, and memory to create complex applications. While LangChain started with single-agent tool-using scenarios, it has evolved to support multi-agent interactions as well. *LangChain's Agents* abstraction allows an LLM to decide which tool or other agent to invoke at each step, based on the conversation context. A recent extension called LangChain Hub/LangGraph enables defining workflows as graphs where nodes are agents or tools and edges represent information flows. This graph-based approach is well suited to multi-agent systems, as it can represent loops, branches, and asynchronous interactions beyond simple linear sequences. For example, using LangChain or LangGraph, one could implement a workflow where a "Researcher" agent, a "Writer" agent, and an "Editor" agent collaborate on writing a report—sharing context via the graph's memory and calling external APIs for data. LangChain's modular design is powerful for prototyping but faces production trade-offs: it centralizes orchestration logic often in application code, which increases integration complexity when scaling multiple distinct agents; memory and state consistency across graph nodes require custom engineering; and it lacks standardized performance claims across versions, so reported latencies depend heavily on choice of LLM and hosting. Where available, benchmark reports show rising per-task latency as workflow graphs deepen (author reports and community benchmarks; systematic studies are still nascent) (Tran et al., 2025).

Beyond these, many specialized frameworks have emerged to target multi-agent applications. Crew (or CrewAI) is one such framework focused on *production-ready* multi-agent systems. It allows developers to define a "crew" of agents, each with distinct roles and abilities, that work together toward a goal. CrewAI emphasizes clean integration and practical deployments, providing infrastructure for agent communication and life-cycle management in real-world use cases (its creators even offer courses on multi-agent system design in enterprise settings). Meanwhile, research prototypes such as HuggingGPT (Shen et al., 2023) demonstrated how an LLM (ChatGPT) could act as a *controller* that delegates subtasks to a pool of specialist models (e.g., a vision model, a speech model) available on platforms such as Hugging Face. This can be viewed as a multi-agent orchestration where a central agent manages an ensemble of expert AI models—effectively treating other models as collaborative agents through a natural language interface. Similarly, the CAMEL framework (Li et al., 2023) explored role-playing agents that communicate to solve problems (e.g., an AI user and an AI assistant discussing a task) to boost performance via collaboration. Production-oriented frameworks such as CrewAI prioritize operational concerns but may introduce vendor dependencies and opinionated life cycles (e.g., logging/telemetry schemas, runtime APIs). Such choices speed internal adoption at the cost of cross-vendor portability unless the framework exposes standard protocol adapters (MCP/ACP/A2A). This again motivates standard protocol adoption for cross-cutting interoperability (Tran et al., 2025).

Generative agents (Park et al., 2023) represent another fascinating direction: autonomous agents with long-term memory and dynamic behaviors interacting in a simulated environment. In that work, multiple agents (each powered by an LLM and given a backstory and objectives) lived out day-to-day activities in a sandbox "town," speaking to each other and remembering past interactions. While more of a simulation than a utility-driven workflow, generative agents' research highlights the importance of memory, persistence, and emergent social behaviors in multi-agent settings—attributes that more utilitarian platforms may also need to incorporate for complex collaboration. Table 1 summarizes a few prominent agentic AI frameworks and their characteristics.

Each approach contributes different strengths—from practical tool integration to advanced research on agent autonomy and interaction. This is not an exhaustive list; numerous other frameworks (e.g., BabyAGI, MetaGPT, IBM Watson

Table 1: Examples of frameworks and projects enabling agentic workflows and multi-agent systems

Framework	Developer/ Source	Key Features	Multi-Agent Capabilities
AutoGen	Microsoft (2023)	Open-source library for LLM-based agents; agents converse via natural language or code. Flexible conversation patterns programmable via Python.	Yes—designed for composing multiple agents in conversation (e.g., Assistant and Developer agents solving tasks collaboratively).
LangChain	Open Source (2022)	"LEGO block" modular toolkit for LLM apps. Provides prompt templates, memory, tool integrations, and agent classes.	Partially—primarily single-agent tool use, but supports calling other agents as tools. Extended by LangGraph for complex agent workflows.
CrewAI	CrewAI (Startup, 2024)	Framework for deploying teams of agents with defined roles. Emphasizes robust production use (logging, error handling).	Yes—built around concept of a coordinated crew of specialized agents working together. Communication handled via the framework's runtime.
HuggingGPT	Microsoft Research (2023)	Research prototype/ controller agent (uses OpenAI GPT-4) that delegates tasks to expert models on Hugging Face. Natural language used as a universal interface.	Yes—though not multiple LLM agents chatting, it involves one agent managing many AI models (treated as tool-agents). Demonstrates dynamic orchestration of heterogeneous models.
Generative agents	Stanford (2023)	Simulated multiple agents with long-term memory and emergent social behavior. Agents prompted to behave like characters, remembering and evolving over time.	Yes—many agents co-exist and communicate in a sandbox environment (e.g., a virtual town). Focus on social interaction and memory.

Orchestrate, etc.) and libraries are part of the rapidly evolving ecosystem. The unifying trend is moving from single, all-purpose AI agents toward assemblies of specialized agents that can cooperate.

Despite their promise, building multi-agent workflows remains nontrivial. Challenges include: coordinating agent communications, avoiding redundant or conflicting actions, managing shared memory or context, and ensuring security (so that autonomous agents do not perform unsafe operations). Moreover, different frameworks often used proprietary or ad-hoc methods for agent communication, making it hard to mix agents from different vendors. To address these issues, the community has increasingly turned attention to standardizing the "language" that agents use to talk to each other and to external tools. This is where recent agent-to-agent communication protocols come into play, as discussed in the following text (Ehtesham, et al., 2025).

Emerging Protocols for Agent-to-Agent Communication

Model Context Protocol (MCP)

Introduced by Anthropic in late 2024, the Model Context Protocol (MCP) represents a foundational shift in how AI systems acquire and operationalize external context. Rather than focusing on inter-agent dialogue, MCP addresses a more fundamental challenge: enabling agents to reliably connect to data repositories, developer tools, and application environments through a standardized interface. In doing so, MCP lays the groundwork not for collaboration, but for capability; it ensures an agent is no longer confined to its internal model but can act upon the real world (Anthropic, 2024).

Purpose: The primary purpose of MCP is to eliminate brittle, bespoke integrations by providing a unified protocol through which agents can query or invoke external tools. In practical terms, it formalizes agent-to-tool interoperability, allowing a language model to perform actions such as file access, code execution, or database search without hard-coded API logic. MCP is thus best understood as an infrastructural protocol that supplies agents with situational awareness and operational reach (Anthropic, 2024).

Technical Basis: MCP defines a client–server interface (JSON/RPC (JavaScript Object Notation Remote Procedure Call)-like semantics) allowing MCP clients (agents/LLM apps) to discover and call MCP servers (data sources, tool endpoints). Transport includes HTTP (Hypertext Transfer Protocol) and STDIO (standard input/output) with optional streaming via server-sent events; schema and capability manifests standardize tool descriptions (Anthropic, 2024).

Adoption Status: Anthropic published MCP in November 2024 and MCP demonstrators appeared in the Claude desktop client; developer platforms such as Replit and Codeium have pilot integrations and major OS vendors (e.g., Microsoft Windows AI Foundry) have signalled platform-level support. Broader enterprise adoption is underway but still early (The Verge, 2024, 2025).

Security & Privacy: MCP places identity, authorization, and auditing at the MCP server boundary, and therefore relies on implementers to enforce least-privilege, scopes, and sandboxing; prompt/tool injection remains a recognized risk that requires hardened server-side validation and runtime monitoring (Anthropic, 2024; The Verge, 2025).

Critical Perspective/Contrast: MCP solves the integration problem (agent↔tool/context), not the collaboration problem (agent↔agent). In practice, it complements higher-level protocols: an agent retrieves data or executes actions via MCP, then collaborates with peer agents through an inter-agent protocol (e.g., A2A). This "stacked" view is consistent with long-standing multi-agent system designs where tool invocation and inter-agent coordination are layered concerns (Finin et al., 1994; FIPA, 2002).

Agent-to-Agent (A2A) Protocol

Google's Agent-to-Agent (A2A) protocol (co-announced with a large partner coalition) targets the missing piece that MCP does not: a standard way for heterogeneous agents to discover one another, exchange tasks, and coordinate across enterprise application estates. Where MCP focuses on an agent's reach into tools, A2A focuses on an agent team's ability to form and execute a workflow (Google, 2025a; Surapaneni et al., 2025).

Purpose: A2A enables agents to securely communicate and coordinate actions across vendors and platforms, allowing organizations to stitch together domain-specific agents (e.g., HR, finance, developer productivity, customer support) into end-to-end processes. It is explicitly positioned to complement MCP: agents talk to tools via MCP and to each other via A2A. This protocol provides a standardized task life cycle and discovery framework to allow heterogeneous agents to discover capabilities, negotiate tasks, exchange progress updates and artifacts, and coordinate end-to-end workflows in enterprise settings (Google, 2025a).

Technical Basis: A2A defines a peer model in which agents publish Agent Cards (self-descriptions of capabilities/endpoints) for discovery. Interactions follow a task life cycle model (request → progress → result), multimodal payload support, and transport/serialization conventions (JSON over HTTP with negotiation primitives). Integration tooling (Google Cloud) assists building A2A-compliant agents (Google, 2025a).

Adoption Status: Announced in April 2025 with a 50+ partner coalition and Cloud tooling support, A2A has since received upgrades and tooling via Google Cloud to help teams build, deploy, evaluate, and commercialize A2A-compliant agents—an indicator of sustained enterprise momentum (Google, 2025b).

Security & Privacy: A2A assumes an enterprise governance model—capability access is policy-gated at the agent boundary, and communications are authenticated/authorized within organizational IAM (Identity and Access Management) and audit frameworks. The task life cycle and artifact exchange provide audit points, policy enforcement at agent boundaries, and durable traces for compliance (Google, 2025a).

Critical Perspective/Contrast: A2A prioritizes governed interoperability across enterprise SaaS ecosystems—sharp contrast to decentralized protocols (e.g., Agent Network Protocol (ANP)/Coral) that emphasize permissionless discovery and cross-domain identity without central authorities. Practically, A2A + MCP forms a high-control, high-assurance stack for enterprise agentic workflows. This "teams of agents" orientation aligns with recent evidence that coordinated multi-agent planning can outperform single-agent baselines on complex tasks when communication is structured. Interoperability with MCP (for tool access) and ACP (transport) is expected in practice (Park et al., 2023; Wu et al., 2024).

Agent Communication Protocol (ACP)

IBM Research's Agent Communication Protocol (ACP), developed under the Linux Foundation with a reference implementation in BeeAI, proposes a lightweight, HTTP-native "wire format" so agents—regardless of framework—can interoperate via a simple REST (Representational State Transfer) API. Rather than prescribing workflow semantics (as A2A does) or tool schemas (as MCP does), ACP focuses on transport-level interoperability that existing stacks can adopt quickly (IBM Research, 2025; Agent Communication Protocol, 2025; Koundinya et al 2025).

Purpose: ACP provides a minimal, framework-agnostic messaging substrate (transport layer) so agents implemented in different stacks can exchange messages and streaming updates over standard web primitives (Agent Communication Protocol, 2025).

Technical Basis: ACP defines standardized REST endpoints for sending/receiving multimodal messages, supporting synchronous calls, and streaming updates. It is intentionally "curl/Postman-friendly" and ships adapters to bridge into other ecosystems (e.g., calling MCP-exposed tools) (Agent Communication Protocol, 2025).

Adoption Status: ACP is being advanced under the Linux Foundation with BeeAI as a production-grade reference; community communications describe alignment with broader interop efforts (including A2A) to reduce fragmentation (IBM Research, 2025).

Security & Privacy: ACP relies on web security primitives (TLS, OAuth/JWT, service authorization) and leaves fine-grained data governance to the agent service and underlying connectors—an explicit design choice to keep the protocol thin and composable (Agent Communication Protocol, 2025).

Critical Perspective/Contrast: ACP's strength is adoptability as a transport layer that can undergo higher-level protocols such as A2A (workflow/task life cycle) or interface with MCP (tools). Its strength is adoptability; its limitation is that discovery, policy, and life cycle semantics remain out of scope. This separation of concerns echoes earlier MAS standards—FIPA-ACL (Foundation for Intelligent Physical Agents – Agent Communication Language)-specified communicative acts and content languages while leaving many architectural choices to implementers (FIPA, 2002).

Agent Network Protocol (ANP)

The Agent Network Protocol (ANP) advances a decentralized vision: an "HTTP of agents" for permissionless networks where independently owned agents can establish identity, discover one another, and collaborate without centralized authorities. In contrast to A2A's enterprise orientation, ANP foregrounds open discovery, cross-domain semantics, and cryptographic trust as first-class citizens (Agent Network Protocol, 2025a, 2025b).

Purpose: ANP provides a decentralized, permissionless stack enabling agent identity, discovery, and inter-agent collaboration across organizational boundaries using decentralized identity primitives (Agent Network Protocol, 2025a).

Technical Basis: ANP layers—(i) Identity and encrypted communication based on W3C decentralized identifiers (DIDs); (ii) a meta-protocol for dynamic negotiation of formats/ontologies; and (iii) an application layer using JSON-LD/RDF (with schema.org-aligned profiles) for agent descriptions and messages. Discovery blends distributed directories and searchable JSON-LD descriptions; all discovery info conforms to an agent description profile (ADP) (Agent Network Protocol, 2025a, 2025b).

Adoption Status: ANP provides specifications, white papers, and open-source assets oriented to decentralized identity and discovery. Its early community is building DID methods and tooling to support cross-organizational collaboration and agent marketplaces (Agent Network Protocol, 2025a).

Security & Privacy: Security is foundational by design—DID-based authentication, verifiable credentials, and end-to-end encryption support private channels without central mediators. This privacy posture shifts operational complexity to implementers for key custody, rotation, and revocation registries (Agent Network Protocol, 2025a).

Critical Perspective/Contrast: ANP maximizes openness and cross-boundary interoperability at the cost of higher operational complexity; it stands in sharp relief to A2A's curated, policy-driven enterprise model. Where A2A assumes centralized IAM and governance, ANP assumes cryptographic identity and decentralized semantics. This reflects a broader design fork in MAS deployments between governed intranets and permissionless internets (Finin et al., 1994; FIPA, 2002).

Coral Protocol

Coral positions itself as an open, decentralized collaboration infrastructure that not only standardizes agent communication and coordination but also builds in trust and payments, aiming to underpin an economic layer for an Internet of

Agents. If ANP emphasizes identity and semantics, Coral adds mechanisms for secure team formation and value exchange, pushing beyond communication into accountability and incentive alignment (Georgio et al., 2025).

Purpose: Coral provides an open substrate for agent communication plus trust and programmable payments, so agents can form teams, exchange value, and complete multi-party workflows in permissionless settings (Georgio et al., 2025).

Technical Basis: Coral introduces standardized message formats and orchestration primitives, plus trust mechanisms (reputation/attestations) and transaction rails for agent-to-agent payments. Its architecture targets broad compatibility and vendor neutrality and is discussed in the agent-oriented software engineering community as a candidate building block for secure team formation (Georgio et al., 2025; Rodriguez et al., 2026).

Adoption Status: As of 2025, Coral is documented in an academic white paper with an emerging OSS (Open Source Software) community; demonstrators emphasize Internet-scale collaboration scenarios and economic alignment via programmable payments (Georgio et al., 2025).

Security & Privacy: By coupling communication with reputation/attestation and payments, Coral aims to enhance accountability (e.g., audit trails, reduced Sybil/spam risk). Concrete guarantees depend on deployments (key management, custody, attestations) and on how Coral integrates with identity layers (DIDs, enterprise IAM) (Georgio et al., 2025).

Critical Perspective/Contrast: Coral complements ANP's identity/semantics with economic trust—a design choice that may be unnecessary in closed enterprises (A2A settings) but crucial in open ecosystems where incentives and accountability must be protocolized. This aligns with evidence from MAS literature that reliable cooperation in open systems often requires incentive and reputation mechanisms alongside secure communication (Park et al., 2023).

Comparative Synthesis & Analysis

Scope & Layering: In synthesis, MCP standardizes agent↔tool access (Anthropic, 2024), while A2A standardizes agent↔agent workflow life cycles (discovery, task life cycle, multimodal artifacts) (Google, 2025a, 2025b). ACP functions as a thin transport substrate (IBM Research, 2025). ANP and Coral emphasize permissionless discovery, decentralized identity, and economic trust (Georgio et al., 2025). In enterprise practice, A2A + MCP provides a cohesive stack for governed collaboration with deep tool reach; ACP can serve as a thin wire format beneath. In open ecosystems, ANP/Coral supply the primitives for permissionless discovery and economically accountable cooperation (Anthropic, 2024; Google, 2025a, 2025b; Agent Communication Protocol, 2025; Agent Network Protocol, 2025a; Georgio et al., 2025). The need for standard benchmarks and formal evaluations is highlighted in recent surveys (Tran et al.,

2025), and MCP security analyses further underline outstanding hard problems in runtime governance.

Enterprise vs. Decentralized: A2A's design centers on organizational policy, IAM, and auditability, offering predictability across vendor estates. ANP/Coral reject centralized control, substituting cryptographic identity (DIDs), verifiable credentials, and reputation/payment to establish trust across unknown counterparties. The trade-off is governance clarity and tooling maturity (A2A) versus openness and cross-boundary composability (ANP/Coral) (Google, 2025a, 2025b; Agent Network Protocol, 2025a; Georgio et al., 2025).

Security & Privacy Postures: As agentic AI systems gain deeper access to enterprise infrastructure and sensitive data, the security and privacy posture of their underlying communication protocols becomes critical. The following emerging protocols illustrate diverse approaches to identity, authorization, and secure collaboration.

- **MCP:** Delegates identity and authorization to the server boundary using a client–server architecture. While MCP facilitates tool access via JSON/RPC, it lacks built-in identity life-cycle management, making least-privilege enforcement and sandboxing essential as agents gain direct system access (Anthropic, 2024; The Verge, 2025).
- **A2A:** Embeds policy-gated capability access and authenticated task/artifact exchange within enterprise IAM frameworks. It supports secure multi-agent workflows through signed Agent Cards, gRPC (Google Remote Procedure Call)-based communication, and integration with existing cloud-native architectures (Google, 2025a).
- **ACP:** Leverages web security primitives such as TLS, OAuth, and JWT for agent interoperability. While ACP provides a lightweight RESTful interface, it defers fine-grained data minimization and access control to service implementers, raising concerns about consistency across deployments (Agent Communication Protocol, 2025; IBM Research, 2025).
- **ANP:** Implements DID-based authentication and end-to-end encryption, shifting the operational burden to key life-cycle management. ANP supports decentralized agent discovery and secure communication, aligning with privacy-by-design principles for open agentic networks (Agent Network Protocol, 2025a).
- **Coral:** Augments open agent networks with reputation scoring, attestations, and on-chain payments, enhancing accountability and deterring abuse. Coral's infrastructure supports multi-agent orchestration and monetized collaboration, positioning it as a foundational layer for the emerging Internet of Agents (Georgio et al., 2025).

Maturity & Trajectory: MCP is moving from developer pilots to OS/platform integrations; A2A shows growing enterprise backing and tooling; ACP is stabilizing as a Linux Foundation effort aligned with interop; ANP/Coral are earlier but articulate the architectural requirements for an Internet of Agents. Given their complementary scopes, convergence is likely—for example, A2A for enterprise workflows, implemented over HTTP transports (ACP-like), with MCP adapters for tools—while gateways bridge enterprise "islands" to decentralized networks (ANP/Coral) when open collaboration or marketplace dynamics are required (The Verge, 2024, 2025; Google, 2025a, 2025b; Agent Communication Protocol, 2025; Agent Network Protocol, 2025a; Georgio et al., 2025). Peer-reviewed MAS work supports this layered view: earlier agent communication languages (KQML (Knowledge Query and Manipulation Language), FIPA-ACL) provided semantics for messages, while practical adoption required surrounding infrastructure and governance—much as today's protocols require identity, security, and discovery layers to succeed (Finin et al., 1994; FIPA, 2002; Park et al., 2023; Wu et al., 2024).

Discussion and Outlook

The rapid evolution of agentic frameworks and protocols indicates that AI is moving toward more modular, networked intelligence. Instead of monolithic AI systems, we are likely to see ecosystems of AI services—*agents*—each specialized but able to communicate and coordinate. This recalls the trajectory of software in general (from mainframes to microservices), suggesting that breaking tasks into interoperating parts yields greater scalability and flexibility. By standardizing how agents talk to each other and to tools, the AI community is laying the groundwork for an "*Internet of Agents*" where knowledge and capabilities can be dynamically composed across boundaries.

However, realizing this vision comes with many challenges, security is paramount—as agents gain autonomy and the ability to take actions (especially with protocols such as MCP that interface with real data and systems), robust safeguards, and permissioning are needed to prevent errors or misuse. Protocols such as ANP's use of DIDs and Coral's trust layer are initial steps, but questions remain on how to authenticate agents, prevent malicious actors, and handle failure modes. Standard maturity is another issue—as of 2025, A2A and ACP are still drafts, and it will take time for consensus to form on their final form. Competing interests of big tech companies versus open communities need to be balanced in governance. We may see some convergence or bridging—for example, future updates might align ACP's message format with A2A's task schema, or define interoperability between A2A and ANP (perhaps via gateway agents that speak both).

As agentic AI systems continue to evolve toward greater autonomy, adaptability, and collaborative intelligence, several critical research avenues warrant focused exploration (Tran et al., 2025):

1. **Standardized Benchmarks for Agent Orchestration:** There is a pressing need for comprehensive benchmarking frameworks that evaluate agent

orchestration across key dimensions such as latency, operational cost, and fault tolerance. These benchmarks should support reproducibility and enable comparative analysis across heterogeneous agentic architectures and deployment contexts.

2. **Failure-Mode Taxonomies and Automated Recovery Patterns:** Multi-agent workflows are inherently complex and prone to diverse failure modes. Developing practical taxonomies of these failure scenarios—alongside automated recovery mechanisms informed by agent introspection, redundancy, and inter-agent signaling—will be essential for building robust and resilient agentic systems.
3. **Lightweight Protocols for Edge and IoT Agents:** As agentic intelligence extends to edge and IoT environments, there is a growing demand for ultra-lightweight coordination and communication protocols. These protocols must accommodate low computational overhead and minimal latency, enabling real-time decision-making and decentralized control in resource-constrained settings.
4. **Privacy-Preserving Multi-Agent Collaboration:** Ensuring privacy in multi-agent interactions is paramount, particularly in domains involving sensitive data. Research should focus on federated learning, secure multi-party computation (SMPC), and other privacy-preserving techniques that allow agents to collaborate and learn jointly without compromising data confidentiality or user trust.
5. **Governance Models for Accountability and Certification:** As agentic systems become increasingly embedded in critical infrastructure and decision-making processes, robust governance frameworks are needed to ensure accountability, transparency, and ethical compliance. This includes mechanisms for agent certification, auditability, and alignment with regulatory standards across sectors such as healthcare, finance, and public services.

As agentic AI systems increasingly assume operational control over business processes and facilitate cross-boundary data exchange, regulatory frameworks must evolve to ensure responsible deployment and oversight. Drawing parallels from algorithmic trading and high-frequency systems, regulators should consider implementing policy mechanisms that promote transparency, accountability, and safety in agentic environments. These include:

1. Mandatory audit logging to enable traceability of agent decisions and interactions, supporting post-hoc analysis and forensic investigation in the event of system failures or unintended outcomes.
2. Defined fallback behaviors that ensure predictable and safe system responses under uncertainty or failure conditions, particularly in safety-critical domains.
3. Operator accountability frameworks that delineate human oversight responsibilities, clarify liability boundaries, and support informed intervention when necessary.

4. Certification protocols for agentic stacks, encompassing both technical validation and ethical compliance, to ensure agents meet minimum standards before deployment in sensitive or regulated sectors.

Such policy levers will become increasingly essential as agentic systems transition from experimental prototypes to integral components of enterprise infrastructure, public services, and autonomous decision-making pipelines.

Another consideration is performance and scaling. Agentic workflows can be resource-intensive; multiple LLMs chatting can incur high compute costs and latencies. Research into more efficient communication (e.g., sending compressed context or using specialized lighter models for inter-agent dialogue) will be important. There is also the question of emergent behavior—with many autonomous agents interacting, unexpected behaviors (positive or negative) could emerge, echoing prior multi-agent phenomena such as cooperation, competition, or collusion. Common failure modes include agent unresponsiveness, inconsistent shared state, message loss, and cascading task failure. Practically, these require (i) monitoring and heartbeat semantics, (ii) agent re-assignment and idempotent task design, and (iii) circuit-breaker patterns to halt runaway costs. Compute and latency scale roughly with the number of LLM invocations and size of shared context; preliminary community benchmarks indicate nonlinear latency increases as agent graph depth and parallelism grow, reinforcing the need for lightweight inter-agent encodings (e.g., compressed summaries, distilled intermediary models) for large deployments. This requires careful design and possibly new evaluation methodologies, such as the survey by Tran et al. (2025) provides a framework for analyzing multi-agent collaboration mechanisms.

From an application standpoint, we are already seeing practical deployments of these ideas. For example, in software engineering, one can imagine an AI pair programmer agent working with a testing agent and a documentation agent—some early tools already attempt this kind of trio. In business processes, an "order processing" agent might coordinate with an "inventory" agent and a "shipping" agent, all via A2A or ACP, to handle an e-commerce order end-to-end with minimal human oversight. Each agent could be built by a different vendor (one by SAP, one by Oracle, etc.), yet they can speak a common language. This interoperability is precisely what A2A's industry backers are aiming for. On the open internet, one can envision agent marketplaces where, say, a small business's AI agent can hire a search engine optimization agent or a graphic design agent on the fly, pay it via Coral, and integrate the results—essentially AI-as-a-service interactions happening autonomously.

Looking ahead, the adoption of agentic systems is poised to unfold in overlapping phases, guided by the maturation of standards and the resolution of key technical and governance challenges. In the near term (2025–2027), we anticipate a period of consolidation and early enterprise integration, where protocols such as MCP and A2A will mature from pilot demonstrations into foundational elements of corporate AI infrastructure, supported by growing tooling and vendor compliance. This will be followed by a cross-boundary expansion phase (2027–2029), as gateways and interoperability layers bridge enterprise-centric stacks (A2A/MCP) with decentralized networks (ANP/Coral), enabling dynamic agent teams to form across organizational silos and the open web. By the end of the decade (2030+), these developments are likely to culminate in the emergence of a pervasive Internet of Agents, a fluid ecosystem where certified, specialized agents from diverse providers can securely discover, negotiate, and collaborate on complex goals with minimal human oversight, ultimately realizing the vision of a scalable, economically aligned artificial collective intelligence.

In conclusion, the landscape of agentic AI in 2025 is defined by two parallel trends: the creation of sophisticated agents and multi-agent workflows that push the envelope of what AI can do autonomously, and the development of standards and protocols that ensure these agents can work together across ecosystems. This review has outlined how frameworks such as AutoGen, LangChain, and others allow us to build the agents and workflows, while protocols such as MCP, A2A, ACP, ANP, and Coral provide the communication fabric to connect everything. It is an exciting, formative time—the analogies to the early days of the internet or the PC industry are apt. Many ideas are being tested; some will become foundational. For AI researchers and engineers, staying abreast of these emerging standards is crucial, as today's experiments could become tomorrow's infrastructure for AI collaboration. The ultimate promise is an *agent-based computing paradigm* where distributed AI systems seamlessly cooperate to tackle complex goals, effectively functioning as an artificial collective intelligence. Realizing that promise will require continued innovation in design, rigorous evaluation, and above all, collaboration—appropriately, the very thing these multi-agent technologies are designed to enhance.

References

Agent Communication Protocol. (2025). Agent Communication Protocol: Welcome. https://agentcommunicationprotocol.dev/

Agent Network Protocol. (2025a, July 24). ANP Technical White Paper. https://agent-network-protocol.com/specs/white-paper.html

Agent Network Protocol. (2025b). ANP Technical White Paper (English page). https://agentnetworkprotocol.com/en/specs/01-agentnetworkprotocol-technical-white-paper/

Anthropic. (2024, November 25). Introducing the Model Context Protocol (MCP). https://www.anthropic.com/news/model-context-protocol

Ehtesham, A., Singh, A., Gupta, G. K. & Kumar, S. (2025). A Survey of Agent Interoperability Protocols: MCP, ACP, A2A, and ANP. arXiv:2505.02279 [cs.AI]

Finin, T., Fritzson, R., McKay, D. & McEntire, R. (1994). KQML as an agent communication language. *Proceedings of the Third International Conference on Information and Knowledge Management* (CIKM '94) (pp. 456–463). ACM. https://doi.org/10.1145/191246.191322

FIPA. (2002). FIPA ACL message structure specification (Standard No. SC00061G). http://www.fipa.org/specs/fipa00061/SC00061G.pdf.

Georgio, R. J., Forder, C., Deb, S., Carroll, P. & Gürcan, Ö. (2025). The Coral Protocol: Open infrastructure connecting the Internet of Agents (arXiv:2505.00749 [cs.MA]). arXiv. https://arxiv.org/abs/2505.00749

Google. (2025a, April 9). Announcing the Agent2Agent protocol (A2A). Google Developers Blog. https://developers.googleblog.com/en/a2a-a-new-era-of-agent-interoperability/

Google. (2025b, August). Agent2Agent protocol (A2A) is getting an upgrade. Google Cloud Blog. https://cloud.google.com/blog/products/ai-machine-learning/agent2agent-protocol-is-getting-an-upgrade

IBM. (2023). What are agentic workflows? IBM Think Blog.

IBM Research. (2025, May 28). An open-source protocol for AI agents to interact. https://research.ibm.com/blog/agent-communication-protocol-ai

Koundinya, S. (2025, March 11). IBM's ACP is an "Extension" of Anthropic's MCP. *Analytics India Magazine* (AI News).

Li, Guohao, Hasan Abed Al Kader Hammoud, Hani Itani et al. (2023). Camel: Communicative agents for "mind" exploration of large language model society. *Advances in Neural Information Processing Systems*, 36, 51991–52008.

Ng, A. (2024). Four AI agent strategies that improve GPT-4 and GPT-3.5. The Batch, deeplearning.ai. Available at: https://www.deeplearning.ai/the-batch/how-agents-can-improve-llm-performance/

Park, J. S., O'Brien, J. C., Cai, C. J., Morris, M. R., Goodman, N., Shah, S., … Bernstein, M. S. (2023). Generative agents: Interactive simulacra of human behavior. *Proceedings of the 36th Annual ACM Symposium on User Interface Software and Technology* (UIST '23) (pp. 146–158). https://doi.org/10.1145/3586183.3606774.

Rodriguez, S., Bairy, A., Baldoni, M., Benjamin, P., Blessing, C., Brandstetter, N., ... Yang, Y. (2026). Engineering the next generation of multi-agent systems: A community roadmap from EMAS 2025. *In: Engineering Multi-Agent Systems* (*EMAS 2025*) (Vol. 16407, pp. 238–258). Springer Cham. Advance online publication. https://doi.org/10.1007/978-3-032-18011-7_14

Shen, Y., Song, K., Tan, X., Li, D., Lu, W. & Zhuang, Y. (2023). HuggingGPT: Solving AI tasks with ChatGPT and its friends in hugging face. *Proceedings of NeurIPS 2023.*

Shinn, N., Cassano, F., Berman, E., Gopinath, A., Narasimhan, K., & Yao, S. (2023). Reflexion: Language Agents with Verbal Reinforcement Learning. arXiv preprint arXiv:2303.11366. DOI: https://doi.org/10.48550/arXiv.2303.11366

Surapaneni, R., Jha, M., Vakoc, M., & Segal, T. (2025, April 9). Announcing the Agent2Agent Protocol (A2A). Google Developers Blog.

The Verge. (2024, November 25). Anthropic launches tool to connect AI systems directly to datasets. https://www.theverge.com/2024/11/25/24305774/anthropic-model-context-protocol-data-sources

The Verge. (2025, May). Windows is getting support for the "USB-C of AI apps". https://www.theverge.com/news/669298/microsoft-windows-ai-foundry-mcp-support

Tran, K. T., Dao, D., Nguyen, M. D., Pham, Q. V., O'Sullivan, B., & Nguyen, H. D. (2025). Multi-agent Collaboration Mechanisms: A Survey of LLMs. arXiv preprint arXiv:2501.06322.

Völker, D., & Oberhauser, M. (2024). Agentic AI: The Next Frontier in Autonomous Enterprise Systems (ESCP Impact Paper). SSRN. https://doi.org/10.2139/ssrn.5467412

Wang, L., Ma, C., Feng, X., Zhang, Z., Yang, H., Zhang, J., … Wen, J.-R. (2024). A survey on large language model based autonomous agents. *Frontiers of Computer Science*, 18(6), 186345. https://doi.org/10.1007/s11704-024-40231-1

Wu, Q., Bansal, G., Zhang, J., Wu, Y., Li, B., Zhu, E., … Burger, D. (2024). AutoGen: Enabling next-gen LLM applications via multi-agent conversation. *Proceedings of the Conference on Language Model 2024* (LLM Agents Workshop at ICLR 2024). https://arxiv.org/abs/2308.12241

Xi, Z., Chen, W., Guo, X. et al. (2025). The rise and potential of large language model based agents: A survey. *Sci. China Inf. Sci.*, 68, 121101. https://doi.org/10.1007/s11432-024-4222-0

CHAPTER

4

Artificial-Neural-Network-Based Correlation for Predicting Oil Formation Volume Factor Using White-Box Technique

Caspar Daniel Adenutsi[1*], Tahiru Abdul-Mumin[2], Jeffery Godwin Ackah[2], Joshua Nsiah Turkson[3,4] and Lian Wange[5,6]

[1] Department of Petroleum Engineering, Faculty of Civil and Geo-Engineering, Kwame Nkrumah University of Science and Technology, Kumasi, Ghana

[2] Department of Chemical Engineering, Faculty of Chemical and Mechanical Engineering, Kwame Nkrumah University of Science and Technology, Kumasi, Ghana

[3] Petroleum Engineering Department, Faculty of Geoscience and Petroleum Engineering, Universiti Teknologi PETRONAS, 32610 Bandar Seri Iskandar, Perak Darul Ridzuan, Malaysia

[4] Centre of Reservoir Dynamics (COReD), Institute of Sustainable Energy & Resources (ISER), Universiti Teknologi PETRONAS, 32610 Bandar Seri Iskandar, Perak Darul Ridzuan, Malaysia

[5] Zhenhua Oil Co., Ltd., Beijing 100031, China

[6] State Key Laboratory of Oil and Gas Reservoir Geology and Exploitation, Southwest Petroleum University, Chengdu 610500, China

Introduction

Oil formation volume factor (B_o or oil FVF) is an important crude oil property that relates crude oil volume in the subsurface to crude oil volume at the surface. This crude oil property is needed in many reservoir-engineering applications such as well testing, reservoir simulation, material balance equation, and production engineering—as production facility design and future production performance prediction. Particularly, oil FVF is needed to estimate oil-in-place and surface production rates accurately. Inaccurate determination of oil FVF

*Corresponding author: casdanad@hotmail.com

DOI: 10.1201/9781003506348-4

would result in errors in the computation of surface volumes of oil-in-place and errors in the estimation of production rates. This would have dire economic consequences for field operators in the petroleum industry. As a result, field operators go to great lengths to sample representative crude oil samples for careful measurement of this property in the laboratory (Elsharkawy, 2003). However, laboratory measurement of oil FVF is expensive, arduous, and can be limited by unrepresentative fluid samples.

To overcome these challenges, empirical correlations were developed for estimating oil FVF (Ayoub et al., 2022; Mahdiani & Norouzi, 2018; Rashidi et al., 2021). However, many of these correlations are based on region-specific data and often fail to generalize accurately for crude oils from other regions. For instance, with a mean absolute percentage error (MAPE) of 1.17%, Standing (1947) provided graphical correlations for determining the oil FVF using 105 experimental data points on 22 hydrocarbon systems in North America (California). The experiment was conducted in a laboratory using oil samples with an API gravity ranging from 16.5 to 63.8. Similarly, using linear and nonlinear regressions, a correlation for the oil FVF was developed by Glaso (1980) using 41 datasets from the North Sea. He adjusted the correlation to account for the impact of nonhydrocarbon components. The average percent deviation was reported to be –0.41%. Al-Marhoun (1988) also developed an empirical correlation to predict oil FVF by employing nonlinear multiple regression on 160 oil samples from the Middle East with an API gravity range of 19.40–44.6. A mean absolute error (MAE) of 0.88% was reported. The correlation coefficient of the model was 0.997, with a MAPE of 3.66%. An oil FVF correlation based on 119 data points was reported by Abdul-Majeed & Salman (1988). Except for the newly computed coefficients, their model is similar to that of Al-Marhoun (1988). The authors employed a nonlinear multiple regression technique to obtain the correlation and reported an MAE of 1.40%. A new empirical correlation was also produced by Kartoatmodjo & Schmidt (1991) utilizing a substantial quantity of data from reservoirs worldwide, including those in Indonesia, North America, the Middle East, and Latin America. They developed the correlation using data from two distinct databases. While the second database, which contained 998 data points, was especially utilized to validate the correlation, the first database with 740 distinct crude oil samples (5392 data points) was primarily used to create correlations. The Vazquez & Beggs (1980) oil FVF correlation served as the foundation for their correlation development. The range of API gravity of the samples was 14.4–59.0, and the authors reported a MAPE of 2.025%. Dokla & Osman (1992) proposed a correlation for estimating the new coefficients for oil FVF for crude oil (with API gravity of 28.21–40.3) from the United Arab Emirates using 51 data points. The reported absolute average percentage relative error (AAPRE) was 1.225%. The correlation of Dokla & Osman (1992) outperformed the pre-existing correlations (Al-Marhoun, 1988; Glaso, 1980; Standing, 1947). Petrosky Jr & Farshad (1993) provided a novel correlation for calculating the oil FVF utilizing 90 datasets with 128 data points from laboratory analysis.

They employed a nonlinear regression model to match measured crude oil date from the Gulf of Mexico hydrocarbon system. An average percent deviation of –0.01% and MAE of 0.64% were reported. An algorithm for predicting the oil FVF utilizing 93 datasets from Malaysian oil reservoirs was published by Omar & Todd (1993). The API gravity was 26.6–53.2. Their empirical model was similar to that of Standing (1947). However, the proposed correlation was enhanced by newly computed constants using nonlinear and linear regression analysis. The developed correlation yielded an AAPRE of 1.44% and an MAE of 1.44%. Sulaimon et al. (2014) reported a correlation for calculating oil FVF using the group method of data handling technique as a family of inductive algorithms. They employed 93 pressure–volume–temperature (PVT) data points from Malaysian crude oil. The authors reported an MAE of 0.976%. There are several empirical correlations developed and the discussions in this work are not exhaustive. These correlations can be referred to in the classical review by Al-Shammasi (2001). Moreover, empirical correlations reported after 2001 can be found in the works of Ayoub et al. (2022) and Rashidi et al. (2021). It is evident that many of the empirical correlations were developed based on data from crude oil samples from specific geographical areas and tend to perform unsatisfactorily when applied to crude oils from different geographical locations. Due to this, few scholars attempted to develop correlations based on data from crude oil samples worldwide (Abdul-Majeed & Salman, 1988; Al-Shammasi, 1999, 2001; Ayoub et al., 2022; Mahdiani & Norouzi, 2018).

Due to the superb performance of machine learning in solving various tasks in the petroleum industry (Hassaan et al., 2024; Onwuchekwa, 2018; Otchere et al., 2021; Sulaimon et al., 2024; Wang et al., 2023; Wang et al., 2021; Wang et al., 2022), the technique has also been employed to estimate oil FVF to surmount the challenges posed by the laboratory and empirical methods. Oil FVF has been predicted by several algorithms such as artificial neural network (ANN) (Gharbi et al., 1999; Moghadam et al., 2011), Elman neural network (Kohzadvand et al., 2024), alternating conditional expectation (Malallah et al., 2006), and support vector machine (El-Sebakhy, 2009). Other techniques employed in estimating oil FVF include multilayer perceptron (MLP), radial basis function, least square support vector machine, committee machine intelligent system (Khoukhi, 2012), and gradient boosting decision tree (Kharazi Esfahani et al., 2023). Some scholars also used hybrid models such as functional network coupled with particle swarm optimization (FN-PSO), fuzzy C-mean adaptive neuro-fuzzy inference system, and grid partition adaptive neuro-fuzzy inference system among others (Salehinia et al., 2016; Seyyedattar et al., 2020; Tariq et al., 2021). The machine learning models discussed in the earlier text performed very well with a coefficient of determination (R^2) ranging from 0.95 to 0.99. However, these models employed the black-box technique.

Several researchers employed ANN as a stand-alone model in the estimation of oil FVF (Al-Marhoun & Osman, 2002; Boukadi et al., 1999; Dutta & Gupta, 2010; Elkatatny & Mahmoud, 2018; Gharbi et al., 1999; Goda et al., 2003;

Moghadam et al., 2011). Boukadi et al. (1999) proposed a correlation based on 92 data points for crude oil samples from Oman. The proposed correlation recorded an R^2 of 0.95 and an absolute average percent deviation (AAPD) of 1.6%. In the same year, Gharbi et al. (1999) developed a correlation based on 5200 data points acquired worldwide. Their correlation outperformed that of Boukadi et al. (1999) in terms of R^2. They reported an R^2 of 0.9875 and AAPD of 1.97%. Al-Marhoun & Osman (2002) brought forth a correlation for a Saudi crude using 282 data points and reported an AAPD of 0.5116%. Based on 163 data points from the Middle East, Goda et al. (2003) also proposed a correlation that recorded an R^2 and AAPD of 0.9981 and 0.03070%, respectively. Dutta & Gupta (2010) used 1852 data points from India to develop a correlation that yielded an R^2 of 0.9890 and an AAPD of 1.779%. Using 218 data points from Iran, Moghadam et al. (2011) developed a correlation and reported an R^2 of 0.997 and AAPD of 0.53%. From the performance metrics reported by these authors, it can be deduced that these ANN models performed extremely well. However, these models (Al-Marhoun & Osman, 2002; Dutta & Gupta, 2010; Gharbi et al., 1999; Moghadam et al., 2011) are black-box models and the empirical equations are not available for comparative studies.

Despite the remarkable accuracy reported, the developed models suffer from transparency and interpretability due to their "black-box" nature. Black-box models operate as closed systems, making it challenging to derive meaningful physical insights or validate the results against established theoretical principles. Furthermore, black-box approaches are often prone to overfitting, especially when the training dataset is not sufficiently representative of the underlying physical phenomena. To overcome the roadblock, subsequent studies developed "white-box" models based on ANN to predict oil FVF but the models were limited to bubble point pressure (Elkatatny & Mahmoud, 2018; Gouda & Attia, 2022; Rammay & Abdulraheem, 2017; Wood & Choubineh, 2019). White-box models address the issues faced by their counterparts by offering greater explainability and interpretability, enabling stakeholders to uncover physical correlations within the data and improve the robustness of predictions. For instance, Rammay & Abdulraheem (2017) used the ANN white-box technique to develop a correlation based on 166 data points for crude oil from Pakistan to predict oil FVF at the bubble point. The authors reported an average absolute percentage error (AAPE) of 0.4975% and R^2 of 0.997. Elkatatny & Mahmoud (2018) collected 760 published data points on different crude oils worldwide to propose a white-box ANN-based empirical correlation to predict oil FVF at bubble point. The authors reported an R^2 of 0.99 and an MAE of 0.93%. Wood & Choubineh (2019) also employed the transparent open-box learning network algorithm to predict oil FVF at bubble points using published data in the literature. They reported a root mean square error (RMSE) of 0.03 and R^2 of 0.95 for the more evenly distributed dataset. Gouda & Attia (2022) used 600 datasets from Egyptian crude oils to predict oil FVF at the bubble point pressure by employing ANN white-box method. They reported an R^2 value of 0.974.

To address the key existing limitations of the current techniques, we introduce a novel empirical model for predicting the oil formation volume factor (B_o or oil FVF) using an ANN white-box technique, considering not only pressures at the bubble point but also pressures below the bubble point. As mentioned earlier, traditional empirical correlations, often developed with region-specific data, lack generalizability and may perform inconsistently across different crude oil types. Additionally, many machine learning models function as black boxes, offering limited interpretability, which can hinder their practical application in understanding physical relationships. The few white-box models available are also limited to FVF at bubble point.

Our study aims to overcome these limitations by developing a novel white-box ANN model using 221 experimental data points from crude oils across the Mediterranean, Africa, the Persian Gulf, the North Sea, and Malaysia to ensure its broader applicability. The ANN model developed in this study extracts weights and biases, enhancing transparency and allowing for better insights into the factors influencing oil FVF. The detailed development process is discussed, and the performance of the developed model is rigorously compared with widely used industry correlations. The proposed model offers improved accuracy and generalizability in estimating oil FVF at and below the bubble point pressure, providing more reliable results for reservoir and production engineering applications.

Methodology

The methodology used in this study is divided into three steps: data acquisition, features selection, and ANN model development.

Data Acquisition and Description

A total of 223 PVT datasets were sourced from various published works (De Ghetto et al., 1994; Omar & Todd, 1993; Soliman et al., 2020). The build-in functions "*rmmissing*" and "*rmoutliers*" in MATLAB (R2021a) were used clean (remove missing data entries and outliers, respectively). The "*rmoutliers*" function uses median absolute deviation (MAD) method to detect any outliers present in the data. This reduced the total number of datasets to 221. Of these, 128 datasets originate from the Mediterranean Basins, Africa, the Persian Gulf, and the North Sea (De Ghetto et al., 1994; Soliman et al., 2020), while the remaining 93 datasets pertain to Malaysian crude oil, as reported by Omar & Todd (1993). The dataset encompasses key variables such as gas solubility, R_s (scf/STB); reservoir temperature, T (°F); API oil gravity (°AIP); pressure, P (psia); gas gravity, γ_g; the oil-specific gravity, γ_o; and oil FVF, B_o (bbl/STB). The distribution of these variables is illustrated through histogram plots in Figure 1.

Out of the 221 datasets, 155 were allocated for training the ANN model, 33 were reserved for testing, and the remaining 33 were used for cross-validation

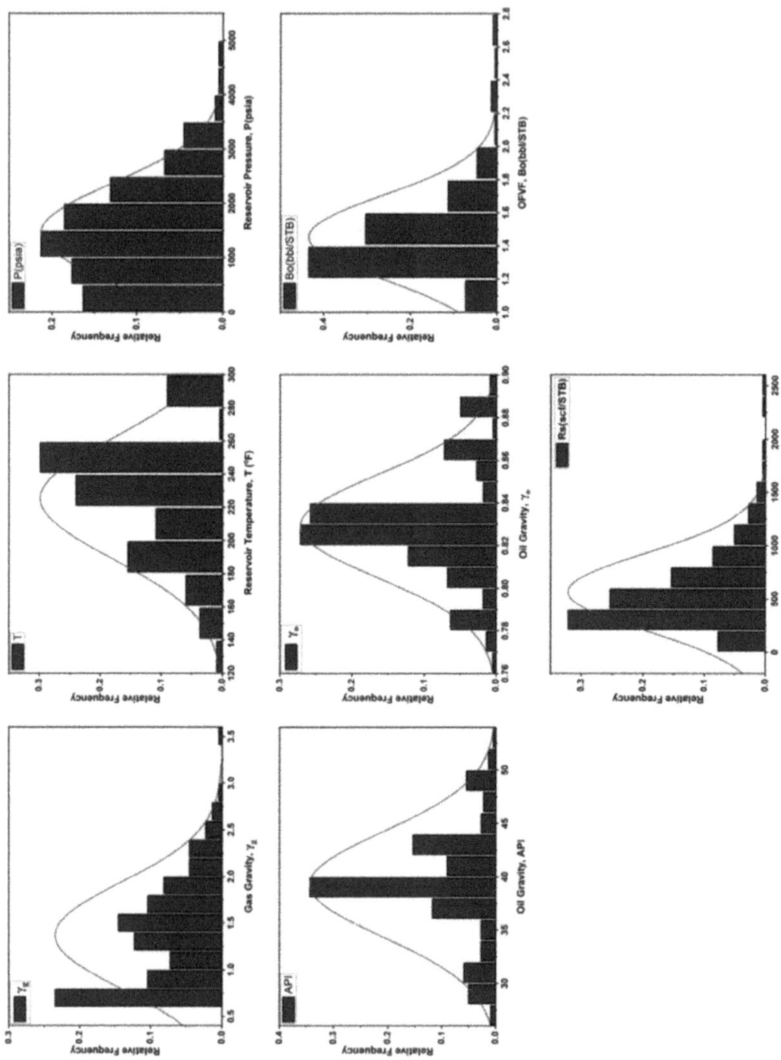

Figure 1: Frequency histograms of the dataset (221 data points).

to verify the relationships established during the training process. Statistical descriptions of the input data and the target variable, including the maximum, minimum, average, range, and standard deviation (st. dev.), are presented in Table 1.

Table 1: Statistical description of all the data

PVT Properties	Minimum	Maximum	Average	Range	St. Dev.	Skewness	Kurtosis
Pressure	55	4975	1452.70	4920	916.2202	0.6788	3.4001
Oil FVF	0.2150	2.7130	1.4463	2.4980	0.2655	1.3778	9.6405
Temperature	125.00	296.00	225.3077	171.00	36.0621	–0.2959	2.8663
Gas gravity	0.612	3.4445	1.3670	2.8325	0.5647	0.6550	3.0423
API oil gravity	26.600	53.200	39.2747	26.600	4.9501	–0.0604	3.3739
Gas solubility	24.000	2496.0	563.9548	2472	365.420	2.0274	9.2671

Features Selection

A critical factor that influences the performance of a machine learning model is the selection of features used during model training, ensuring that the most relevant features are chosen is essential for optimizing model accuracy and performance. Dimensionality reduction techniques are therefore considered a crucial step in data preprocessing. They help eliminate features that could negatively impact model outcomes (Trunk, 1979). Given that this study addresses a supervised machine learning problem, supervised feature selection techniques were employed. Typically, three approaches to supervised feature selection are available: filter, wrapper, and embedded methods. In this study, the filter method was applied, utilizing Pearson correlation (Figure 2) and Spearman's rho correlation (Figure 3) to visualize relationships and covariance between the predictor variables and the target variable (B_0). These correlations were used to assess the relationships between B_0 and variables such as R_s, γ_g, γ_o, P, T, and API, as illustrated in Figures 2 and 3. Based on the results of these analyses, the optimal features were selected for model training.

From Figure 2, a correlation value close to +1 indicates a strong positive relationship, –1 indicates a strong inverse relationship, and 0 suggests no relationship between the variables. In this study, two criteria were employed for feature selection: (1) feature–target correlation coefficients and (2) feature–feature correlation coefficients. The first criterion involved selecting features that exhibited a correlation value of 0.5 or higher with the target variable (B_0). Based on this, (R_s) and (T) were chosen, as they had correlation values of 0.90 and 0.57, respectively, with (B_0).

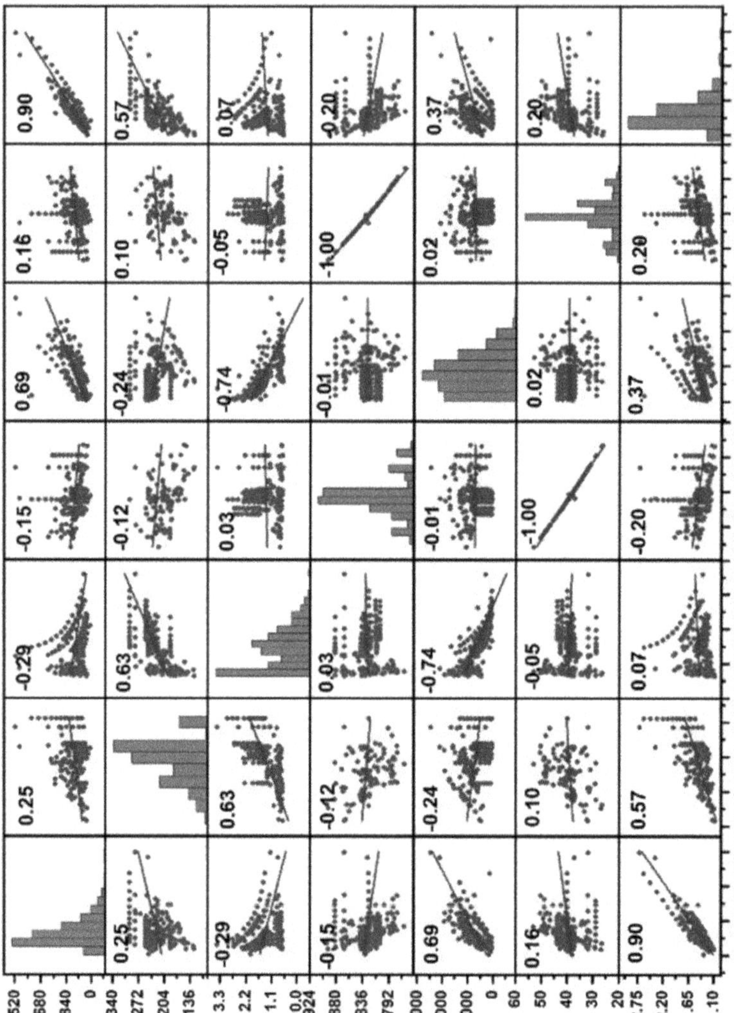

Figure 2: Pair plot distribution of the features and oil FVF.

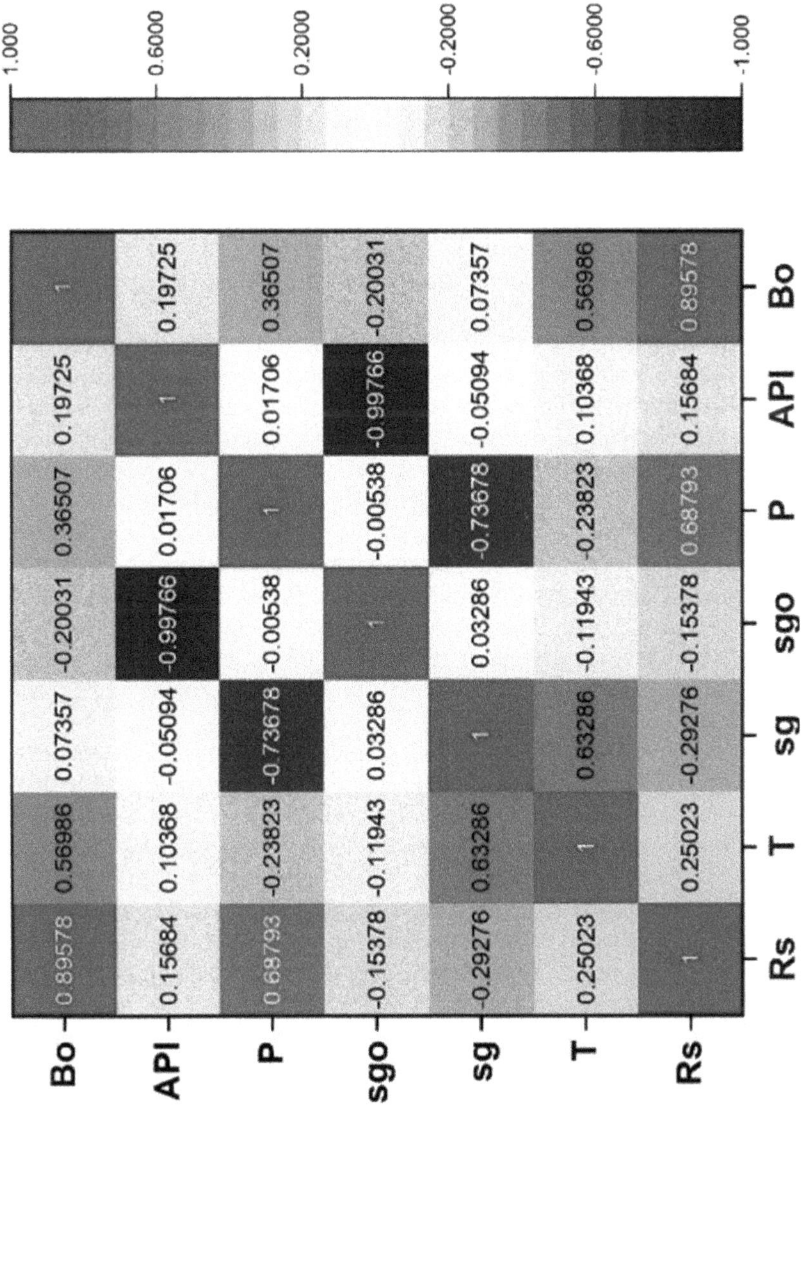

Figure 3: Heatmap plot of the Spearman rho correlation for all the data.

The second criterion was implemented to manage multicollinearity, which occurs when two or more predictor variables in a regression model are highly correlated. This situation can obscure the individual effects of independent variables on the dependent variable, making interpretation challenging. To address this, features were selected when their "feature-to-feature" correlation was above 0.7 or below –0.7. For instance, (γ_g) (denoted as *sg* in Figures 2 and 3) had a correlation value of –0.74 with *P* and was selected over *P*, particularly because R_s, which had a strong linear relationship with *P*, was already selected. Similarly, specific oil gravity γ_o (denoted as *sgo* in Figures 2 and 3) exhibited a strong inverse correlation with API, with a value of –1, and was selected over API. While multicollinearity may not significantly affect model accuracy, it can complicate model interpretation by making it difficult to discern the individual impact of independent features on the target variable.

The Pearson correlation coefficient and the coefficient of determination can be used to measure the strength of the linear relationship between variables that are normally distributed (Rashidi et al., 2021; Tabaaza et al., 2023; Turkson et al., 2024b). However, it will be impractical to assume that the variables involved have a linear relationship or are normally distributed. Hence Spearman's rho correlation method, which is a non-parametric statistical method that measures the nonlinear relationships between the independent features and the oil FVF, was employed. The Spearman rho correlation covariance matrix plotted as a heatmap (Figure 3) depicts the intensity and direction of correlation between the individual features. The significance of each correlation depends on the *p*-values which range from –1 to 1. Based on the "feature-to-target" criterion, R_s and *T* passed the selection criterion. Similarly, γ_o and γ_g passed the "feature-to-feature" criterion. Again, γ_o and γ_g were selected over *API* and *P*, respectively, due to the same reasons stated earlier. Thus, the Spearman rho correlation corroborates the results of the earlier feature selection method.

While the correlation-based feature selection or filter methods employed in this study provide valuable insights into linear and monotonic relationships, they may not fully capture intricate interdependencies between features in high-dimensional data. These interdependencies can mask subtle patterns that are crucial for understanding complex systems such as reservoir properties. Incorporating advanced feature selection techniques approaches such as embedded (e.g., lasso and ridge regression), intrinsic, wrapper (e.g., recursive and backward feature elimination), univariate methods (e.g., SelectKBest), etc. (Otchere et al., 2022a; Otchere et al., 2022b; Tabaaza et al., 2023; Turkson et al., 2024a) could further enhance the robustness of feature selection and model performance in future studies.

Development of the ANN Architecture and Empirical Correlation

MATLAB's Neural Net Fitting tool was used to create, train, validate, and test the networks. It was also used to generate the parameters' goodness of fit, which includes correlation coefficients and slope for the predicted output (B_o). The dataset was randomly divided into three 3 parts, that is, 70% for training, 15% for testing, and the remaining 15% for validation. The training process of the ANN model was done using the training data to generate the predicted results. During the training process, the error estimated between the results and the target aided the network in adjusting the weights of each input by the feedforward backpropagation algorithm. The feedforward backpropagation algorithm returns the error to the network and modifies the weights until calibration is achieved (Otchere et al., 2021). The testing dataset was used to measure how well a network has generalized and to halt training when generalization no longer improves. Finally, the performance of the network was assessed using the validation dataset.

The network comprises the input, hidden, and output layers. The gas solubility (R_s), gas-specific gravity (γ_g), oil-specific gravity (γ_o), and temperature ($\underline{T}$) were the independent variables fed to the input layer, while the output layer has the dependent variable oil FVF (B_o). Engelbrecht (2007) stated that a feedforward network with one hidden layer containing enough neurons can fit any finite input–output mapping problem. Based on that, a single hidden layer feedforward and back propagation network was employed in this study. Again, Li et al. (1995) proposed an equation named "seed neurons' number" that relates the number of data points used to train the model and the optimum number of neurons in a hidden layer. This equation has been accepted and adopted by several researchers (Agatonovic-Kustrin & Beresford, 2000; Ocampo et al., 2021) due to its performance. The "seed neurons' number" was used in this work to select the optimum number of neurons in the hidden layer, which was found to be 20.

The number of neurons in the hidden layer was selected based on the seed neurons' number as shown in Eq. 1 (Agatonovic-Kustrin & Beresford, 2000; Li et al., 1995):

$$\text{Number of neurons} = \frac{\sqrt{1+8n}-1}{2} \tag{1}$$

where n is the number of samples (221 for this study), resulting in 20.52 neurons, which was rounded down to 20. The architecture of the ANN model is shown in Figure 4.

The layers in the ANN architecture are shown in [X–N1–Y]. The input layer is represented by "X," whose value indicates the number of independent variables; the hidden layer is represented by "N1," whose number shows the number of neurons in that layer. "Y" denotes the output layer and its number denotes the number of the dependent variables. In this work, 4 independent variables were fed to the input layer, with 20 neurons in the hidden layer and 1 dependent variable at the output layer; thus, the ANN configuration for this work is [4–20–1]. For calculations between the input and the hidden layer, the hyperbolic tangent sigmoid transfer function (tansig) was used, while the pure

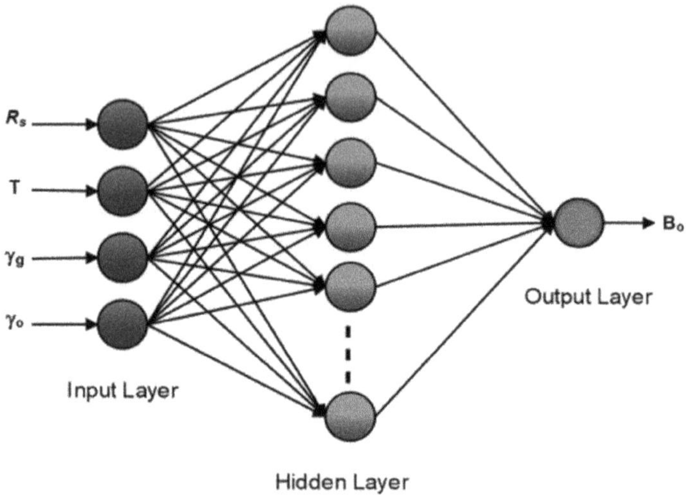

Figure 4: Architecture of the ANN model.

linear transfer function (purelin) was employed to calculate the output from the hidden layer to the output layer. These tansig and purelin transfer functions are shown in Eqs. 2 and 3, respectively (Akhgar et al., 2019):

$$f(x) = \frac{2}{1-e^{-2x}} - 1 \tag{2}$$

$$f(x) = x \tag{3}$$

Additional hyperparameters known as training algorithms have an impact on how well the ANN model learns and performs during training (Ehret et al., 2015). Because the Levenberg–Marquardt function is a widely used training function and one of the algorithms with profound learning capability and training performance (Abidoye et al., 2018), it was utilized to train the model along with the backpropagation technique. The backpropagation neural network technique was chosen because of its exceptional processing capabilities while handling massive amounts of data (Ehret et al., 2015). Its method, which also applies to complicated data, involves iteratively adjusting the weights and biases in response to the error value between the expected and desired outputs (Abidoye et al., 2018). It is also capable of analyzing and learning different types of embedded relationships found in the training set.

The ANN was a single hidden layer feedforward and backpropagation neural network. In backpropagation, the error vector is computed by taking the difference between the actual and target values. This error vector at the output layer is defined by a quadratic cost function shown in Eq. 4:

$$E_r = \frac{1}{2}\sum_{1}^{n}(y_{\mathrm{exp}} - y_{\mathrm{pred}})^2 \quad (4)$$

In Eq. 4, E_r is the error vector; y_{exp} represents the experimental values of the target variable, and y_{pred} is the predicted value. The estimated error (E_r) is then used with the weights, biases, and the momentum parameter to iteratively adjust the weights and biases between the actual and target outputs using Eq. 5 (Avseth & Mukerji, 2002):

$$\Delta w(t) = \eta \nabla E_r + \alpha \Delta w(t-1) \quad (5)$$

In Eq. 5, Δw is the updated weight at iteration t; $\Delta w(t-1)$ is the weight increment at iteration $(t-1)$, E_r is the estimated error between the predicted and actual output, α is the momentum parameter (<1), and η is the learning rate. Note that the momentum parameter α helps overcome slow convergence problems by allowing the iteration process to accelerate the movement toward consistent gradients, as this momentum can result in models that more effectively generalize to blind data (Chun et al., 2020). It also lessens oscillations and unpredictable behavior of the training process, leading to smoother paths toward the global minima (Chun et al., 2020).

Each iteration process updates the weights and the biases, which then improves the model by reducing the mean square error (MSE). This procedure continues until it reaches an epoch (iteration) where the error tolerance is satisfied, at which point it stops and logs the result as the best validation performance point along with the associated MSE. Figure 5 shows how the errors are looped back through the network to tune the weight. In the feedforward step, inputs are transmitted through neurons to estimate the output targets.

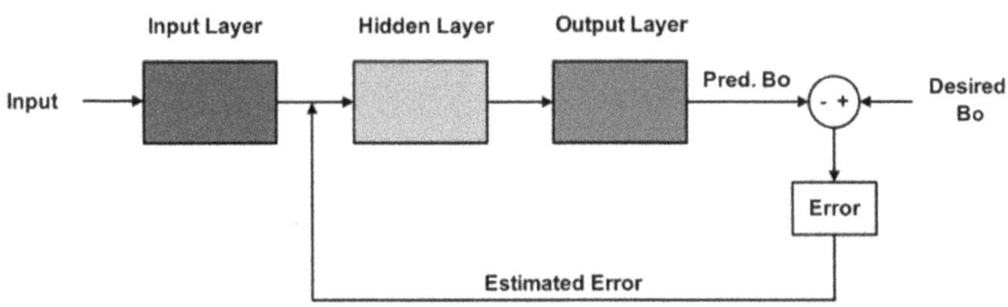

Figure 5: A backpropagation method showing how the error is looped back through the network for weight adjustment.

The number of error tolerance and iterations were regulated during the training process using goals and epochs as the stopping criteria, respectively. Epoch refers to the number of times all training sets are given to the network and goal is the maximum error tolerance that the produced network must meet. As a result, whenever the error goal or the maximum number of epochs is attained, training ends.

Steps for Developing the ANN-Based Correlation from Weights and Biases

Extracting the base equation using the weights and biases generated from the trained network is essential in model development. This comes after an ANN model has been successfully trained to acquire the weights and biases for a target variable. The process for extracting the equation is illustrated in the following steps.

Step 1. Normalization of the input data at the input layer: Normalization is a data scaling method that is typically used in neural network training. The output of a neural network can be significantly enhanced by using the right normalization procedure. Normalization scales the input and the target variables to have the lowest and highest values, respectively, mostly in the range of –1 to 1. In this study, the min–max normalization method was used where the input variables (R_s, γ_g, γ_o and T) were normalized at the input layer. Equation 6 displays the mapminmax mathematical expression (Abidoye et al., 2018)

$$y = \frac{(y_{max} - y_{min})\,(x - x_{min})}{(x_{max} - x_{min})} + y_{min} \tag{6}$$

In Eq. 6, y is the normalized value of the input variables; x is the input variable to be normalized which are R_s, γ_g, γ_o and T; x_{min} is the minimum value of the input variable; x_{max} is the maximum value of the input variable; y_{max} is the maximum normalized value which is equal to 1 and y_{min} is the minimum normalized value which is equal to –1.

Using Eq. 6, the normalized values of the input variables x are expressed in the following text:

$$R_{s(N)} = 0.000809061488673\ R_s - 1.019417475728155 \tag{7}$$

$$\gamma_{g(N)} = 0.706090026478376\ \gamma_g + 1.432127096204766 \tag{8}$$

$$\gamma_{o(N)} = 15.5163899147161\ \gamma_o - 12.887218045112775 \tag{9}$$

$$T_{(N)} = 0.011695906432748\ T - 2.461988304093563 \tag{10}$$

Step 2. Assignments of weights and biases at the hidden layer: Normalization of the input variables is followed by the assignment of weights and biases to the normalized variables. W and b in Table 2 represent the weights and biases that the network automatically generated. Using the following equation, the relationship is formed linearly:

$$E_{I,i,n} = \sum_{i=1}^{j} \left(W_{I,i,R_s} R_{s(N)} + W_{I,i,\gamma_g} \gamma_{g(N)} + W_{I,i,\gamma_o} \gamma_{o(N)} + W_{I,i,T}\, T_{(N)}\right) + b_{I,i} \tag{11}$$

In Eq. 11, W_{I,i,R_s}, W_{I,i,γ_g}, W_{I,i,γ_o} and $W_{I,i,T}$ represent the weights allocated, respectively, to R_s, γ_g, γ_o and T in association with a particular neuron (i) at the hidden layer (I); $b_{I,i}$ is the bias at a particular hidden layer (I) assigned to the neuron (i); $R_{s(N)}$, $\gamma_{g(N)}$, $\gamma_{o(N)}$ and $T_{(N)}$ are the nth normalized values of R_s, γ_g, γ_o and T, respectively. $E_{I,i,n}$ refers to the nth sum of the weighted normalized values, based on the weight assignment in association with i^{th} neuron at a particular hidden layer (I); j is the final neuron count in the network. The weights and biases of the ANN model generated from this study are displayed in Table 2.

Table 2: Weights and the biases of the ANN model in this study

Neurons s(*i..j*)	Weights and Biases of the Input Layer					Hidden Layer Weights
	W_{I,i,R_s}	W_{I,i,γ_g}	W_{I,i,γ_o}	$W_{I,i,T}$	$b_{I,i}$	$W_{o,i}$
1	2.8605	−1.5782	−0.3324	−0.1083	−2.9158	−0.5285
2	1.1629	−0.5131	1.0839	−2.4517	−2.3184	−0.3380
3	−0.4431	0.3111	2.1040	1.8214	2.2403	0.0121
4	3.4798	0.3638	2.8263	−1.4343	−1.6174	−1.0344
5	−2.3006	−0.9198	−1.6993	−0.3951	1.3495	−0.2057
6	0.1183	2.3582	1.0873	−1.0833	−0.8932	−0.7900
7	0.7489	−1.3447	−1.1019	−2.1037	−1.0511	0.0368
8	2.3679	−0.6043	−1.0511	0.2182	−1.0852	0.0929
9	3.4553	1.1799	0.8559	−1.9318	−0.3919	1.4544
10	−1.5186	−1.7151	−1.3278	1.7128	0.2388	−1.3798
11	1.7416	−2.1886	−0.7271	0.9237	0.5531	0.1322
12	−1.2792	1.1561	−2.7749	−0.7555	0.1952	−0.1460
13	−2.4427	−1.1594	0.7517	−0.5188	−0.5644	−0.2000
14	−0.7630	0.0304	−1.5823	0.8316	−0.3401	0.2611
15	−1.6891	1.5989	−0.9908	1.1452	−0.9367	0.1310
16	0.3559	−1.3222	1.5324	1.8756	2.3373	0.2034
17	−2.2950	−0.9203	−0.3174	2.1212	−2.3553	−0.6117
18	−1.7585	−1.0869	−0.0124	2.7669	−2.1098	0.4386
19	−1.9137	1.5226	0.3003	1.1600	−2.8272	0.2195
20	2.5270	1.6128	−1.2110	−0.3049	2.5730	0.0574

Step 3. Evaluation of the transfer function at the hidden layer: The development of the ANN-based model necessitates a mathematical operation employing the tansig transfer function in the last stage. For a single-hidden-layer model, the function can be stated mathematically as:

$$\text{Tansig}\ (E_{I,i,n}) = F_{I,i,n} = \left(\frac{2}{(1 + \exp\ (-2\ E_{I,i,n})}\right) - 1 \tag{12}$$

In Eq. 12, $E_{I,i,n}$ is the nth sum of the weighted normalized variables calculated in Step 2. It is worth mentioning that when dealing with n number of hidden layers, the tansig transfer function will operate on the n number of layers separately.

Step 4. Assignments of weights and purelin transfer function at the output layer: The output layer of the ANN architecture in Figure 4 demands the application of the pure linear or "purelin" transfer function after the "tansig" transfer function in Step 3, which is preceded by the assignment of weights and biases in Step 2. At this stage, the weights and biases generated by the ANN at the output layer are assigned to the already calculated variable $F_{I,i,n}$ in Step 3. Equations 13 and 14 illustrate how the weights and biases were assigned to the preceding variable $F_{I,i,n}$ using the "purelin" function:

$$a_o = \sum_{i=1}^{j} (W_{o,i}\ F_{I,i,n}) + b_o \tag{13}$$

$$b_o = -0.1274$$

The values of $W_{o,i}$ are reported in Table 2. Using Eq. 13, the value of a_o is expressed as

$$\begin{aligned} a_o = {} & (-0.5285F_1 - 0.3380F_2 + 0.0121F_3 - 1.0344F_4 - 0.2057F_5 - 0.7900F_6 \\ & + 0.0368F_7 + 0.0929F_8 + 0.4544F_9 - 1.3798F_{10} + 0.1322F_{11} - 0.1460F_{12} \\ & - 0.1100F_{13} + 0.2611F_{14} + 0.1310F_{15} + 0.2034F_{16} - 0.6117_{17} + 0.4386F_{18} \\ & + 0.2195F_{19} + 0.0574F_{20} - 0.1274) \end{aligned} \tag{14}$$

In Eq. 14, a_o is the standardized final output; $W_{o,i}$ is the weight at the output layer (o) that corresponds to each neuron (i); b_o denotes the bias at the output layer and $F_{I,i,n}$ is the previously established variable of the tansig-transformed in Step 3.

Step 5. Denormalization of the normalized output: It is important to denormalize the output value found in Eq. 13 or Eq. 14 in order to retrieve the true value of the output, as the input variables were normalized. Herein, Eq. 15 was used to denormalize the normalized final output, a_o. This was accomplished by making x the subject of Eq. 6 since x is the desired outcome rather than y. It should be noted that a_o is used in place of the normalized input variable y; and x is the target variable which is the predicted oil FVF.

$$x = \frac{(a_o - y_{min})\ (x_{max} - x_{min})}{(y_{max} - y_{min})} + x_{min} \tag{15}$$

The parameters a_o, y_{min} and x_{min} are already defined in Eqs. 6 and 13. Using the above equation, the final empirical correlation for the oil FVF is expressed as:

$$B_o = \left(\frac{a_o + 1}{1.22850122850123} \right) + 1.084099 \qquad (16)$$

The entire development of this ANN-based empirical correlation is displayed in Figure 6.

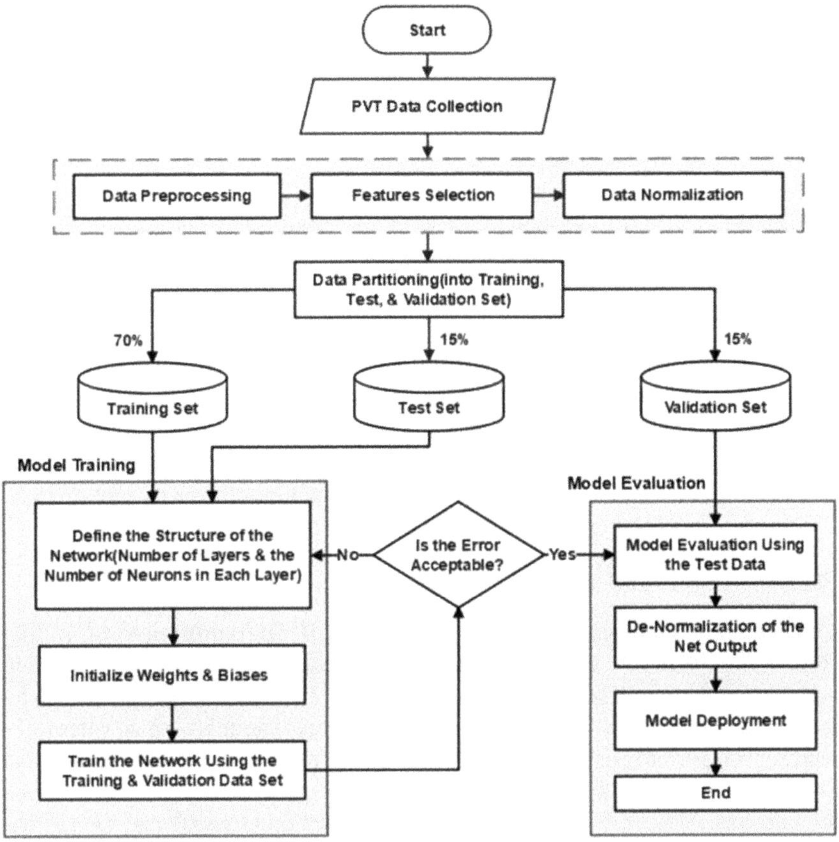

Figure 6: Flowchart of the ANN development process.

Verification of the ANN-Based Empirical Correlation

To verify the proposed correlation, 15% of the total dataset constituting 33 datasets were used to test the correlation. The left panel of Figure 7 shows the cross-plot of the measured oil FVF against the predicted oil FVF, while the right panel of Figure 7 shows the test results for the proposed ANN correlation. It had a correlation coefficient (R) of 0.99289 which is very close to 1 (predicted equals target). Based on the results shown in Figure 7, it is reasonable to conclude that there is a strong relationship between the measured oil FVF values and the predicted oil FVF values from the proposed correlation.

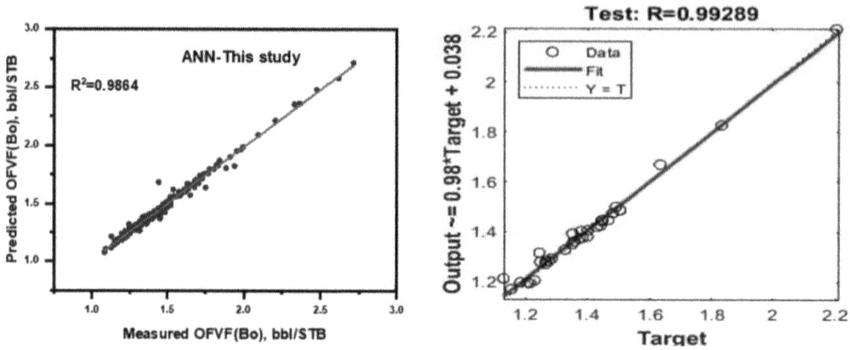

Figure 7: (Left) Cross plot of measured oil FVF against the predicted oil FVF. (Right) Test results for the model.

Performance Assessment of the ANN Model

Following the successful training of the ANN model, its performance was evaluated with different statistical and graphical methods including the correlation coefficient (R), MSE, AAPRE, st. dev., RMSE, and minimum and maximum absolute error. The performance and accuracy of the correlation are visualized using graphical error analysis, which includes the cross-plot method. The proposed correlation was then compared with existing empirical correlations.

Statistical Error Analysis

Correlation Coefficient (R) and Coefficient of Determination (R^2): The correlation coefficient is a statistical measurement of the relationship between variables and their association with each other. It is normally represented by the letter R and ranges from 1 to –1. Between two variables, an R value of 1 represents a perfect positive association while –1 represents the perfect negative association between those variables. If there is no correlation at all between the two variables, R is 0. The square of the correlation coefficient (R^2) is the coefficient of determination. The mathematical expression of the correlation coefficient is:

$$R = \frac{\sum (x_i - \underline{x})\,(y_i - \underline{y})}{\sqrt{\sum (x_i - \underline{x})^2\,(y_i - \underline{y})^2}} \tag{17}$$

R is the correlation coefficient, x_i are values of the x-variable in a sample, $\underline{x}$ is the mean of the values of the x-variables, y_i is the values of the y-variable in a sample, $\underline{y}$ is the mean of the values of the y-variables.

Mean Square Error (MSE): The MSE measures the average of the squares of the errors between the experimental value ($y_{i,\text{exp}}$) and the predicted or estimated value ($y_{i,\text{pred}}$). It shows how close a regression line is to the measured data. The closer the MSE value is to 0, the better the model and it is given by Wood & Choubineh (2019):

$$\text{MSE} = \frac{1}{N}\sum_{n=1}^{N}(y_{i,\exp} - y_{i,\text{pred}})^2 \tag{18}$$

where $y_{i,\exp}$ is the experimental value; $y_{i,\text{pred}}$ is the predicted or estimated value; n is the sequential data count; i is the sequential neuron count; and N is the number of data points.

Average Absolute Percent Relative Error (AAPRE): This is the average of the relative deviation in the prediction of a particular variable, and it is expressed as a percentage. Lower values of AAPRE indicate better model performance. It can be computed as follows:

$$\text{AAPRE} = \frac{1}{N}\sum_{n=1}^{N}\left|\frac{y_{i,\exp} - y_{i,\text{pred}}}{y_{i,\exp}}\right| \times 100 \tag{19}$$

Standard Deviation: Within a set of data, this metric shows the degree of variation or dispersion. The accuracy of the model under consideration increases with a decreasing standard deviation. A lower value indicates less scatter. It is given as:

$$\text{St. Dev.} = \sqrt{\frac{1}{n-1}\sum_{i=1}^{n}\left(\frac{y_{i,\exp} - y_{i,\text{pred}}}{y_{i,\exp}}\right)^2} \tag{20}$$

Root Mean Square Error (RMSE): This parameter determines how the data points are distributed with respect to zero deviation. Its value would decrease as the model fits the experimental data more closely. It can be computed as follows:

$$\text{RMSE} = \sqrt{\frac{1}{N}\sum_{n=1}^{N}(y_{i,\exp} - y_{i,\text{pred}})^2} \tag{21}$$

Minimum Absolute Error ($Er_{\min}$): The minimum error is given as:

$$Er_{\min} = \min_{i=1}^{n}\left[\frac{1}{n}\sum_{i=1}^{n}\left|\frac{y_{i,\exp} - y_{i,\text{pred}}}{y_{i,\exp}}\right|\right] \tag{22}$$

Maximum Absolute Error ($Er_{\max}$): The maximum error is given as:

$$Er_{\max} = \max_{i=1}^{n}\left[\frac{1}{n}\sum_{i=1}^{n}\left|\frac{y_{i,\exp} - y_{i,\text{pred}}}{y_{i,\exp}}\right|\right] \tag{23}$$

Graphical Error Analysis

Graphical analysis techniques help in visualizing the accuracy of a model. In this study, two of these techniques were employed. These include cross plot and the residual plot. With the cross plot, measured values are plotted against the predicted values with a line of best fit normally drawn from the origin at an angle of 45°, which indicates that the measured values are equal to the predicted results. As a result, the closer the plotted data is to this line, the higher the performance and accuracy of the model. The residual plot is obtained by plotting the residual (measured minus predicted) values on the vertical axis and the independent variables (data points) on the horizontal axis.

Results and Discussion

Training Performance Plot

One of the measures utilized to assess the performance of the ANN was an analysis of the training performance plot. Figure 8 displays the generated training performance curve of the proposed model. It is evident from Figure 8 that the MSE values for the train, validation, and test datasets were high at the start of the training process and declined with each epoch. The best training performance point for the three data groups is shown by the intersection of the two dotted lines and it was reached when the MSE values approached the most ideal value. At the 12th epoch in this investigation, the best validation performance was reported as 0.00096043, and this was when the optimal MSE value was obtained. When all of the curves are identically formed, it indicates that the network responds

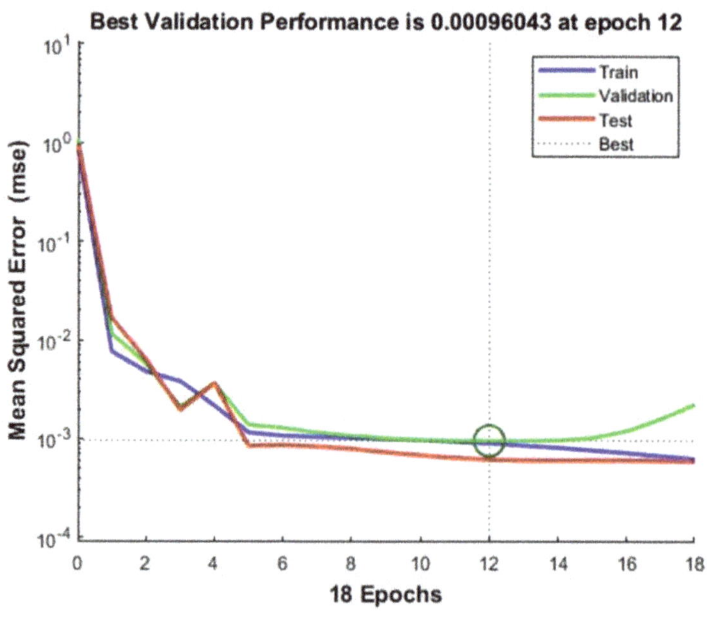

Figure 8: Performance plot of the ANN model in terms of MSE.

similarly to training data as well as to the test and validation data (Watanabe & Yamana, 2021). As a result, the likelihood of overfitting is extremely low, but it still cannot be completely ruled out. As can be observed from Figure 8, the difference between the validation and test curves is insignificant. If, however, the test curve deviates considerably from the validation curve, then there might have been an instance of overfitting (Hawkins, 2004; Nilizadeh et al., 2021). Thus, there are no significant issues with the performance plot of the proposed model. Furthermore, it is observed that the test curve is lower than the validation curve from the fifth epoch, indicating that the test phase of the ANN model is finished with the best possible performance values. As a result, the trained network performs better when using test data than when using data that is used in the training, indicating exceptional model generalizability. The best validation performance of the model was observed at epoch number 12, which corresponds to the lowest MSE of 0.00096043.

Training State Performance Plot

Figure 9 shows the training state performance plot depicting the best validation performance of the model. With a validation check of 6, the gradient was recorded as 0.0010103 with a momentum (Mu) of 0.00001 at the 18th epoch. This shows that the errors were repeated 6 times after the 12th epoch, and the process terminated at the 18th epoch. The performance parameters were also

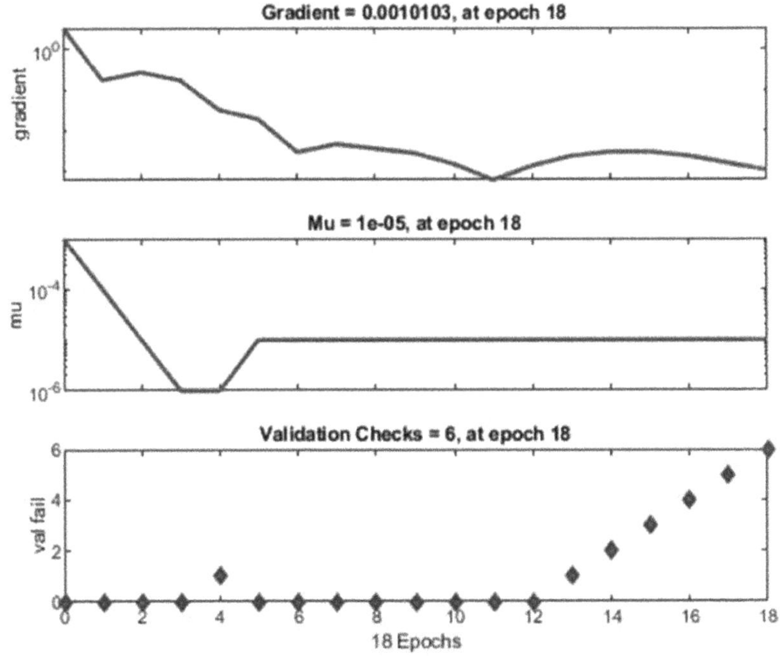

Figure 9: Training state performance plot.

calculated and evaluated for the ANN model. The calculated *R*-value for the training state of the neural network was 0.992593 with an MSE of 0.000899964. The *R*-value for the validation state was 0.995395 with an MSE of 0.00096042, and the *R*-value for the testing state was 0.992892 with an MSE of 0.000619068. Since the *R* values are remarkably close to one with exceedingly small values of MSEs, the accuracy of the prediction of the ANN model was high and reliable.

Regression Plot

Another technique for validating the network is the regression plot. It illustrates the relationship between the outputs of the network and their related objectives. In numerical form, it is represented by the regression factor or correlation coefficient, *R*. The first three plots in Figure 10 are for the training, validation, and testing dataset. These three were then combined to give the fourth plot (All). The solid line is the linear regression line of best fit between outputs and targets. The dashed line in each plot stands for the perfect result (measured = predicted). The trend shown by the plot indicates that the network learns more effectively, and the linear regression fits well to the data points, which aligns the expected

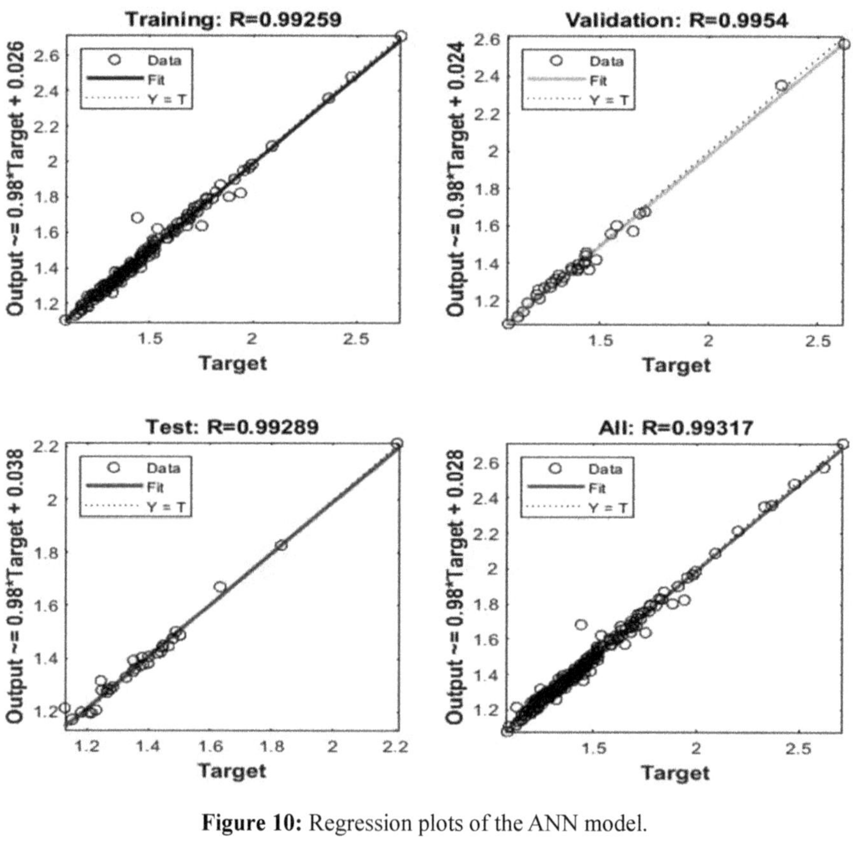

Figure 10: Regression plots of the ANN model.

output with the real target. It is clear from the correlation coefficient (R) of 0.99317 that the ANN model has remarkably performed well.

Error Histogram

The error histogram illustrated in Figure 11 also provides more insight into the quality of the trained network. It shows how the residuals are distributed between the targets and the network output. The errors of the model have spread around the zero-error value showing that the ANN model has a high degree of reliability. It additionally demonstrates that the overall error of the neural network spans from –0.2325 (the leftmost bin) to 0.1072 (the rightmost bin) with most errors falling between –0.01797 and 0.01779. There is a zero-error line and an error value of -9.20×10^{-5} in the bin with the highest number of errors for the training, validation, and test. This indicates that most of the prediction errors are close to zero (target = predicted), and that, there are fewer deviations (more reliable). Bins are the number of vertical bars that are visible on the graph. Each vertical bar represents the number of samples in the dataset that fall into a specific bin. The width the bars stand for the most common type of error, and this was evaluated using Eq. 24 (Ocampo et al., 2021):

$$\text{Width of bin} = \frac{\text{Maximum error} - \text{Minimum error}}{\text{Number of bins}} \tag{24}$$

In this study, the maximum error is 0.1072; the minimum error is –0.2325 while the number of bins is 20, and thus the width of each bin is 0.016985.

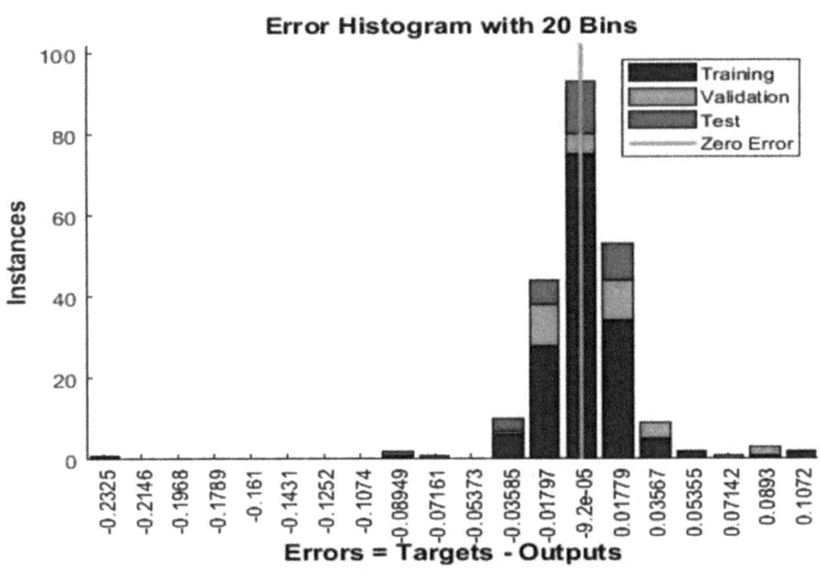

Figure 11: Error histogram of the ANN model.

Performance of the Proposed ANN Empirical Correlation

In this study, ten prominent and widely used empirical correlations for estimating oil FVF of crude oil samples were examined (Abdul-Majeed & Salman, 1988; Al-Marhoun, 1988; Dokla & Osman, 1992; Glaso, 1980; Kartoatmodjo & Schmidt, 1991; Mazandarani & Asghari, 2007; Omar & Todd, 1993; Petrosky Jr & Farshad, 1993; Standing, 1947; Sulaimon et al., 2014). Afterward, the accuracy and performance of the proposed ANN model were assessed in reference to the empirical correlations provided in the earlier text. Table 3 displays the performance metrics for the empirical correlations including the proposed ANN model. The table indicates that the majority of the empirical correlations analyzed were unable to offer a satisfactory level of accuracy for the oil FVF predictions. The AAPRE values of the empirical correlations mentioned earlier and that of the proposed ANN model are depicted in Figure 12. Figure 13 displays the RMSE of the empirical correlations mentioned in the earlier text, as well as the ANN model proposed in this study. Figure 14 is the correlation coefficient plot of the empirical correlations as well as the proposed ANN.

From Figure 12, it is seen that the worst performance was exhibited by the correlation of Dokla & Osman (1992) with an AAPRE of 49.02%. This model also exhibited the highest RMSE of 0.6940. This is followed by Al-Marhoun (1988), Standing (1947), and Mazandarani & Asghari (2007) with their corresponding AAPREs being 29.27%, 18.83%, and 14.28%, respectively. Again, their respective RMSEs are 0.4252, 0.3501, and 0.2832. The correlations of Glaso (1980), Petrosky Jr & Farshad (1993), Omar & Todd (1993), Sulaimon et al. (2014), and Abdul-Majeed & Salman (1988) had similar performance of an acceptable degree with their AAPRE ranging from 3.59% to 6.11%. The performance of the correlation proposed by Kartoatmodjo & Schmidt (1991) was exceptional as compared to all of the aforementioned correlations. The AAPRE and RMSE for their correlation were 2.49% and 0.0547, respectively. Nonetheless, the proposed ANN model in this study outperformed the pre-existing correlations with an AAPRE value of 1.19% and an RMSE of 0.0294.

To visually support the claim that the proposed model outperformed existing models, confidence interval bands were incorporated around the predicted values to help communicate the uncertainty in predictions. This provides a statistical measure of the confidence in each prediction, giving a better understanding of the reliability of the ANN model in different regions of the data. Figure 15 shows the regression plot with the confidence interval bands (95% confidence and prediction bands). This band indicates the range within which the true relationship between the predicted and measured values lies with 95% confidence. With 95% confidence the R^2 and AAPRE were 0.9864 and 1.19%, respectively. The narrow confidence band around the regression line as seen in the plot suggests that the model is consistent and provides reliable predictions, with little uncertainty in the estimated relationship. The prediction band further ensures that future predictions will fall within an acceptable range of uncertainty, reinforcing the model's robustness.

Table 3: Statistical comparison of the existing correlations and the ANN model

Correlations	Er_{min}	Er_{max}	MSE	R	St. Dev.	AAPRE (%)	RMSE
Standing (1947)	0.0420	1.2611	0.12254	0.97427	0.1923	18.83	0.3501
Petrosky Jr & Farshad (1993)	−0.4682	0.2377	0.00805	0.97306	0.0734	4.31	0.0897
Glaso (1980)	−0.4249	0.2436	0.00426	0.97910	0.0518	3.59	0.0653
Dokla & Osman (1992)	0.2821	0.8637	0.48170	0.98390	0.0529	49.02	0.6940
Al-Marhoun (1988)	0.1090	0.6707	0.18082	0.98421	0.0601	29.27	0.4252
Omar & Todd (1993)	−0.2609	0.3446	0.01706	0.90848	0.1056	6.46	0.1306
Sulaimon et al. (2014)	−0.2790	0.5881	0.01932	0.89615	0.1385	6.53	0.1390
Abdul-Majeed & Salman (1988)	−0.1028	0.7608	0.02273	0.88322	0.1300	6.11	0.1508
Mazandarani & Asghari (2007)	0.0103	1.0506	0.08021	0.96080	0.1718	14.28	0.2832
Kartoatmodjo & Schmidt (1991)	−0.3105	0.2087	0.00300	0.97872	0.0545	2.49	0.0547
ANN—this study	−0.2415	0.1161	0.00087	0.99317	0.0295	1.19	0.0294

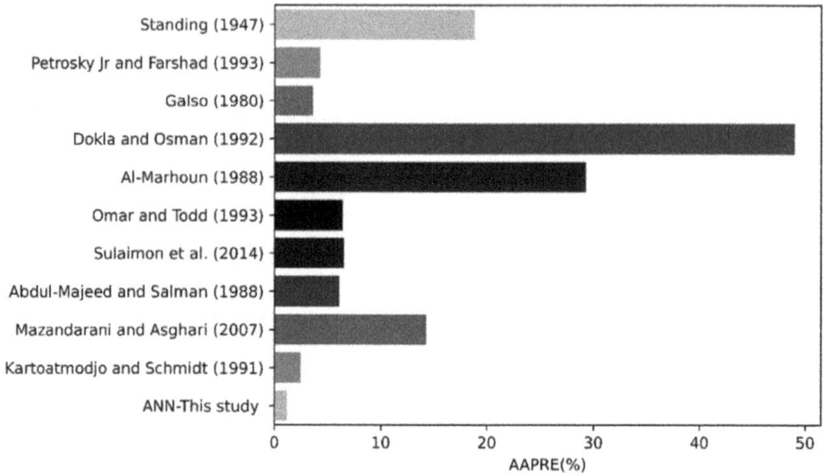

Figure 12: Comparison between average absolute percent relative error of empirical correlations and the proposed model.

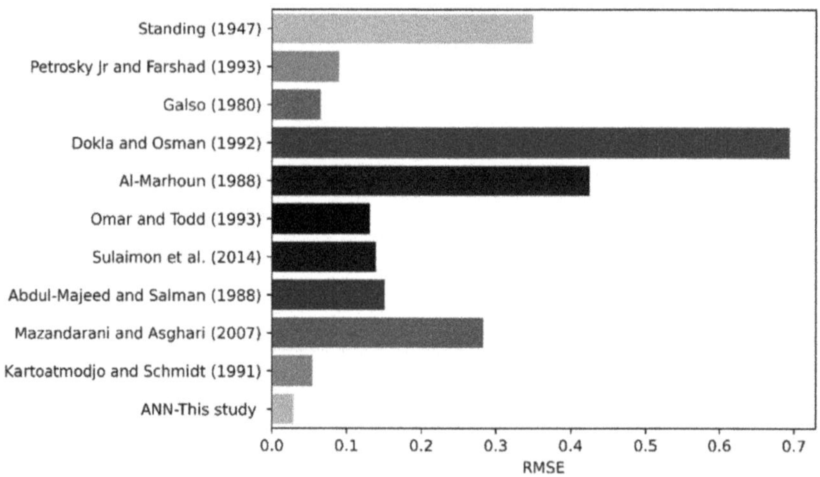

Figure 13: RMSE of empirical correlations and the proposed model.

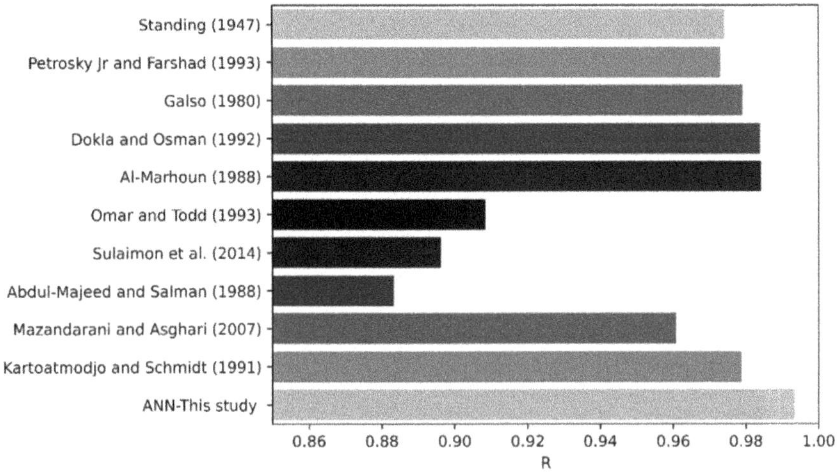

Figure 14: Plot of the correlation coefficient (R).

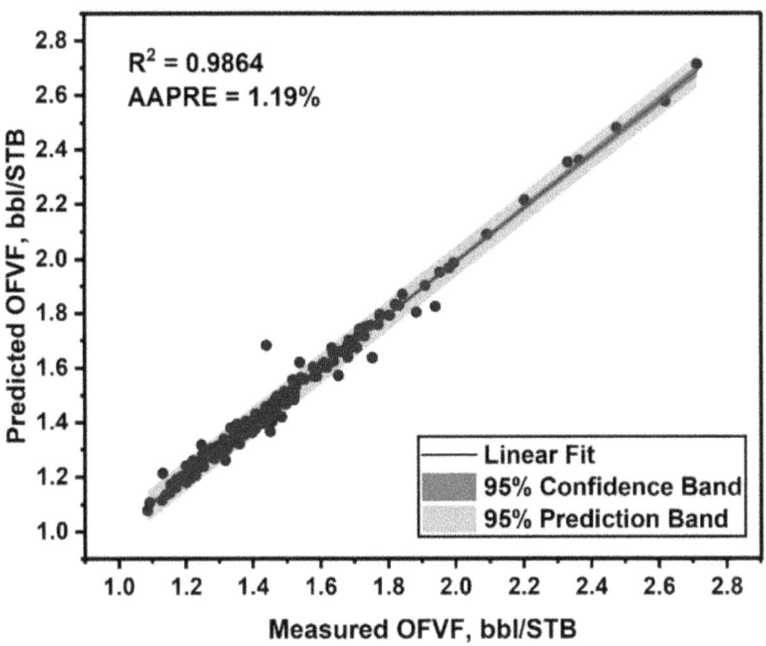

Figure 15: A plot of measured OFVF versus predicted OFVF with 95% confidence interval band.

Thus, the proposed ANN model in this study had the best performance in predicting the oil FVF. The correlation coefficient provides further evidence that the proposed ANN model performed excellently. The correlation coefficient was determined to be 0.99317, a value close to 1. Moreover, most of the predicted values fell within a confidence and prediction bands of 95%. As a result, in terms of oil FVF prediction, the proposed ANN model outperformed the existing empirical correlations in terms of accuracy and capability.

Figure 16 illustrates the cross-plots of oil FVF for the published empirical correlations and the proposed ANN model. It shows that the ANN model had more data points falling along the 45° line, proving that there is a strong relationship between the predicted values from the ANN model and the measured data.

Residual Plot

Residual is defined as the vertical difference between the actual and the predicted value. In other words, it measures how much a regression line vertically misses a target data point. The residual plot is obtained by plotting the residual values on the vertical axis against the independent variables (data points) on the horizontal axis. A good linear model is expected to have its residual values scattered equally and randomly spaced around the horizontal axis (residual = 0 line) (Feng et al., 2020). However, in some cases, the residual values may show no pattern and will be randomly dispersed, or even clustered in a U-shaped pattern (Santos et al., 2021). In that case, it can be reasonably concluded that the linear model is not good and reliable. The residual plot for the oil FVF by the published correlations and the proposed ANN model in this study is shown in Figure 17.

From the plot, the correlation of Dokla & Osman (1992) appeared to be the least accurate with most of its residuals falling in the range of 0.6 to 0.8. This is followed by the correlation of Al-Marhoun (1988) with most of its residuals falling in the range of 0.2 to 0.6. The Standing (1947) correlation shows similar performance as that of Mazandarani & Asghari (2007), with most of their residuals falling in the range of 0 to 0.4. Again, the correlations of Petrosky Jr & Farshad (1993), Glaso (1980), Omar & Todd (1993), Abdul-Majeed & Salman (1988), and Sulaimon et al. (2014) also appeared to have a similar performance having most of their residuals in the range of –0.2 to 0.1, but showed better performance over the correlations of Dokla & Osman (1992), Mazandarani & Asghari (2007), Al-Marhoun (1988), and Standing (1947). The correlation of Kartoatmodjo & Schmidt (1991) emerged as the best-performing correlation after the proposed ANN correlation developed in this study.

Physical Verification of the Proposed Model via Trend Analysis

To relate the developed white-box model to the physical features, trend analysis was performed aiming to see how well the model can accurately represent

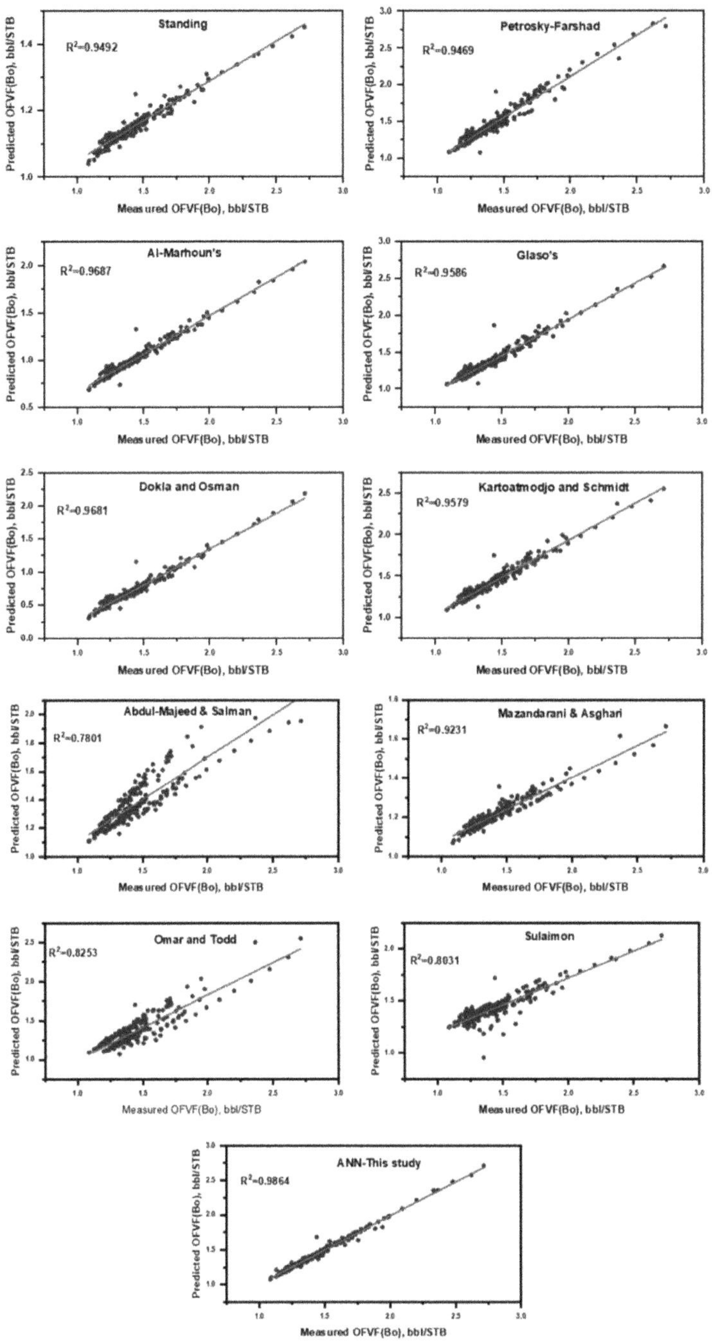

Figure 16: Cross-plots of oil formation volume factor for 10 selected correlations and ANN model.

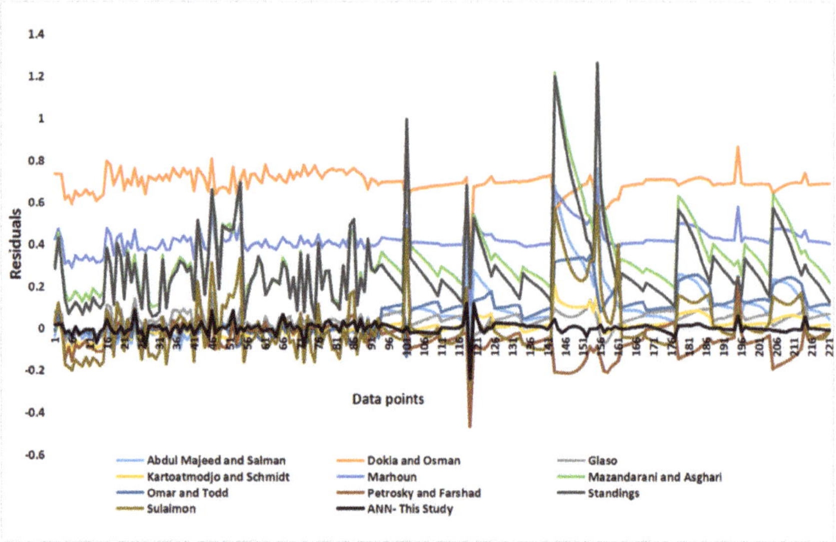

Figure 17: Residual plot for the oil FVF by the well-established correlations and the proposed ANN model.

the underlying physical phenomenon in real-world applications. The analysis involved generating synthetic datasets where each time, only one input parameter was varied while all other parameters were held constant at their average values [R_s = 563.9548 (scf/STB), γ_g = 1.3670, γ_o = 0.8290, T = 225.3077 (°F)]. This was done to examine the effect of each input variable on OFVF.

Figure 18 shows the plot in each case. Figure 18(a) shows the effect of gas solubility (R_s) on OFVF. It is observed that OFVF increases with R_s. The increase in OFVF can be attributed to the higher gas dissolution at higher R_s. As R_s increases, the volume of oil with its dissolved gas also increases in the reservoir and this increases OFVF. Figure 18(b) shows the variation of oil-specific gravity (γ_o) with OFVF. The graph shows that OFVF decreases as γ_o increases, which is physically valid. Heavier oils (with higher specific gravity) generally have less dissolved gas. Therefore, if a heavy oil is brought to standard conditions, it loses a minimal amount of gas, thereby resulting in high volumes of oil. This results in a lower OFVF. This behavior aligns with empirical observations and fluid expansion characteristics in real reservoirs. The effect of reservoir temperature (T) on OFVF is also depicted by Figure 18(c). It shows OFVF increasing with T, which is physically valid. Higher temperatures cause oil to expand and allow more gas to dissolve in the oil at reservoir conditions. This increases the volume of oil in the reservoir resulting in higher OFVF. However, the magnitude of this increase depends on the specific reservoir fluid properties. Similar trend was observed in the case of gas gravity (γ_g) shown in Figure 18(d). A higher gas gravity implies that the gas is heavy and has heavier components. Therefore, below the bubble

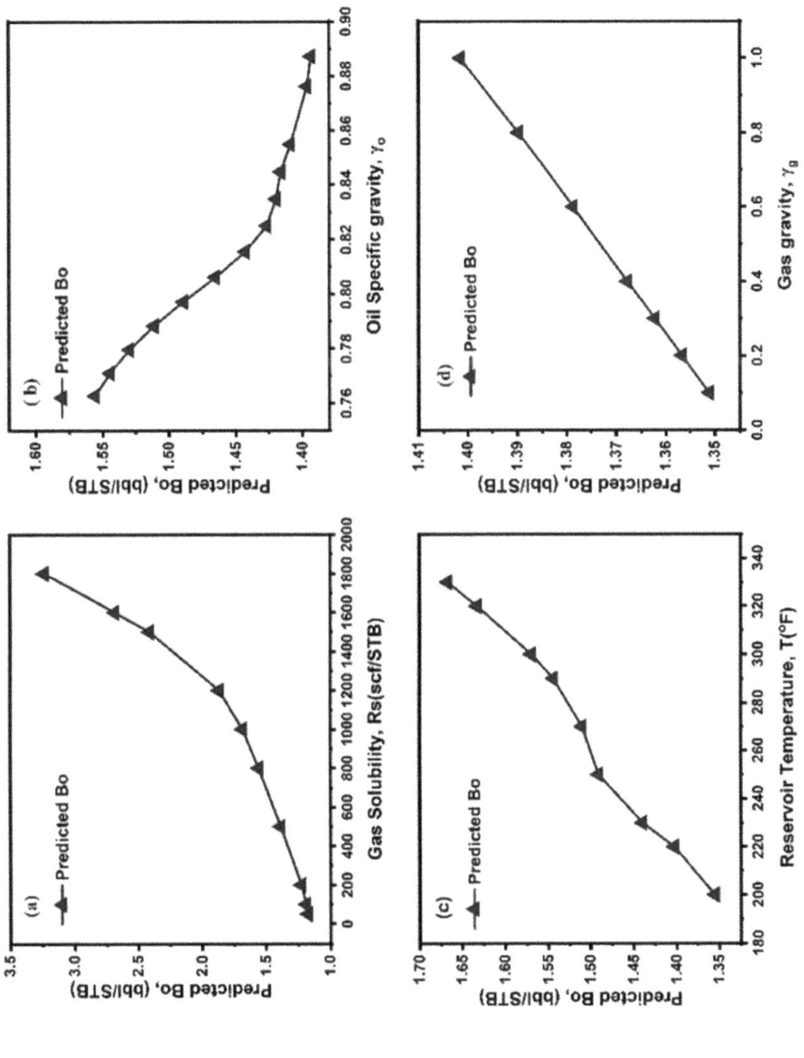

Figure 18: Effect of (a) gas solubility, (b) oil gravity, (c) reservoir temperature, and (d) gas gravity on the OFVF.

point, less gas will come out of solution compared to a system with lower gas gravity (light gas). Since less gas comes out of solution, the volume of oil and its dissolved gas in the reservoir remains high for a gas with high gas gravity. This leads to increase in OFVF as γ_g increases.

The verification of the proposed model which was done by examining the dependence of the correlation on its input variables showed that all the observed trends make physical sense. Consequently, the proposed correlation in this study is physically accurate.

Conclusion

We proposed a new empirical correlation to predict the oil FVF at the bubble point and pressures below the bubble point using the ANN white-box technique. Data on crude oil samples were collected worldwide; including data on Malaysian crude oil, and gas-saturated and undersaturated oils in the Mediterranean Basins, Africa, the Persian Gulf, and the North Sea. The proposed correlation was analyzed and compared to ten well-established empirical correlations in the literature. Key insights from the rigorous analysis include:

1. The results from the feature selection techniques revealed that the solution gas–oil ratio (R_s) is the most influential variable on oil FVF. This is followed by temperature (T), oil gravity (γ_o), and API gravity, with gas gravity (γ_g) as the least influential variable. The oil gravity (γ_o) exhibited a strong inverse relation with API gravity and an explanation of this is appreciated with the formulaic relationship between API and γ_o.
2. The accuracy and reliability of the trained model were examined by graphical tools including training performance plot, error histogram, and residual plot. The training performance plot showed that the best validation performance of the model was observed at the 12th epoch, which corresponds to the lowest MSE of 0.00096043. The error histogram showed that an error value of -9.20×10^{-5} was in the bin with the highest number of errors for the training, testing, and validation; and this means there are fewer deviations, making the model reliable. The residual plot revealed that the residuals for the ANN model proposed in this study lay close to the zero residual line demonstrating the high accuracy of the model.
3. The root mean square error, mean square error, average absolute percent relative error, and correlation coefficient recorded for the proposed ANN model are 0.0294, 0.00087, 1.19%, and 0.99317, respectively.
4. The proposed empirical ANN correlation outperformed ten popular and widely used correlations in the industry.

The proposed ANN-based empirical correlation stands out due to its simplicity and practical utility. Its ability to be executed in a spreadsheet without requiring specialized software significantly enhances its accessibility for field professionals. This ease of use facilitates real-time decision-making in field applications, such as adjusting production parameters or estimating reserves more quickly and accurately. By eliminating the reliance on proprietary software, the model reduces operational bottlenecks, enabling efficient workflows. Moreover, the transparency of the white-box technique ensures that the correlation is reproducible and auditable, promoting confidence in its predictions. These features collectively make the proposed correlation a valuable tool for industry professionals, providing greater efficiency in reservoir management and improving the adaptability of decision-making processes.

Acknowledgment

The authors are grateful to the College of Engineering of the Kwame Nkrumah University of Science and Technology (KNUST), Kumasi, for providing opportunities to undertake this research. We also thank Mr. Amedeka Emperor (Teaching Assistant at the Department of Petroleum Engineering, KNUST) for his assistance in this study. Joshua Nsiah Turkson acknowledges Universiti Teknologi PETRONAS for supporting his graduate studies through the Graduate Assistantship program.

Disclosure Statement

We declare that we do not have any known competing financial interests or personal relationships that could have appeared to influence our work.

Data Availability Statement

Data will be made available upon reasonable request.

References

Abdul-Majeed, G. H. & Salman, N. H. (1988). An empirical correlation for oil FVF prediction. *J. Can. Pet. Technol.*, 27(06). https://doi.org/10.2118/88-06-10.

Abidoye, L., Mahdi, F., Idris, M., Alabi, O. & Wahab, A. (2018). ANN-derived equation and ITS application in the prediction of dielectric properties of pure and impure CO_2. *J. Cleaner Prod.*, 175, 123–132. https://doi.org/10.1016/j.jclepro.2017.12.013.

Agatonovic-Kustrin, S. & Beresford, R. (2000). Basic concepts of artificial neural network (ANN) modeling and its application in pharmaceutical research. *J. Pharm. Biomed. Anal.*, 22(5), 717–727. https://doi.org/10.1016/S0731-7085(99)00272-1.

Akhgar, A., Toghraie, D., Sina, N. & Afrand, M. (2019). Developing dissimilar artificial neural networks (ANNs) to prediction the thermal conductivity of MWCNT–TiO_2/ water–ethylene glycol hybrid nanofluid. *Powder Technol.*, 355, 602–610. https://doi.org/10.1016/j.powtec.2019.07.086.

Al-Marhoun, M. & Osman, E. (2002). Using artificial neural networks to develop new PVT correlations for Saudi crude oils. *Abu Dhabi International Petroleum Exhibition and Conference*. https://doi.org/10.2118/78592-MS.

Al-Marhoun, M. A. (1988). PVT correlations for Middle East crude oils. *Journal of Petroleum Technology*, 40(05), 650–666. https://doi.org/10.2118/13718-PA.

Al-Shammasi, A. (1999). Bubble point pressure and oil formation volume factor correlations. *SPE Middle East Oil and Gas Show and Conference*. https://doi.org/10.2118/53185-MS.

Al-Shammasi, A. (2001). A review of bubble point pressure and oil formation volume factor correlations. *SPE Reservoir Eval. Eng.*, 4(02), 146–160. https://doi.org/10.2118/71302-PA.

Avseth, P. & Mukerji, T. (2002). Seismic lithofacies classification from well logs using statistical rock physics. *Petrophysics – The SPWLA Journal of Formation Evaluation and Reservoir Description*, 43(02).

Ayoub, M. A., Elhadi, A., Fatherlhman, D., Saleh, M., Alakbari, F. S. & Mohyaldinn, M. E. (2022). A new correlation for accurate prediction of oil formation volume factor at the bubble point pressure using Group Method of Data Handling approach. *J. Pet. Sci. Eng.*, 208, 109410. https://doi.org/10.1016/j.petrol.2021.109410.

Boukadi, F., Al-Alawi, S., Al-Bemani, A. & Al-Qassabi, S. (1999). Establishing PVT correlations for Omani oils. *Petroleum Science and Technology*, 17(5–6), 637–662. https://doi.org/10.1080/10916469908949738.

Chun, I. Y., Huang, Z., Lim, H. & Fessler, J. A. (2020). Momentum-Net: Fast and convergent iterative neural network for inverse problems. *IEEE Transactions on Pattern Analysis and Machine Intelligence*, 45(4), 4915–4931. https://doi.org/10.1109/TPAMI.2020.3012955.

De Ghetto, G., Paone, F. & Villa, M. (1994). Reliability analysis on PVT correlations. *SPE Europec Featured at EAGE Conference and Exhibition.*

Dokla, M. E. & Osman, M. E. (1992). Correlation of PVT properties for UAE crudes. *SPE Form. Eval.*, 7(01), 41–46. https://doi.org/10.2118/20989-PA.

Dutta, S. & Gupta, J. (2010). PVT correlations for Indian crude using artificial neural networks. *J. Pet. Sci. Eng.*, 72(1–2), 93–109. https://doi.org/10.1016/j.petrol.2010.03.007.

Ehret, A., Hochstuhl, D., Gianola, D. & Thaller, G. (2015). Application of neural networks with back-propagation to genome-enabled prediction of complex traits in Holstein–Friesian and German Fleckvieh cattle. *Genet. Sel. Evol.*, 47, 1–9. https://doi.org/10.1186/s12711-015-0097-5.

El-Sebakhy, E. A. (2009). Forecasting PVT properties of crude oil systems based on support vector machines modeling scheme. *J. Pet. Sci. Eng.*, 64(1–4), 25–34. https://doi.org/10.1016/j.petrol.2008.12.006.

Elkatatny, S. & Mahmoud, M. (2018). Development of new correlations for the oil formation volume factor in oil reservoirs using artificial intelligent white box technique. *Petroleum*, 4(2), 178–186. https://doi.org/10.1016/j.petlm.2017.09.009.

Elsharkawy, A. M. (2003). An empirical model for estimating the saturation pressures of crude oils. *J. Pet. Sci. Eng.*, 38(1-2), 57–77. https://doi.org/10.1016/S0920-4105(03)00035-4.

Engelbrecht, A. P. (2007). *Computational Intelligence: An Introduction* (2nd ed.). John Wiley & Sons.

Feng, C., Li, L. & Sadeghpour, A. (2020). A comparison of residual diagnosis tools for diagnosing regression models for count data. *BMC Med. Res. Methodol.*, 20, 1–21. https://doi.org/10.1186/s12874-020-01055-2.

Gharbi, R. B., Elsharkawy, A. M. & Karkoub, M. (1999). Universal neural-network-based model for estimating the PVT properties of crude oil systems. *Energy Fuels*, 13(2), 454–458. https://doi.org/10.1021/ef980143v.

Glaso, O. (1980). Generalized pressure–volume–temperature correlations. *Journal of Petroleum Technology*, 32(05), 785–795. https://doi.org/10.2118/8016-PA.

Goda, H. M., El-M Shokir, E. M., Fattah, K. A. & Sayyouh, M. H. (2003). Prediction of the PVT data using neural network computing theory. *SPE Nigeria Annual International Conference and Exhibition*. https://doi.org/10.2118/85650-MS.

Gouda, A. & Attia, A. M. (2022). Development of a new approach using an artificial neural network for estimating oil formation volume factor at bubble point pressure of Egyptian crude oil. *J. King Saud Univ., Eng. Sci.* https://doi.org/10.1016/j.jksues.2022.08.001.

Hassaan, S., Mohamed, A., Ibrahim, A. F. & Elkatatny, S. (2024). Real-time prediction of petrophysical properties using machine learning based on drilling parameters. *ACS Omega*, 9(15), 17066–17075. https://doi.org/10.1021/acsomega.3c08795.

Hawkins, D. M. (2004). The problem of overfitting. *J. Chem. Inf. Comput. Sci.*, 44(1), 1–12. https://doi.org/10.1021/ci0342472.

Kartoatmodjo, T. & Schmidt, Z. (1991). New correlations for crude oil physical properties. *Paper SPE*, 23556.

Kharazi Esfahani, P., Peiro Ahmady Langeroudy, K. & Khorsand Movaghar, M. R. (2023). Enhanced machine learning—Ensemble method for estimation of oil formation volume factor at reservoir conditions. *Sci. Rep.*, 13(1), 15199. https://doi.org/10.1038/s41598-023-42469-4.

Khoukhi, A. (2012). Hybrid soft computing systems for reservoir PVT properties prediction. *Computers & Geosciences*, 44, 109–119. https://doi.org/10.1016/j.cageo.2012.03.016.

Kohzadvand, K., Mahmoudi Kouhi, M., Ghasemi, M. & Shafiei, A. (2024). Novel robust Elman neural network-based predictive models for bubble point oil formation volume factor and solution gas–oil ratio using experimental data. *Neural Computing and Applications*, 1–24. https://doi.org/10.1007/s00521-024-09821-9.

Li, J.-Y., Chow, T. W. & Yu, Y.-L. (1995). The estimation theory and optimization algorithm for the number of hidden units in the higher-order feedforward neural network. *Proceedings of ICNN'95-International Conference on Neural Networks*. https://doi.org/10.1109/ICNN.1995.487330.

Mahdiani, M. R. & Norouzi, M. (2018). A new heuristic model for estimating the oil formation volume factor. *Petroleum*, 4(3), 300–308. https://doi.org/10.1016/j.petlm.2018.03.006.

Malallah, A., Gharbi, R. & Algharaib, M. (2006). Accurate estimation of the world crude oil PVT properties using graphical alternating conditional expectation. *Energy Fuels*, 20(2), 688–698. https://doi.org/10.1021/ef0501750.

Mazandarani, M. T. & Asghari, S. M. (2007). Correlations for predicting solution gas–oil ratio, bubble point pressure and oil formation volume factor at bubble point of Iran crude oils. *European Congress of Chemical Engineering*, Copenhagen. https://folk.ntnu.no/skoge/prost/proceedings/ecce6_sep07/upload/2429.pdf.

Moghadam, J. N., Salahshoor, K. & Kharrat, R. (2011). Introducing a new method for predicting PVT properties of Iranian crude oils by applying artificial neural networks. *Petroleum Science and Technology*, 29(10), 1066–1079. https://doi.org/10.1080/10916460903551040.

Nilizadeh, A., Leavens, G. T., Le, X.-B. D., Păsăreanu, C. S. & Cok, D. R. (2021). Exploring true test overfitting in dynamic automated program repair using formal methods. *14th*

IEEE Conference on Software Testing, Verification and Validation (*ICST*), https://doi.org/10.1109/ICST49551.2021.00033.

Ocampo, I., López, R. R., Camacho-León, S., Nerguizian, V. & Stiharu, I. (2021). Comparative evaluation of artificial neural networks and data analysis in predicting liposome size in a periodic disturbance micromixer. *Micromachines*, 12(10), 1164. https://doi.org/10.3390/mi12101164.

Omar, M. & Todd, A. (1993). Development of new modified black oil correlations for Malaysian crudes. *SPE Asia Pacific Oil and Gas Conference and Exhibition*. https://doi.org/10.2118/25338-MS.

Onwuchekwa, C. (2018). Application of machine learning ideas to reservoir fluid properties estimation. *SPE Nigeria Annual International Conference and Exhibition*. https://doi.org/10.2118/193461-MS.

Otchere, D. A., Ganat, T. O. A., Gholami, R. & Ridha, S. (2021). Application of supervised machine learning paradigms in the prediction of petroleum reservoir properties: Comparative analysis of ANN and SVM models. *J. Pet. Sci. Eng.*, 200, 108182. https://doi.org/10.1016/j.petrol.2020.108182.

Otchere, D. A., Ganat, T. O. A., Ojero, J. O., Tackie-Otoo, B. N. & Taki, M. Y. (2022a). Application of gradient boosting regression model for the evaluation of feature selection techniques in improving reservoir characterisation predictions. *J. Pet. Sci. Eng.*, 208, 109244. https://doi.org/10.1016/j.petrol.2021.109244.

Otchere, D. A., Mohammed, M. A. A., Al-Hadrami, H. & Boakye, T. B. (2022b). Enhancing drilling fluid lost-circulation prediction: Using model agnostic and supervised machine learning. *In: Data Science and Machine Learning Applications in Subsurface Engineering* (pp. 6–32). CRC Press. https://doi.org/10.1201/9781003366980

Petrosky Jr, G. & Farshad, F. (1993). Pressure–volume–temperature correlations for Gulf of Mexico crude oils. *SPE Annual Technical Conference and Exhibition*.

Rammay, M. H. & Abdulraheem, A. (2017). *PVT Correlations for Pakistani Crude Oils Using Artificial Neural Network*. Springer.

Rashidi, S., Mehrad, M., Ghorbani, H., Wood, D. A., Mohamadian, N., Moghadasi, J. & Davoodi, S. (2021). Determination of bubble point pressure & oil formation volume factor of crude oils applying multiple hidden layers extreme learning machine algorithms. *J. Pet. Sci. Eng.*, 202, 108425. https://doi.org/10.1016/j.petrol.2021.108425.

Salehinia, S., Salehinia, Y., Alimadadi, F. & Sadati, S. H. (2016). Forecasting density, oil formation volume factor and bubble point pressure of crude oil systems based on nonlinear system identification approach. *J. Pet. Sci. Eng.*, 147, 47–55. https://doi.org/10.1016/j.petrol.2016.05.008.

Santos, J., Firmiano, A. & Wendland, E. (2021). Numerical results on the residual error indicator for a contaminant transport model. *Trends in Computational and Applied Mathematics*, 22, 341–358.

Seyyedattar, M., Ghiasi, M. M., Zendehboudi, S. & Butt, S. (2020). Determination of bubble point pressure and oil formation volume factor: Extra trees compared with LSSVM-CSA hybrid and ANFIS models. *Fuel*, 269, 116834. https://doi.org/10.1016/j.fuel.2019.116834.

Soliman, A., Attia, A. & Abdelwahab, M. (2020). Development of new models for predicting crude oil bubble point pressure, oil formation volume factor, and solution gas–oil ratio using genetic algorithm. *J. Pet. Min. Eng.*, 22(2), 17–39. https://doi.org/10.21608/jpme.2020.31955.1035.

Standing, M. B. (1947). A pressure–volume–temperature correlation for mixtures of California oils and gases. *Drilling and Production Practice*, API, 275–287.

Sulaimon, A., Ramli, N., Adeyemi, B. & Saaid, I. (2014). New correlation for oil formation volume factor. *SPE Nigeria Annual International Conference and Exhibition*. https://doi.org/10.2118/172396-MS.

Sulaimon, A. A., Turkson, J., Umar, A. A., Abdulwasiu, U., Rehan, A. D. C. & Yusof, M. A. M. (2024). Asphaltene stability prediction using hybrid artificial neural network modeling approach. *SPE Nigeria Annual International Conference and Exhibition*, https://doi.org/10.2118/221598-MS.

Tabaaza, G. A., Tackie-Otoo, B. N., Zaini, D. B., Otchere, D. A. & Lal, B. (2023). Application of machine learning models to predict cytotoxicity of ionic liquids using VolSurf principal properties. *Comput. Toxicol.*, 26, 100266. https://doi.org/10.1016/j.comtox.2023.100266.

Tariq, Z., Mahmoud, M. & Abdulraheem, A. (2021). Machine learning-based improved pressure–volume–temperature correlations for black oil reservoirs. *J. Energy Resour. Technol.*, 143(11), 113003. https://doi.org/10.1115/1.4050579.

Trunk, G. V. (1979). A problem of dimensionality: A simple example. *IEEE Transactions on Pattern Analysis and Machine Intelligence*, 3, 306–307. https://doi.org/10.1109/TPAMI.1979.4766926.

Turkson, J. N., Md Yusof, M. A., Fjelde, I., Sokama-Neuyam, Y. A. & Darkwah-Owusu, V. (2024a). Estimating oil recovery efficiency of carbonated water injection with supervised machine learning paradigms and implications for uncertainty analysis. *SSRN*. https://doi.org/10.2139/ssrn.4815241.

Turkson, J. N., Md Yusof, M. A., Fjelde, I., Sokama-Neuyam, Y. A., Darkwah-Owusu, V. & Tackie-Otoo, B. N. (2024b). Harnessing ensemble learning techniques for accurate interfacial tension estimation in aqueous CO_2 systems. *SPE Gas & Oil Technology Showcase and Conference*. https://doi.org/10.2118/219176-MS.

Vazquez, M. & Beggs, H. D. (1980). Correlations for fluid physical property prediction. *Journal of Petroleum Technology*, 32(06), 968–970, Article SPE-6719-PA. https://doi.org/https://doi.org/10.2118/6719-PA.

Wang, L., Yao, Y., Luo, X., Adenutsi, C. D., Zhao, G. & Lai, F. (2023). A critical review on intelligent optimization algorithms and surrogate models for conventional and unconventional reservoir production optimization. *Fuel*, 350, 128826. https://doi.org/10.1016/j.fuel.2023.128826.

Wang, L., Yao, Y., Wang, K., Adenutsi, C. D., Zhao, G. & Lai, F. (2021). A novel surrogate-assisted multi-objective optimization method for well control parameters based on tri-training. *Natural Resources Research*, 30(6), 4825–4841.

Wang, L., Yao, Y., Wang, K., Adenutsi, C. D., Zhao, G. & Lai, F. J. (2022). Data-driven multi-objective optimization design method for shale gas fracturing parameters. *Journal of Natural Gas Science and Engineering*, 99, 104420. https://doi.org/10.1016/j.jngse.2022.104420.

Watanabe, S. & Yamana, H. (2021). Overfitting measurement of deep neural networks using no data. *2021 IEEE 8th International Conference on Data Science and Advanced Analytics* (*DSAA*). https://doi.org/10.1109/DSAA53316.2021.9564119.

Wood, D. A. & Choubineh, A. (2019). Reliable predictions of oil formation volume factor based on transparent and auditable machine learning approaches. *Adv. Geo-Energy Res.*, 3(3). https://doi.org/10.26804/ager.2019.03.01.

CHAPTER
5

CitricNet: Citrus Quality Grading Using Computer Vision

Yohanes Nuwara[1*], Robertus Theodore[1*], Bayu Dwi Apri Nugroho[2*] and Deny Rahardjo[3]

[1] Agrari, Jakarta, Indonesia
[2] Faculty of Agricultural Technology, Gadjah Mada University, Yogyakarta, Indonesia
[3] Independent, Singapore

Introduction

Recent advancements in computer vision and machine learning have significantly impacted the agricultural sector, especially in quality grading of fruits before or after harvest. A variety of studies have been conducted to develop innovative methods for citrus fruit detection, grading, yield estimation, and disease identification. This literature review highlights key contributions in these areas, showcasing the evolution of techniques from traditional approaches to more sophisticated deep-learning and AI-based systems.

Chen et al. (2018) proposed an automatic orange grading detection method based on computer vision, with a grading accuracy of 94.38%. Chen et al. (2020) proposed a rapid citrus recognition method for detecting different growth periods of citrus in a natural environment, with an average accuracy of 96.04%. Chen et al. (2021) presented a deep-learning-based vision system for fast on-line sorting of defective citrus fruits, with a detection precision of 93.6%. Huang et al. (2022) presented a citrus detection task, combining UAV (Unmanned Aerial Vehicle) data collection, AI embedded device, and target detection algorithm, with a detection accuracy of 93.32%. Malik et al. (2016) conducted research to develop a technique

*Corresponding authors: yohanes@agrari.id; robertus@agrari.id; bayu@tep.ugm.ac.id

DOI: 10.1201/9781003506348-5

to estimate citrus fruit yield from tree images, with a strong correlation between the automated and manual counts, with coefficients of determination R^2 up to 0.99. Patel et al. (2019) conducted research on nondestructive quality assessment of fruits, proposing a new orange size and bacteria spot defect detection and classification system. Saha (2020) proposed a deep learning approach to recognize orange fruit diseases at the initial phase of production, with an overall accuracy of 93.21%. Vijayakumar et al. (2023) investigated tree-level citrus yield prediction using three machine learning (ML) based models, with the second model chosen as the best model due to its low prediction error value and data collection simplicity. Zhang et al. (2022) researched a method of estimating fruit yield using computer vision and deep learning, with an mAP (Mean Average Precision) of 0.957 and mean absolute error (MAE) of 0.081. The research of Mirhaji et al. (2021) examines a method of estimating fruit yield in an orchard with hundreds of trees, with YOLO ((You Only Look Once)v4 model having precision, recall, F1-score, and mAP scores of 91.23%, 92.8%, 92%, and 90.8%, respectively. Finally, Cubero et al. (2016) reviewed recent works that use color and nonstandard computer vision systems for the automated inspection of citrus, such as yield prediction, robotic harvesting, and early detection of dangerous diseases, as well as postharvest handling applications including automated external quality inspection and sorting into commercial categories at high speed. Hyperspectral imaging has enabled the detection of not only external defects, but also their association to dangerous diseases such as citrus cancer, black spot, and citrus Huanglongbing (HLB) disease.

The gap that we discovered from previous works is that the yield estimation directly based on photo is still not implemented. Converting the detection result by model into market information, such as the yield and grade of orange, is important. We have structured this study into two main parts. In the first part, we proposed a new computer vision model called CitricNet, which is used to detect and classify oranges based on coloration and smoothness. In the second part, we developed a calculation workflow to estimate the yield of orange divided by grades based on the detection result by CitricNet.

Methods

Data Collection and Annotation

We photographed 30 orange trees across *Citrus sinensis* L. plantation in Simalungun Regency, Sumatra, Indonesia. The trees are of varying ages: 21 of them are 3 years old, 4 of them are 4 years old, 4 of them are 5 years old, and 1 of them is 8 years old. For each tree, photos were taken from four different directions—north, south, west, and east—at a distance of 3 meters. This resulted in a total of 120 photos for the study, as each tree had four images taken. The camera specifications used to capture these photographs are detailed in Table 1.

Next, we annotated every single orange fruit in the photo based on two criteria, namely, color level and smoothness level. The color level is defined by

Table 1: Imaging system specification

Attribute	Value
Device name	Samsung Galaxy Note 9
Display	6.4″ Super AMOLED
Rear camera effective pixels (megapixels)	12
Rear camera focal length (mm)	26
Rear camera aperture	1.5
Pixel density (ppi)	556

the coloration level of orange differentiated into five classes—0%, 25%, 50%, 75%, and 100%. The smoothness level is defined by the existence level of silver scurfs on the orange surface differentiated into three classes—smooth, mild silver scurfs, and severe silver scurfs. The classification detection is based on the combination of color and smoothness level. To achieve this objective, we create a convention for each of the classes. The naming of annotation is shown inside the brackets in Table 2. For example, an orange that has a perfect orange color (100% orange level) and perfect smoothness is identified as "pw100_mulus," while an orange that has 50% orange color, or in other words appears greenish, and has mild silver scurfs on its surface is identified as "pw50_KH." Figure 1 also shows sample of labelled oranges for every combination of color level (0%, 25%, 50%, 75%, and 100%) and smoothness level (smooth and silver scurfs). In addition, we also annotated small-sized oranges which are not fully grown.

We can notice the difference between two color levels, as it gradually changes its color from green to orange. Also notice the appearance of silver scurfs oranges as they typically have rough dark blotches on their surfaces. Another observation from these images is variability that affects how the appearance of some oranges in the photograph is different from other oranges. There are at least three aspects of variability we can observe. The first aspect, brightness, makes some images look brighter or darker due to sunlight exposure or shadow than other images. The second aspect, resolution, makes some oranges appear more blurred than the others due to the distance of mobile phone camera to the orange trees during photography. The third aspect is disturbance. Some oranges are partly covered by the tree leaves.

Model Development and Architecture

The images are divided into 71 training, 30 validation, and 19 testing images. After annotating and splitting the photos for training, we found out that our data suffers from a significant class imbalance. Class imbalance is a condition where the instance of some classes outnumbers the instance of other classes, or majority over minority classes. The most imbalanced class or the majority class in this dataset is the "pw0_mulus" orange with 198 training instances. Recalling

Category	0% orangeness (pw0)	25% orangeness (pw25)	50% orangeness (pw50)	75% orangeness (pw75)	100% orangeness (pw100)
Smooth (mulus)		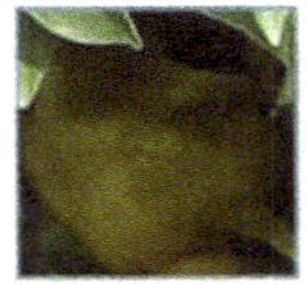	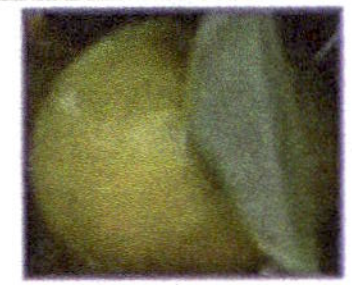	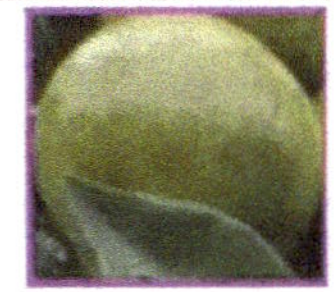	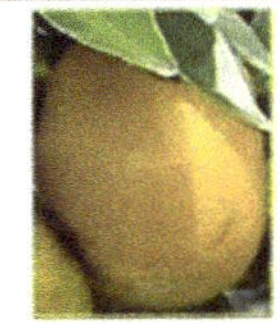
Mild silver scurf (KH)	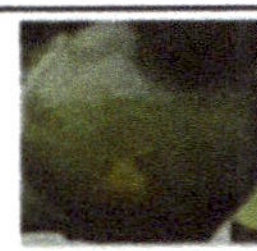	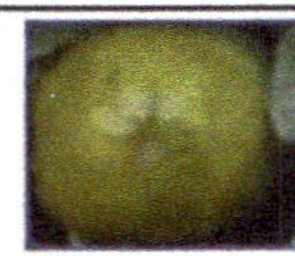	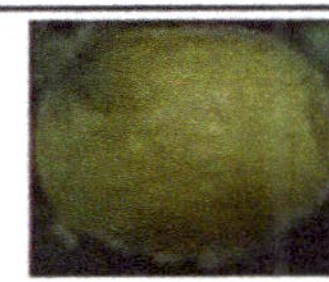	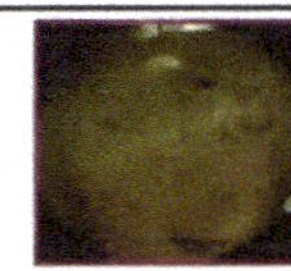	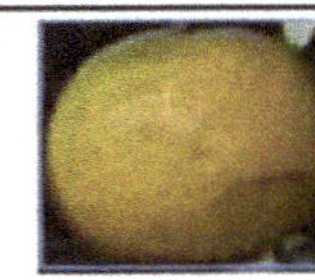
Severe silver scurf (KK)		No sample	No sample	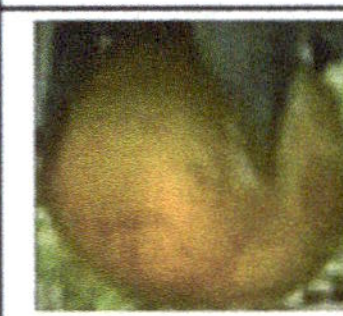	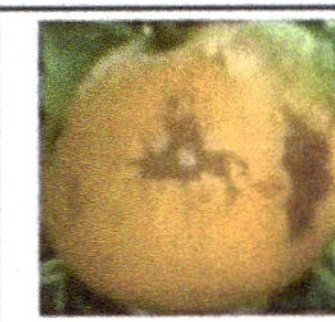

Figure 1: Quality categories of orange based on its color and smoothness, each with sample image. The annotation name is inside brackets.

the previous definition of this class, the "pw0_mulus" is a class of orange that has entirely green color and smooth surface. This is due to the larger population of oranges that are not ripe already for harvesting compared to oranges which are ready for harvesting.

We developed a computer vision model called CitricNet, which consists of two parts. The first part is an object detection model which we used, YOLOv5, and the second part is a classification model, which is built of EfficientNetV2S architecture. YOLOv5 is the fifth generation of YOLO models family released in 2021 by Ultralytics.

A YOLOv8 model is trained on this training sample. We modified the MixUp hyperparameter by increasing the value to 0.4. MixUp is an effective hyperparameter that can deal with the class imbalance by cropping the pictures of fruits from the tree photos, mixing up, and adding them to other random photos such that we increase the number of instances in minority class. We used F1-score as our metric to evaluate the model accuracy on every class.

Quality Grading

The acceptance of orange to the market for preparation and packaging depends on the grade of quality. According to OECD (Organisation for Economic Co-operation and Development) (2010), the quality of citrus is divided into four grades ranked from the highest to lowest, namely, extra, class I, class II, and out of grade. The quality also depends on the appearance of the fruit, basically based on coloration and smoothness. Since we have developed classes for orange, we associated each class with the grade. Figure 2 indicates model architecture for quality grading, while Table 2 displays the grading based on coloration and smoothness.

Table 2: Grade of citrus based on coloration and smoothness category

Orange color percentage (%)	Smooth	Mild scurf	Severe scurf
100	Extra	Class I	Class II
75	Extra	Class I	Class II
50	Class I	Class II	Out of grade
25	Class II	Out of grade	Out of grade
0	Out of grade	Out of grade	Out of grade

Yield Calculation from Computer Vision Result

To calculate the yield or total harvested mass of any fruit, we need to estimate the mass of every orange detected in the photo. The mass of any fruit can be estimated from photo by employing the equation as follows:

$$m = \left(\frac{4}{3}\right)\pi\rho\left(\frac{d}{f_o}\right)^3\left(h_p * \frac{h_s}{h_i}\right)^2 (w_p * w_s/w_i) \quad (1)$$

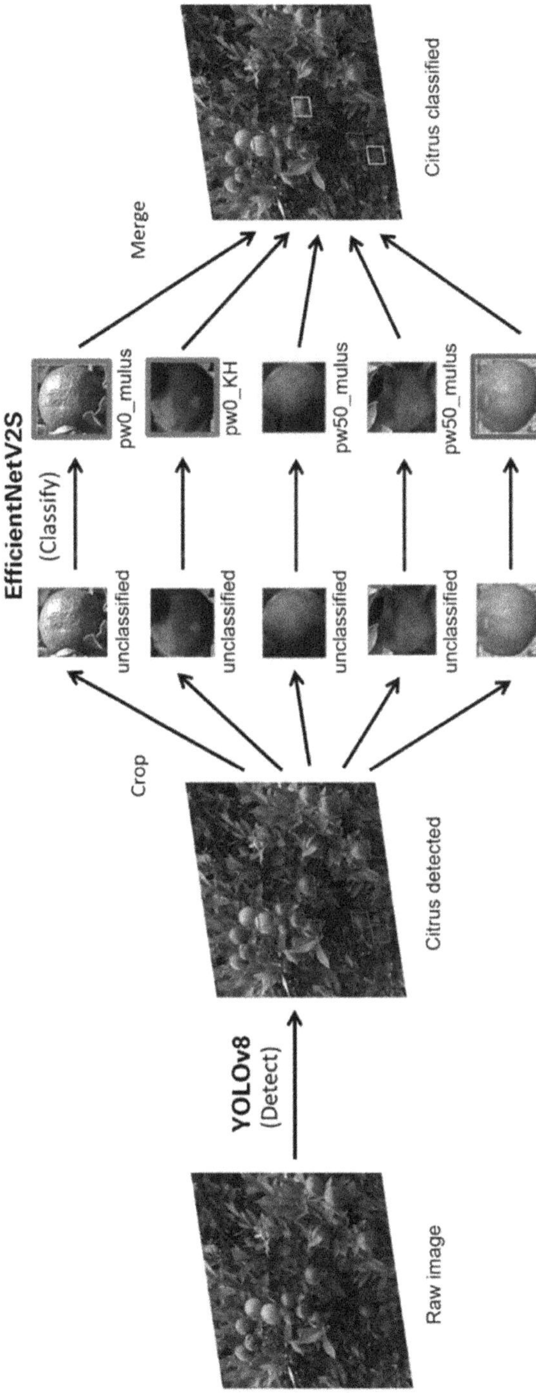

Figure 2: CitricNet model for citrus quality grading.

where ρ is the density of fruit in g/cm^3, d is the distance between the tree (object) and the camera in meters, f_o is the camera focal length in meters, w_p is the object width in pixel, h_p is the object height in pixel, w_i is the photo width in pixel, h_i is the photo height in pixel, w_s is the sensor width in millimeter, and h_s is the sensor height in millimeter. The derivation of this equation can be found in Appendix 1.

The density ρ of a citrus, according to Sharifi et al. (2007), is close to the density of water, which is 1.01 g/cm^3. For a 12-megapixel camera, w_s and h_s are known to be 5.76 mm and 4.29 mm, respectively. The distance d between citrus tree and camera is known to be 3 m, and the f_o of our camera is 26 mm.

By summing all the masses together, we can finally obtain the total yield of oranges from all trees, given the following:

$$Y = \sum_{i=1}^{n} m_i \tag{2}$$

We calculated the yield for each grade of citrus previously mentioned, namely, extra class, class I, class II, and out of grade.

Results

Table 3 shows the performance of our model evaluated by means of precision, recall, and F1-score for every class. In general, the model can accurately detect pw100_mulus oranges with highest accuracy. Note that the model accuracy depends on how much class instance, or number of labels of a class, is present in training set.

Table 3: Classification performance metrics

Class	Number of Instances	Precision	Recall	F1-score
pw100_mulus	23	0.754	0.938	0.836
pw25_mulus	41	0.732	0.935	0.821
pw0_mulus	198	0.711	0.927	0.805
pw75_mulus	11	0.702	0.911	0.793
pw50_mulus	10	0.698	0.906	0.789
jeruk_kecil	463	0.686	0.887	0.774
pw100_KH	3	0.676	0.871	0.761
pw0_KH	4	0.671	0.862	0.755
pw75_KH	3	0.669	0.859	0.752
pw25_KH	8	0.662	0.850	0.744

The class of smooth (or "mulus") has higher accuracy than the class of mild silver scurfs (or "KH") because class "mulus" has more instances than "KH."

We took a sample of tree to present the detection result from our model. Figure 1 shows the result of our orange fruit quality detection model on a single tree captured from all four sides. Note that the annotation names which appear in the photos refer to the description that had been previously discussed. There are also small-sized oranges detected by the model. As it is shown, there are 1 small-sized orange on the north side, 2 on the south side, 2 on the west side, and 4 on the east side of the tree.

Table 4 shows the number of detected orange fruits categorized by the qualities which referred to the detection result in Figure 2. In total, there are 3 oranges with 75% coloration but having mild silver scurfs, 1 orange with 50% coloration but having mild silver scurfs, 1 orange with 25% coloration and smooth, while most oranges with 0% coloration and smooth as much as 19 oranges. With this result and applying the grouping displayed in Table 2, we can infer that this particular tree produced 3 oranges of class I, 2 oranges of class II, and 19 oranges out of grade.

Table 4: Coloration and smoothness across regions

Coloration	Smoothness	North	South	West	East
0%	Smooth	6	3	2	7
25%	Smooth	1	0	0	0
50%	Mild scurf	0	1	0	0
75%	Mild scurf	0	1	2	0

Applying the same grouping of oranges into the three grades for all trees results in the total yield of the crop segregated over grades. Table 5 shows the yield of orange per grade. The best grade of all, extra class, consists of 10% of total yield produced by the crop.

Table 5: Weight distribution by class

Grade	Yield (kg)
Extra class	547
Class I	1094
Class II	2188
Out of grade	1641

Conclusion

In this work, we have proposed a robust computer vision model called CitricNet for quality grading of oranges from citrus plantation crop. This model is able to detect and classify accurately the quality of oranges based on coloration and smoothness with an average F1-score accuracy of 0.785 from 15 different classes.

From this model, we can group all the classes into their quality grades (extra class, class I, class II, and out of grade) and calculate the yield for every grade as shown in Figure 3. We obtained a total of 5,740 kg of oranges, with 10% of total yield belonging to the highest grade, the extra class. We have successfully showed in this study that computer vision can help in quick estimation of total orange yield from any crop and reduce the manual counting done by farmers or agronomists. For future study, this work can be extended by including orange diseases into the model. This application is promising since we can implement this detection for time periods and use for monthly or quarterly inventory.

(a) North side

(b) South side

(c) West side

(d) East side

Figure 3: Result of detected orange fruits appeared in four sides of photo.

Acknowledgments

This was supported by PT Revolusi Agribisnis Berkelanjutan (Agrari), which provided the data needed to build the model and to test it for mock-up implementation.

Appendix A: Derivation of the Mass of Orange Estimation from Photo

First, we need to calculate the real size of every fruit by considering the camera specifications shown in Table 1. The pixel resolution of an image I can be known from the following formulas:

$$w_r = \frac{w_s}{w_I} \tag{3}$$

$$h_r = \frac{h_s}{h_I} \tag{4}$$

where w_r and h_r are the pixel width and height resolution (in mm/pixel), w_s and h_s are the sensor width and height (in mm), and w_I and h_I are the image width and height (in pixels).

Next, we can calculate the size of the object in the photo using:

$$w_o = w_p \cdot w_r \tag{5}$$

$$h_o = h_p \cdot h_r \tag{6}$$

where w_o and h_o are the object width and height on the photo (in mm), and w_p and h_p are the pixel width and height (in pixels).

Accordingly, we can calculate the size of the object using the following formulas:

$$w = w_o \cdot d/f_o \tag{7}$$

$$h = h_o \cdot d/f_o \tag{8}$$

where w and h are the size of the object in the real world (in mm), d is the distance between the object and the camera (in mm), and f_o is the focal length (in mm).

Second, we need to estimate the volume of an orange by assuming it has an ellipsoidal shape. The volume V of an ellipsoid is given by:

$$V = (4/3)\,\pi abc \tag{9}$$

where a, b, and c are the three parameters of the ellipsoid. In this assumption, the a parameter is the width or w (longest diameter), while b and c are the height or h (smallest diameter), which are assumed to be equal. Thus, the volume of an orange can be approximated by:

$$V = (4/3)\ \pi w h^2 \tag{10}$$

Finally, we can derive the estimated mass of an orange by multiplying the volume with the density of an orange, using:

$$m = \rho \cdot V \tag{11}$$

where m is the mass of the orange (in grams), ρ is the density of the orange (in g/cm^3), and V is the volume of the orange (in cm^3).

Substituting the abovementioned equations altogether leads to the final form of equation as follows:

$$m = \left(\frac{4}{3}\right) \pi\rho \left(\frac{d}{f_o}\right)^3 \left(h_p * \frac{h_s}{h_i}\right)^2 (w_p * w_s/w_i) \tag{12}$$

References

Hamzeh Mirhaji, Mohsen Soleymani, Abbas Asakereh & Saman Abdanan Mehdizadeh. (2021). Fruit detection and load estimation of an orange orchard using the YOLO models through simple approaches in different imaging and illumination conditions. *Computers and Electronics in Agriculture*, 191, 106533.

Hardik Patel, Rashmin Prajapati & Milin Patel. (2019). Detection of quality in orange fruit image using SVM classifier. *In: 2019 3rd International Conference on Trends in Electronics and Informatics* (*ICOEI*), pp. 74–78.

Heqing Huang, Tongbin Huang, Zhen Li, Shilei Lyu & Tao Hong. (2022). Design of citrus fruit detection system based on mobile platform and edge computer device. *Sensors*, 22(1).

Mohammad Sharifi, Shahin Rafiee, Alireza Keyhani, Azadeh Jafari, Mobli Hossein, A. Rajabipour & Akram A. (2007). Some physical properties of orange (var. tompson). *International Agrophysics*, 21(01).

OECD. (2010). Citrus fruits. *International Standards for Fruit and Vegetables*. OECD Publishing. Paris, https://doi.org/10.1787/9789264083745-en-fr.

Rupali Saha. (2020). Orange fruit disease classification using deep learning approach. *International Journal of Advanced Trends in Computer Science and Engineering*, 9(2), 2297–2301.

Sergio Cubero, W. S., Lee, N., Aleixos, Francisco Albert & Jose Blasco. (2016). Automated systems based on machine vision for inspecting citrus fruits from the field to postharvest—A review. *Food and Bioprocess Technology*, 9(10).

Vinay Vijayakumar, Yiannis Ampatzidis & Lucas Costa. (2023). Tree-level citrus yield prediction utilizing ground and aerial machine vision and machine learning. *Smart Agricultural Technology*, 3, 100077.

Wenkang Chen, Shenglian Lu, Binghao Liu, Guo Li & Tingting Qian. (2020). Detecting citrus in orchard environment by using improved YOLOv4. *Scientific Programming*, 2020(1), 8859237.

Xiaohua Zhang, Arash Toudeshki, Reza Ehsani, Haoling Li, Wenfeng Zhang & Ruijun Ma. (2022). Yield estimation of citrus fruit using rapid image processing in natural background. *Smart Agricultural Technology*, 2, 100027.

Yaohui Chen, Xiaosong An, Shumin Gao, Shanjun Li & Hanwen Kang. (2021). A deep learning-based vision system combining detection and tracking for fast on-line citrus sorting. *Frontiers in Plant Science*, 12.

Yuwei Chen, Jinghua Wu & Mengtian Cui. (2018). Automatic classification and detection of oranges based on computer vision. *In: 2018 IEEE 4th International Conference on Computer and Communications* (*ICCC*), pp. 1551–1556.

Zeeshan Malik, Sheikh Ziauddin, Ahmad Shahid & Asad Safi. (2016). Detection and counting of on-tree citrus fruit for crop yield estimation. *International Journal of Advanced Computer Science and Applications*, 7(05).

CHAPTER

6

Comparative Analysis of Machine Learning Techniques for Estimating CO_2 Absorption in Ionic Liquids and the Implications for CO_2 Capture

Joshua Nsiah Turkson[1,2,3*], Muhammad Aslam Md Yusof[1,2], Yen Adams Sokama-Neuyam[4], Ingebret Fjelde[5], Caspar Daniel Adenutsi[3], Bennet Nii Tackie-Otoo[6], Victor Darkwah-Owusu[1,2] and Gideon Gyamfi[7]

[1] Petroleum Engineering Department, Faculty of Geoscience and Petroleum Engineering, Universiti Teknologi PETRONAS, 32610 Bandar Seri Iskandar, Perak Darul Ridzuan, Malaysia

[2] Centre of Reservoir Dynamics (COReD), Institute of Sustainable Energy & Resources (ISER), Universiti Teknologi PETRONAS, 32610 Bandar Seri Iskandar, Perak Darul Ridzuan, Malaysia

[3] Department of Chemical Engineering, University of Virginia, 385 McCormick Rd, Charlottesville, VA 22903, USA

[4] Department of Petroleum Engineering, Faculty of Civil and Geo-Engineering, Kwame Nkrumah University of Science and Technology, Kumasi, Ghana

[5] Department of Energy & Technology, NORCE Norwegian Research Centre AS, 4021 Stavanger, Norway

[6] Department of Petroleum Engineering, College of Petroleum Engineering and Geosciences, King Fahd University of Petroleum & Minerals, Dhahran 31261, Saudi Arabia

[7] Chemical and Petroleum Engineering, University of Kansas, Lawrence, Kansas 66045, USA

Introduction

Average global temperature increment, melting of glaciers, displacement, submerging of islands, and heat waves are a couple of consequences of climate change and global warming. It is therefore not surprising that the pair of climate

*Corresponding author: joshua_22009915@utp.edu.my

DOI: 10.1201/9781003506348-6

change and global warming has been described as a threat to global health and security (Sokama-Neuyam et al., 2022; Turkson et al., 2024c). Climate change and global warming are attributed to the increasing atmospheric CO_2 concentrations. Carbon capture and storage (CCS) plays a pivotal role in surmounting the extant challenge of high atmospheric CO_2 concentrations (Darkwah-Owusu et al., 2024a). Other options in the carbon emission reduction portfolio include renewable energy, hydrogen energy, and improvement in energy efficiency (Darkwah-Owusu et al., 2024b; Tackie-Otoo & Haq, 2024; Turkson et al., 2024c). CCS technologies include approaches to capturing CO_2 emissions from processes and facilities, followed by its transport and secure isolation (mostly in geological formations such as saline aquifers, depleted hydrocarbon reservoirs, coal seams, etc.) (Sokama-Neuyam et al., 2022).

Developing efficient and cost-effective methods for CO_2 capture from flue gases and other emission sources is central to the success of CCS technologies, especially considering that CO_2 capture is one of the most expensive steps in the CCS value chain. A recent commentary by energy experts at the International Energy Agency posited that the capture cost varies significantly depending on the CO_2 source, ranging from US$15–25 per ton CO_2 for industrial processes that produce "pure" or highly concentrated CO_2 streams (e.g., ethanol production) to US$40–120 per ton CO_2 for processes with more dilute gas streams, such as cement production and power generation, while direct air capture was pegged at disproportionate US$134–342 per ton CO_2 (IEA, 2021). Hence, exploring novel solvents and sorbents with selective CO_2 capture capability has gained significant attention from researchers (Borhani & Wang, 2019; Kondapi, 2024). Ionic liquids (ILs) have garnered considerable traction in this regard. Ionic liquids are characterized by low volatility, high thermal stability and conductivity, and tunable solubility (Ghazani et al., 2018). These characteristics plus their ability to dissolve organometallic, organic, and inorganic compounds including CO_2 make them attractive candidates for integration into CCS processes. Ionic liquids have demonstrated versatile applications across multiple areas including enhanced oil recovery (Atilhan & Aparicio, 2021; Bera & Belhaj, 2016), metal extraction (Asrami et al., 2021), wastewater treatment (Goutham et al., 2022), etc.

CO_2 absorption or solubility in ionic liquids is paramount during carbon capture. To this end, a good number of experimental studies have investigated this crucial parameter (Henni et al., 2023; Jalili et al., 2019; Kang et al., 2020; Noorani & Mehrdad, 2020; Noorani et al., 2021), but the technique is time-consuming and costly given the vast range of available ionic liquids (Venkatraman & Alsberg, 2017). Predictive models emerged as alternatives to the conventional experimental technique to provide fast and reliable prediction of the physical property. These models encompass quantitative structure–property relationships (QSPR) and classical thermodynamics-based methods such as Universal Quasichemical (UNIQUAC) Functional-Group Activity Coefficients (UNIFAC), non-random-two-liquid (NRTL) model, and molecular dynamics (Venkatraman & Alsberg, 2017). These methods are also relatively complex, involve intricate mathematical operations, computationally intensive, and often require an in-depth understanding of molecular thermodynamics, limiting the scalability of these techniques. Due to the limitations of these conventional techniques, recent studies have also deployed machine learning, a subset of artificial intelligence, to predict the solubility of CO_2 and other acidic gases (H_2S, SO_2, etc.) in different systems because of the significant advantages the latter offer (Aghaie & Zendehboudi, 2020; Amar et al., 2023; Baghban et al., 2017; Feng et al., 2022; Mohammadi et al., 2022; Mohan et al., 2024; Mokarizadeh et al., 2020; Mousavi et al., 2023; Ouaer et al., 2019; Tatar et al., 2016).

Machine learning (ML) models can efficiently process and analyze large datasets, enabling the prediction of CO_2 solubility in a wide array of ionic liquids without the need for extensive experimental trials. These paradigms can also rapidly adapt to different ionic liquid compositions, handle complex nonlinear relationships, and offer a scalable alternative to traditional methods. Additionally, ML models can reduce the dependency on expert-level thermodynamic knowledge, overcoming the inherent limitations of classical thermodynamic models. Recently, Kuroki et al. (2023) employed Gaussian process regression (GPR) to explore CO_2 solubility in over 400,000 ionic liquids. The study aimed to identify the optimal cation–anion combination for CO_2 absorption from 6933 cations and 58 anions. Via the ML-assisted approach, [P_{66614}][PFOS] (trihexyl(tetradecyl) phosphonium perfluorooctanesulfonate) was found to demonstrate higher CO_2 solubility than the documented [P_{66614}][TFSA] (trihexyl(tetradecyl)phosphonium bis(trifluoromethanesulfonyl)imide), which was confirmed in subsequent experiments. Ouaer et al. (2019) proposed two data-driven connectionist models (multilayer perceptron [MLP] and gene expression programming) to predict CO_2 solubility in ionic liquids, honoring 744 experimental data points derived from 13 ionic liquids. The models achieved impressive predictive performance, recording coefficient of determination (R^2) > 0.98 and root mean square error (RMSE) < 0.03. The multilayer perceptron optimized with Levenberg–Marquardt (MLP-LM) algorithm emerged as the outstanding model with R^2 and RMSE of 0.9941 and 0.015, respectively. Moreover, the MLP-LM model outperformed the Peng–Robinson (PR) and Soave–Redlich–Kwong (SRK) equations of state (EoS) in a comparative analysis using [bmim][PF6] (1-butyl-3-methylimidazolium hexafluorophosphate) and [bmim][Tf2N] (1-butyl-3-methylimidazolium bis(trifluoromethanesulfonyl)imide). In contrast, the gene expression programming model was the bottom-placed model with R^2 and RMSE of 0.9854 and 0.0238, respectively. Mousavi et al. (2023) also deployed diverse computational techniques including white-box machine learning, deep learning, and ensemble learning to estimate the solubility of H_2S in 37 ionic liquids. The extreme gradient boosting model was adjudged as the most precise among the four developed models (group method of data handling, genetic programming, and deep belief network). Specifically, the model yielded an R^2, RMSE, and average absolute percent relative error (AAPRE) of 0.99, 0.002, and 1.14%, respectively. The modeling study also revealed an inclined trend between alkyl chain length and H_2S solubility in ionic liquids.

Mohan et al. (2024) developed a physics-driven machine learning model to estimate the CO_2 solubility in 149 chemically reactive deep eutectic solvents. The artificial neural network (ANN) achieved optimal predictive performance with an R^2, RMSE, and mean absolute error (MAE) of 0.989, 0.051, and 0.029, respectively, compared to R^2 of 0.790–0.837, RMSE of 0.182–0.238, and MAE of 0.122–0.154 recorded by random forest (RF), support vector regression (SVR), and gradient boosted tree models collectively. Similarly, Baghban et al. (2017) predicted CO_2 solubility in ionic liquids using MLP, adaptive neuro-fuzzy inference system (ANFIS), least square support vector machine (LSSVM), and radial basis function artificial neural network (RBF-ANN). LSSVM demonstrated exceptional predictive capabilities against its counterparts, obtaining an R^2 and MSE of 0.9942 and 0.00035, respectively. In an earlier study, Tatar et al. (2016) developed four data-driven models (RBF, MLP, LSSVM, and committee machine intelligent system (CMIS)) to estimate CO_2 solubility in 14 ionic liquids. The models provided accurate CO_2 solubility predictions. Notwithstanding, the CMIS model demonstrated superior predictive capabilities (R^2 of 0.9992 and RMSE of 0.0059) over developed models in the study and PR-EoS (R^2 of 0.7323 and RMSE of 0.0520) and SRK-EoS (R^2 of 0.6989 and RMSE of 0.0747).

The extensive research outlined in the earlier text highlights the demand for accurate and efficient models to estimate CO_2 absorption in ionic liquids. The current work therefore aims to make significant advancements in the field. Specifically, the study focuses on developing diverse ML models for predicting equilibrium CO_2 absorption in ionic liquids, leveraging key thermophysical properties (such as temperature and pressure) and thermodynamic properties (such as critical temperature, critical pressure, and acentric factor) as predictors because of their direct influence on CO_2 solubility in ionic liquids and their established relevance in gas absorption phenomena.

Temperature and pressure are known to significantly modulate gas solubility, with temperature affecting the solvation energy and pressure impacting the gas–liquid equilibrium by altering CO_2 partial pressure (Pelaquim et al., 2021). The inclusion of critical temperature, critical pressure, and acentric factor further refines the predictive model. These properties are intrinsic to characterizing phase behavior and fluid compressibility, which are critical to understanding and modeling gas–liquid interactions at various operational conditions. Notably, they capture non-idealities that would not be accounted for by temperature and pressure alone, enhancing the robustness of the model in predicting CO_2 solubility across different IL compositions and conditions.

Given this, we contributed to the existing literature by developing non-parametric singular and ensemble models to estimate the equilibrium CO_2 absorption in ionic liquids using one of the largest databases ever considered in the literature to the best of our knowledge. This extensive dataset and predictor choice enables the models to perform with high accuracy and broad applicability. We also examined the influence of thermophysical conditions and thermodynamic properties on CO_2 absorption in ionic liquids to highlight the underexplored phenomenon from a computational modeling perspective.

Beyond accuracy, the developed paradigms promise considerable cost savings and operational efficiency as reliable validation tools for experimental data, thus enabling rapid testing of new ionic liquids for CO_2 capture. To further demonstrate the robustness and wide applicability of the developed paradigms, we conducted an extensive comparative analysis, benchmarking the outstanding model from our study against models developed by fellow scholars. This analysis showcases the superiority of the developed paradigm over existing ones, emphasizing the paradigm shift introduced by the nonparametric ensemble technique. Finally, we discuss the implications of the developed paradigm for CO_2 capture. Our work sets a new yardstick for predictive capabilities in CO_2 capture, with implications extending to large-scale applications and experimental research validation.

Machine Learning Paradigms and Applications

The study focused on distinct machine learning models including support vector regression, decision tree (DT), random forest, and gradient boosting regression (GBR) owing to their remarkable performance across multiple tasks as outlined in Table 1 (Otchere et al., 2022a; Tabaaza et al., 2023; Turkson et al., 2024a; Turkson et al., 2024b). Moreover, the selected models ensure a robust and comprehensive application of different categories of ML models spanning from singular to parallel and sequential ensemble paradigms. The diverse ML paradigms employed in the current study are elaborated on in the following subsections.

Support Vector Regression (SVR)

SVR is an adaptation of the support vector machine (SVM) tailored for regression analysis. It aims to maximize the margin between support vectors (the closest data points) and a hyperplane that defines the decision boundary. Unlike conventional SVMs that penalize every classification error, SVR minimizes the discrepancy between actual and predicted values within a defined tolerance margin using an epsilon-insensitive loss function (Huang et al., 2021; Mardhatillah et al., 2022). This feature enhances the robustness of SVR against outliers and noise. Additionally, SVR can address nonlinear relationships by transforming input data into higher-dimensional spaces with kernel functions, enabling linear separation. This capability allows SVR to effectively handle complex nonlinear regression problems, such as predicting oil recovery efficiency (Huang et al., 2021; Turkson et al., 2024a).

Decision Tree (DT)

A DT is a supervised learning technique used to predict target values in a dataset through a series of binary decisions (Turkson et al., 2024a). The process begins with a root node that splits the data into smaller subsets based on specific feature values. This splitting continues recursively until certain stopping conditions are met, such as reaching a maximum tree depth, having a minimum number

Table 1: Application of ML paradigms across diverse fields

ML Models	Application	Key Findings/Remarks	Reference
DT, RF, SVR	Estimation of oil recovery efficiency	• Developed paradigms successfully captured the relationships between variables with a considerable level of accuracy, precision, and consistency. • RF outperformed DT and SVR, achieving an R^2, RMSE, and MAE of 0.87, 4.53%, and 3.95%, respectively.	(Turkson et al., 2024a)
GBR, LightGBM	CO_2/brine interfacial tension prediction	• Both models yielded exceptional predictive performance, recording an $R^2 > 0.98$, RMSE < 2.00 mN/m, MAE < 1.2 mN/m, and AAPRE < 1.5%. • GBR outperformed LightGBM and other equally robust ML models in the literature in terms of accuracy, precision, and consistency.	(Turkson et al., 2024b)
GPR, RBF, RF	CO_2/brine interfacial tension modeling	• RF model outperformed GPR and RBF, achieving an R^2 of 0.9907, AAPRE of 1.9156%, and RMSE of 0.7279 mN/m. • Model facilitated in structural trapping capacity assessment of possible CO_2 storage site.	(Safaei-Farouji et al., 2022b)
DT, ET, GBR	Surfactant/ hydrocarbon interfacial tension prediction	• Models precisely predicted the system IFT (interfacial tension) with AAPRE of 2.71%–4.12%. • The parameters with significant influence on the system IFT included phase inversion temperature, surfactant concentration, and hydrophilic–lipophilic balance.	(Rashidi-Khaniabadi et al., 2023)
RF	Reservoir facies classification	• The model facilitated the interpretation process in classifying the reservoir facies with high accuracy (95%). • Integration of ensemble modeling with geostatistical techniques (sequential indicator simulation and indicator-kriging) improved the accuracy of the reservoir model.	(Rahimi & Riahi, 2022)
ANN, CMIS, DT, RF, SVM	Low-salinity water injection performance prediction	• RF emerged as the best-performing model with an R^2, RMSE, and AAPRE of 0.8399, 5.757, and 3.302 on out-of-sample data predictions, respectively. • Model can be beneficial for enhanced oil recovery optimization in carbonate oil reservoirs.	(Shafiei et al., 2022)

(Contd.)

(*Contd.*)

ML Models	Application	Key Findings/Remarks	Reference
MLP, RF, SVR, XGB	Modeling CO_2 storage in deep saline aquifers	• RF and XGB demonstrated impressive CO_2 trapping performance prediction during cross-validation. • The RF model provided crucial insights into the influential parameters such as reservoir monitoring and rock properties on CO_2 trapping performance, highlighting the potential application of the model to optimize CO_2 trapping in deep saline aquifers.	(Khanal & Shahriar, 2022)

ANN: Artificial neural network; CMIS: Committee machine intelligent system; DT: Decision tree; ET: Extremely randomized trees; GBR: Gradient boosting regression; GPR: Gaussian process regression; LightGBM: Light gradient boosting machine; MAE: Mean absolute error; MLP: Multilayer perceptron; RBF: Radial basis function; RF: Random forest; RMSE: Root mean square error; SVR: Support vector regression; XGB: Extreme gradient boosting.

of samples per node, or when further splits do not enhance node purity. Each terminal node, or leaf, represents a final prediction or decision. Decision trees are appreciated for their simplicity and interpretability, as they offer transparent decision-making paths (Pekel, 2020). Despite their tendency to overfit and potential limitations in capturing complex relationships, DTs are extensively used across various fields due to their user-friendliness and effectiveness in many situations (Alanazi et al., 2023; Tabaaza et al., 2023).

Random Forest (RF)

Developed by Breiman (2001), RF is a robust ensemble learning technique that enhances prediction accuracy and reduces overfitting by combining multiple decision trees through bagging/bootstrap aggregation. Each DT in an RF is trained on a random subset of the dataset, which introduces diversity among the trees. This diversity allows the RF model to capture a broad range of patterns and generalize effectively to new data. The final prediction from an RF is obtained by averaging the predictions of all the trees in the case of regression tasks or by majority voting for classification tasks. This method leverages the collective predictive capabilities of the individual trees to provide robust and reliable predictions across various predictive modeling scenarios, making RF a versatile and widely adopted ML technique (Khanal & Shahriar, 2022; Kim et al., 2023; Safaei-Farouji et al., 2022a; Turkson et al., 2024a; Vo Thanh et al., 2023).

Gradient Boosting Regression (GBR)

GBR is a sophisticated boosting method in supervised machine learning proposed by Friedman (2002). GBR focuses on building weak learners sequentially to enhance model performance. Each weak learner, typically a decision tree, is trained to correct the errors made by the previous learners, using a gradient descent approach to minimize the loss function (Otchere et al., 2022a; Tabaaza et al., 2023; Turkson et al., 2024b). This iterative refinement process allows the model to progressively improve its predictive accuracy with each new learner added to the ensemble. The process starts by fitting the initial model to the data, then iteratively adds new models to reduce the residual errors of the combined previous models. The primary advantage of gradient boosting is its ability to handle complex, nonlinear relationships within the data, making it suitable for a wide range of predictive tasks. Each iteration of the gradient boosting algorithm fits a new learner to the residuals of the existing ensemble, ensuring that the model continually focuses on the most challenging aspects of the data. This results in a highly accurate and robust predictive model that can generalize well to new, unseen data. Modern machine learning techniques, such as adaptive, light gradient, and extreme gradient boosting, are built on the principles of gradient boosting. These methods have been tailored to enhance computational efficiency and scalability, allowing them to handle large datasets and complex models with ease (Otchere et al., 2022a; Tabaaza et al., 2023).

Methodology

The modeling workflow commenced with the acquisition of relevant data from the literature (Baghban et al., 2017). The predictors included thermophysical properties (temperature and pressure) and thermodynamic properties (critical temperature, critical pressure, and acentric factor) with solubility as the output target (Eq. 1). Figure 1 shows the detailed modeling workflow for the study.

$$\chi = f(T, P, T_c, P_c, \omega) \tag{1}$$

Dataset Acquisition and Description

A comprehensive and extensive database is required for developing robust and reliable paradigms. The current study adapted the data reported by other scholars (Baghban et al., 2017). The instances spanned 60+ ionic liquids. Extensive information on the data sources can be found elsewhere in the literature (Baghban et al., 2017). Notwithstanding, Table 2 provides the statistical summary of the database.

Data Exploration and Preprocessing

Data preprocessing encompassed identifying and handling missing and duplicate values within the database. Missing data within the database can be handled by imputation with a statistical value (mean, median, or mode) or removed. None of these techniques were implemented due to the completeness of the database. Duplicate rows were also identified and removed from the database.

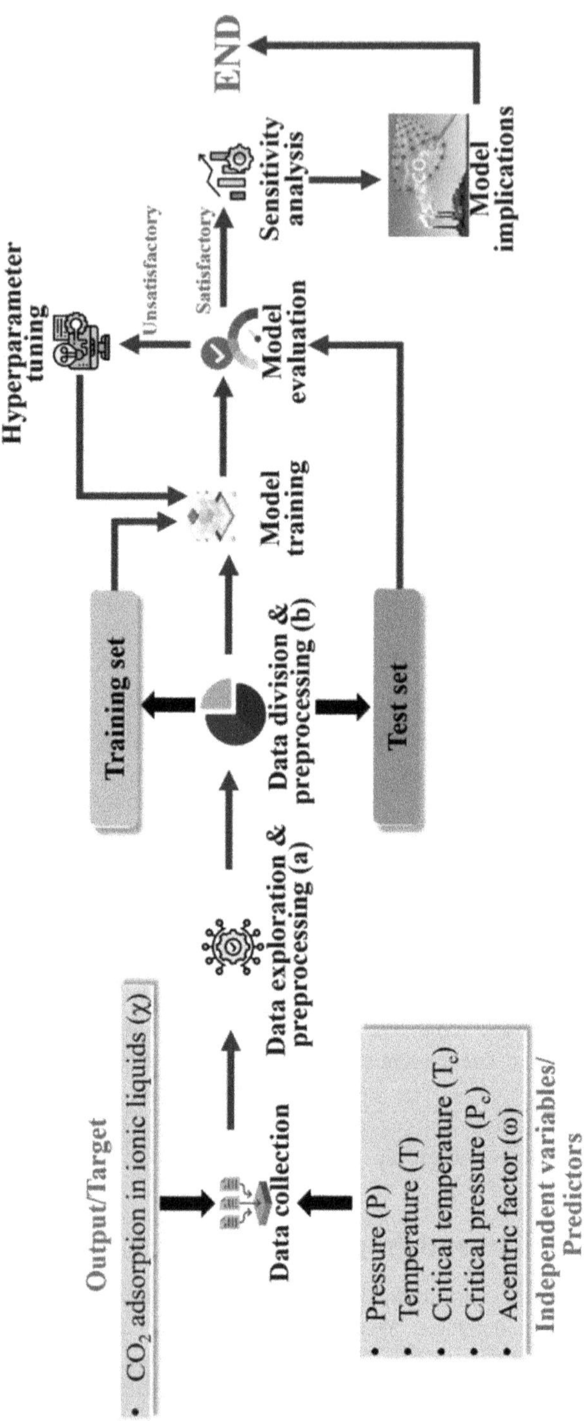

Figure 1: Modeling workflow for estimating CO_2 absorption in ionic liquids.

Table 2: Range of instances for CO_2 solubility in ionic liquids

	P **(MPa)**	***T*** **(K)**	P_c **(MPa)**	T_c **(K)**	ω	χ **(Mole Fraction)**
Count	5223	5223	5223	5223	5223	5223
Mean	330.7	10.0	921.8	3.7	0.9	0.406
Standard Deviation	26.6	15.5	248.4	6.7	0.3	0.244
Minimum	271.1	0.0	585.3	0.8	0.3	0.000
Maximum	453.2	100.1	1604.8	46.3	1.7	0.968
Median	324.2	4.5	831.4	2.2	0.9	0.401
First Quartile	313.2	1.5	753.7	1.8	0.7	0.200
Third Quartile	344.6	10.3	1073.7	2.8	1.2	0.605
Skewness	1.01	2.88	1.17	5.10	0.001	0.071
Kurtosis	2.06	9.09	0.66	26.04	–0.88	–1.114

Here, the first instance of rows of data with the same independent (input) and dependent (target) parameters were retained, and the remaining were removed. Additionally, rows of data with the same independent parameters but different dependent parameters were removed altogether. Consequently, 232 duplicate points were eliminated from the database, reducing the dimensions from (5445, 7) to (5223, 7).

The data was then split into a 70:30 ratio for the training and test sets. The input features varied in scale as depicted in Table 2 and as such required data scaling. However, tree-based models such as random forest and gradient boosting are not influenced by the scale of input features due to their reliance on comparisons between feature thresholds. Instead of considering the distances between data points, these models make decisions by evaluating feature thresholds and partitioning the data at points that optimize certain criteria. In contrast, the support vector regression is scale-dependent, requiring data scaling. Hence, the training and test sets were independently normalized to a standard range of –1 to 1 to prevent data leakage and ensure data integrity.

Multicollinearity is also undesirable in model development. Multicollinearity can distort regression estimates, inflate standard errors, and reduce the reliability of *p*-tests and *t*-tests (Otchere, 2023). Diverse correlation techniques (Spearman and Kendall) were employed to ascertain the feature–feature and feature–target relationships and eliminate possible collinear features. Spearman correlation (Eq. 2) is based on the ranks of the data and is sensitive to the magnitude of the differences between ranks, while Kendall correlation (Eq. 3) focuses solely on the concordance or discordance of the ranks (Tariq et al., 2023). Spearman and Kendall are nonparametric rank correlations independent of normal distribution. The techniques were used concurrently to validate the consistency of the observed relationships to enhance the credibility of the findings. Multiple scholars proposed $-0.8 <$ correlation coefficient < 0.8 (corresponding to a variance inflation factor of 2.78) as the criteria for identifying collinear features (Otchere et al., 2022a; Tabaaza et al., 2023; Turkson et al., 2024a; Turkson et al., 2024b).

$$\rho_{\text{Spearman}} = \frac{\left[\frac{\sum (x_i - \mu_x)(y_i - \mu_y)}{n-1}\right]}{\sigma_x \, \sigma_y} \tag{2}$$

$$\tau_{\text{Kendall}} = \frac{2\,(n_c - n_d)}{k\,(k-1)} \tag{3}$$

where n is the total number of instances, x_i and y_i are the input features and output target (i.e., IFT), μ_x and μ_y are the averages of the input features and output target, σ_x and σ_y are the standard deviation of an input feature and target, n_c and n_d are the number of concordant and discordant pairs.

Model Development

Model training and testing are critical components of the ML framework. In this study, 70% of the data (representing 3,656 instances) was allocated for training, while the remaining 1,567 instances were reserved for testing. To mitigate the risks of overfitting or underfitting, several strategies were implemented throughout the model development process.

The training phase included hyperparameter optimization for each model using the randomized search technique in combination with fivefold cross-validation. For example, SVR parameters such as the regularization parameter (C), epsilon, and kernel type were systematically fine-tuned. Randomized search was chosen due to its ability to explore a wider range of hyperparameters in a computationally efficient manner. Compared to grid search, which can become computationally expensive as the number of hyperparameters, and their possible values, increases, randomized search can achieve comparable or even better results with fewer iterations. Additionally, unlike Bayesian optimization, which involves complex probabilistic models and requires more computational overhead, randomized search offers a good balance of efficiency and effectiveness, making it particularly suitable for models with large hyperparameter spaces (Bergstra & Bengio, 2012). The search range of the hyperparameters for each model and the corresponding optimum option are presented in Table 3.

Additionally, ensemble models such as RF and GBR included mechanisms to address overfitting risks. RF, for instance, inherently reduces overfitting through the averaging of predictions from multiple decision trees trained on random subsets of the data. Nevertheless, RF model hyperparameters such as number of decision trees, features, and maximum tree depth were fine-tuned to avoid overfitting. GBR also leverages sequential learning but requires specific controls to prevent overfitting. To mitigate this, hyperparameters such as the learning rate, the number of boosting stages/trees, and maximum tree depth were fine-tuned as shown in Table 3.

Table 3: Search range of model hyperparameters and optimum option

Model	Hyperparameter	Search Range	Optimum Option
Support vector regression	Kernel	[linear, RBF, polynomial]	RBF
	C	10–10000	1000
	Epsilon	0.01–10	0.01
Decision tree	Maximum number of features	[auto, sqrt, log2]	log2
	Maximum tree depth	3–20	18
	Minimum samples split	2–10	8
	Minimum samples leaf	2–10	2
Random forest	Number of independent trees	50–200	194
	Maximum number of features	[auto, sqrt, log2]	log2
	Maximum tree depth	3–20	17
Gradient boosting regression	Number of independent trees	100–500	441
	Maximum tree depth	3–20	6
	Learning rate	0.01–0.3	0.125

The robustness of the models was further ensured through the use of fivefold cross-validation during training. This approach divides the training data into five subsets, iteratively using one subset for validation while training on the others. This methodology helps assess model performance on unseen data and minimizes the likelihood of overfitting.

After training, the models were tested using 30% of the dataset (out-of-sample data) to evaluate generalization capabilities. To validate further, the models were also applied to the entire dataset, confirming consistency across training, test, and complete datasets. To assess model performance, visualizations and numerical statistical measures including the coefficient of determination (R^2), average absolute error (AAE), and RMSE, described in Eq. 4 through Eq. 6 (Gyamfi et al., 2024; Ouaer et al., 2019; Turkson et al., 2024a), were employed.

$$R^2 = 1 - \frac{\sum_{i=1}^{n} (\chi_{act_i} - \chi_{pred_i})^2}{\sum_{i=1}^{n} (\chi_{act_i} - \chi_{avg})^2} \tag{4}$$

$$\text{AAE} = \frac{1}{n}\sum_{i=1}^{n} \left| \chi_{act_i} - \chi_{pred_i} \right| \tag{5}$$

$$\text{RMSE} = \sqrt{\frac{1}{n}\sum_{i=1}^{n} (\chi_{act_i} - \chi_{pred_i})^2} \tag{6}$$

where n is the number of instances, χ_{act_i} is the observed CO_2 absorption, χ_{pred_i} is the predicted CO_2 absorption, and χ_{avg} is the average CO_2 absorption. The models were implemented in the Jupyter Integrated Development Environment using the scikit-learn library (Pedregosa et al., 2011).

Feature Importance

Feature importance is a critical aspect of ML, since it reveals the significance of input features in determining the model output. This analysis helps identify features with strong influence and vice versa. In this study, diverse techniques including correlations, permutation importance, and SHapley Additive exPlanations (SHAP) were employed for the analysis. Correlation analysis reveals linear relationships between features, while permutation importance assesses the influence of each feature on model performance by randomly shuffling its values. SHAP provides a more detailed view by assigning each feature a contribution value for every individual prediction, offering a clearer understanding of the effect of each feature on the model prediction (Otchere et al., 2022b; Turkson et al., 2024a).

Statistical Validity and Model Reliability Assessment

It is critical to ensure that model results are not unduly affected by a small subset of the data. Therefore, the leverage technique was adapted to assess the statistical validity of the database and model reliability. Leverage signifies the extent of deviation of an independent variable from its mean. High leverage points have a significant impact on the estimation of regression coefficients because they lie far from the center of the data distribution. The leverage values are derived from the hat matrix (H), which is obtained in the context of least squares regression. The hat matrix maps the observed values to the fitted values. The Wiliams plot combines the leverage values with standardized residuals (a measure of the discrepancy between observed and predicted values), which helps identify data categories (Table 4) and instances far from the mean of the predictors and poorly fitted points. For a given set of predictors X, the hat matrix (H) and standardized residuals (SR) are calculated using Eqs. 7 and 8, while the critical leverage is determined using Eq. 9:

$$H = X\,(X^{T} X)^{-1}\,X^{T} \tag{7}$$

$$SR_i = \frac{\chi_{act_i} - \chi_{pred_i}}{\sigma\sqrt{1 - h_i}} \tag{8}$$

$$H^* = \frac{3a + 3}{b} \tag{9}$$

where X is an $a \times b$ matrix, a and b denote the number of instances and input features respectively; χ_{pred_i} and χ_{act_i} are observed and predicted CO_2 absorption

Table 4: Data categories based on the leverage approach (Ouaer et al., 2019; Turkson et al., 2024a).

Category	**Criteria**
Valid data	$0 \leq H \leq H^*$ and $-3 \leq$ SR ≤ 3
Vertical suspect data	$H < H^*$ and SR<−3 or SR > 3
Good high-leverage data	$H > H^*$ and $-3 \leq$ SR ≤ 3
Bad leverage data	$H > H^*$ and SR<−3 or SR > 3

H: Leverage or hat value; H^*: Critical leverage limit; SR: Standardized residuals.

values, respectively; σ represents standard deviation; SR_i and h_i are the standardized residual and leverage value of the *i*th instance; H^* is the critical leverage limit. The ordinary least squares (OLS) regression module was used to obtain the standardized residuals and leverage values.

Results and Discussion

Exploratory Data Analysis

Data analysis provides insights into the data distributions and variable relationships. Pairplot distributions and covariance heatmaps were thus generated to achieve these insights. It is evident that the predictors exhibit varying levels of skewness and peakedness, as depicted in Figure 2. It can also be confirmed from Table 2 that all predictors (except the acentric factor) are right-skewed with skewness values 1.01–5.10. Moreover, pressure, critical pressure, and acentric factor are platykurtic (kurtosis < 3), while temperature and critical temperature are leptokurtic (kurtosis > 3), indicating data is not normally distributed (Verma, 2019). A bell curve demonstrating a symmetrical/normal distribution typically exhibits skewness values between −0.5 and 0.5. The skewness values of a moderately skewed distribution fall within the range of −0.5 to −1.0, while a highly skewed distribution is characterized by values either less than −1.0 or greater than 1.0 (Verma, 2019).

Due to the deviations from normal distribution, the Spearman and Kendall correlation techniques are ideal for analyzing the relationships between variables because of their non-parametricity and nonalignment to data normality. The feature–feature and feature–target correlation coefficients are captured in the covariance heatmap (Figure 3). Positive correlation coefficients represent an inclined relationship while negative correlation coefficients indicate a declining relationship. The stronger the relationship between variables, the closer the correlation coefficient approaches either −1 or 1. In the study, the Spearman and Kendall correlations provide concordant findings regarding the direction and strength of the relationship between variables. It can be ascertained from both heatmaps that temperature, critical temperature, pressure, and the acentric factor positively correlate with CO_2 absorption. On the other hand, critical pressure

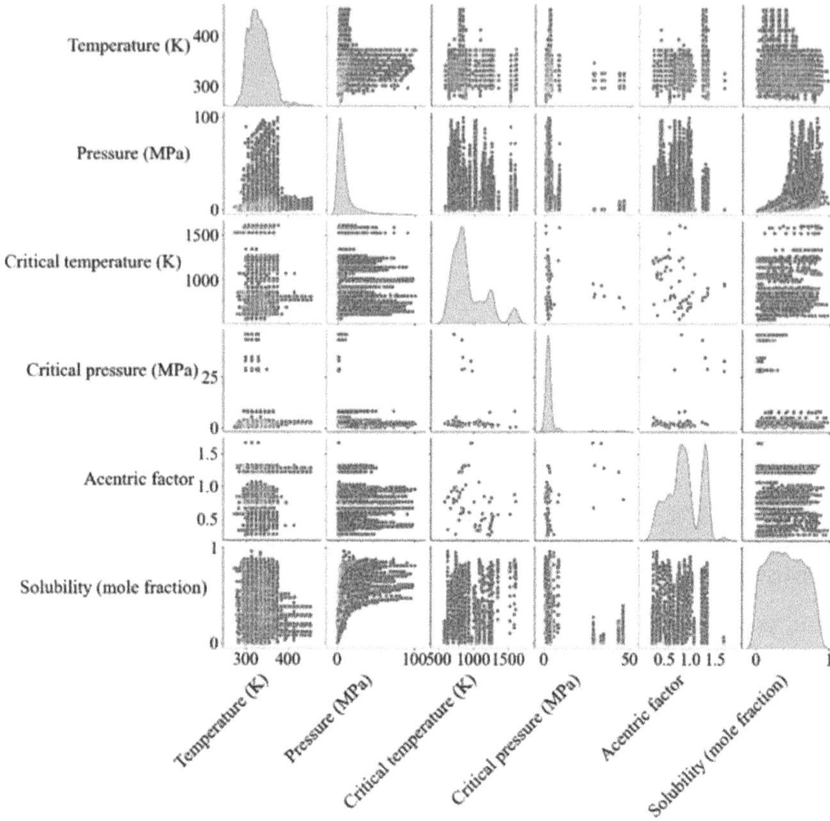

Figure 2: Pairplot distribution for dataset.

negatively correlates with CO_2 absorption. Additionally, pressure shows the highest positive correlation coefficient with CO_2 absorption, while temperature and acentric factor are assigned low correlation coefficients. Furthermore, the database contains no collinear features, since the feature–feature correlation coefficients (−0.24 to 0.33) are well within the proposed multicollinearity cut-off (−0.8 < correlation coefficient < 0.8) (Otchere et al., 2022a; Tabaaza et al., 2023; Turkson et al., 2024a; Turkson et al., 2024b).

Paradigm Performance and Reliability Assessment

The developed models were evaluated on different subsets of the dataset (training, test, and all) to ascertain their robustness and generalization capabilities. The results for the numerical statistical measures for the three subsets are illustrated in Figure 4. The model training performance metrics including an R^2 of 0.90–0.999, AAE of 0.004–0.023, and RMSE of 0.006–0.034, respectively, slightly exceed those of the test data (R^2 of 0.94–0.991, AAE of 0.016–0.039, and RMSE

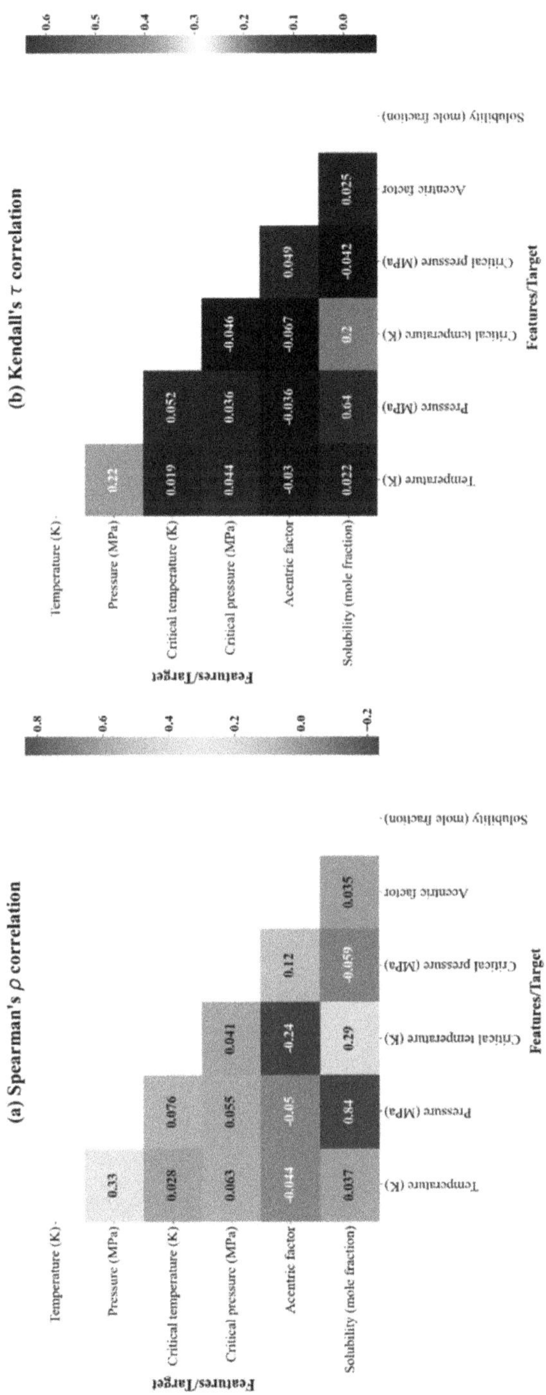

Figure 3: Correlation coefficient heatmaps for Spearman and Kendall techniques.

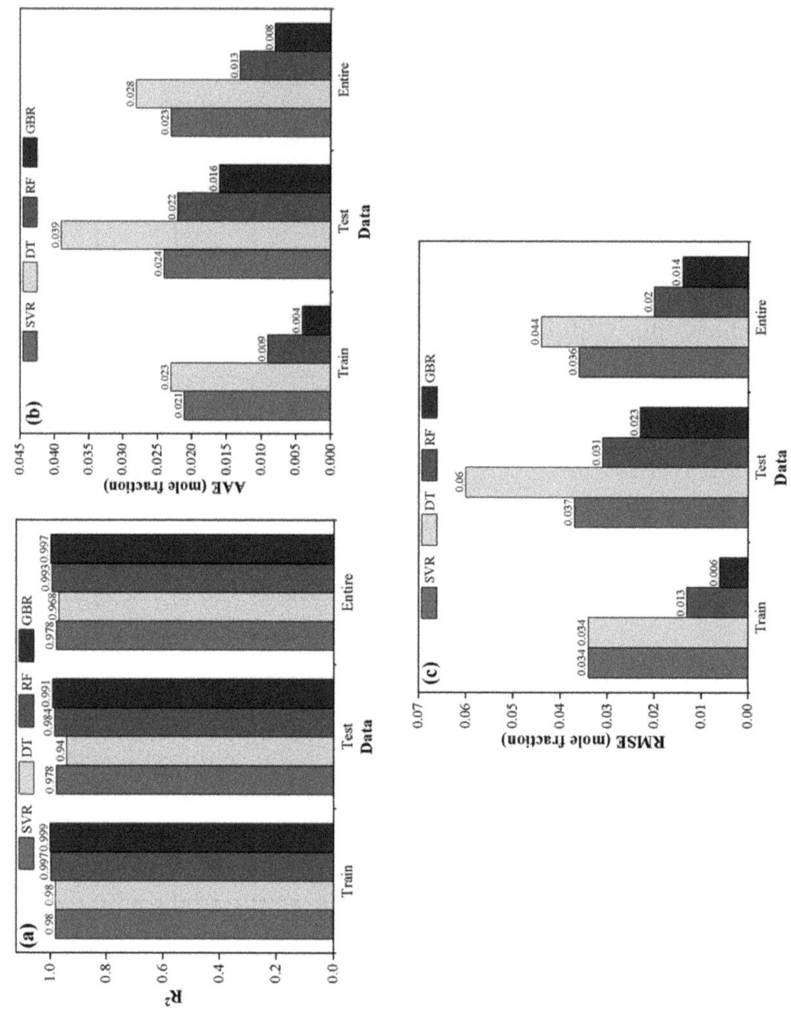

Figure 4: Model performance: (a) R^2, (b) AAE, and (c) RMSE on different subsets of the dataset.

of 0.023–0.060). The observation is attributed to the ability of the models to capture hidden trends and replicate familiar data. Notwithstanding, the test performance is closer to that of the entire dataset, signifying the robustness and exceptional generalizability of the models.

Overall, the models exhibited impressive predictive performance, yielding an $R^2 \geq 0.94$, AAE < 0.04, and RMSE $\leq$ 0.060. That notwithstanding, Figure 4 demonstrates model performance disparities. Across various metrics and datasets, the nonparametric ensemble models (RF and GBR) consistently achieved superior performance with overall R^2 of 0.993–0.997, AAE of 0.008–0.013, and RMSE of 0.014–0.020 over their nonparametric but singular counterparts (DT and SVR) with an overall R^2 of 0.968–0.978, AAE of 0.023–0.028, and RMSE of 0.036–0.044. This superiority was evident in terms of predictive accuracy, robustness to overfitting, and generalization capabilities. In particular, specific strategies were applied to avoid overfitting and underfitting issues that could potentially compromise model performance as explained earlier.

Ensemble learning techniques such as RF and GBR harness the collective capabilities of multiple base learners to make accurate and reliable predictions. They mitigate the limitations inherent in individual models by aggregating the outputs of multiple decision trees. This collective decision-making process enables the ensemble models to capture complex relationships within the data more effectively and achieve higher predictive performance as observed in Figure 4. However, ensemble learners like GBR can be prone to overfitting, especially with complex or smaller datasets (Al-Fakih et al., 2024). Hence, we carefully tuned the hyperparameters using randomized search as shown in Table 3, which efficiently explored a wider range of hyperparameters without excessive computational expense, avoiding overfitting that could result from improper hyperparameter choices. Similarly, optimum hyperparameters for RF model was also derived (Table 3) to avert underfitting or overfitting challenges.

Considering the ensemble learners, GBR outperformed RF, owing to the different underlying frameworks for model development. Specifically, the GBR model recorded an R^2, AAE, and RMSE of 0.991, 0.016, and 0.023, respectively, on out-of-sample data predictions. GBR employs sequential decision trees for ultimate predictions, while RF utilizes parallel decision trees to achieve a similar objective. GBR trains each successive model to rectify errors identified by preceding learners. Each new model in the sequence is equipped with insights into the training data and the deficiencies of its predecessors. The subsequent model leverages this accumulated knowledge to refine its predictions, enhancing the overall accuracy of GBR model output. On the other hand, RF trains multiple DTs on a random part of the data and obtains the final prediction by averaging the individual output from the trees. Unlike GBR, RF might encounter challenges in capturing subtle patterns due to the averaging schemes and random feature selection.

Utilizing joint plots for graphical error analysis offered further insights into the model performance. These visualizations complemented the numerical statistical error measures discussed earlier. By integrating scatter plots, marginal histograms, and kernel density estimation plots, the joint plot provided a comprehensive view of the disparity between predicted and observed CO_2 absorption values, along with their respective distributions as depicted in Figure 5. Notably, the joint plots for all models reveal an insignificant shaded region around the correlation line (X=Y line), signifying their exceptional predictive strengths. However, the predicted CO_2 absorption values of the GBR model closely adhere to the correlation line than those predicted by DT, SVR, and RF, demonstrating the heightened accuracy of GBR compared to alternative models. Conversely, the predictions of the DT

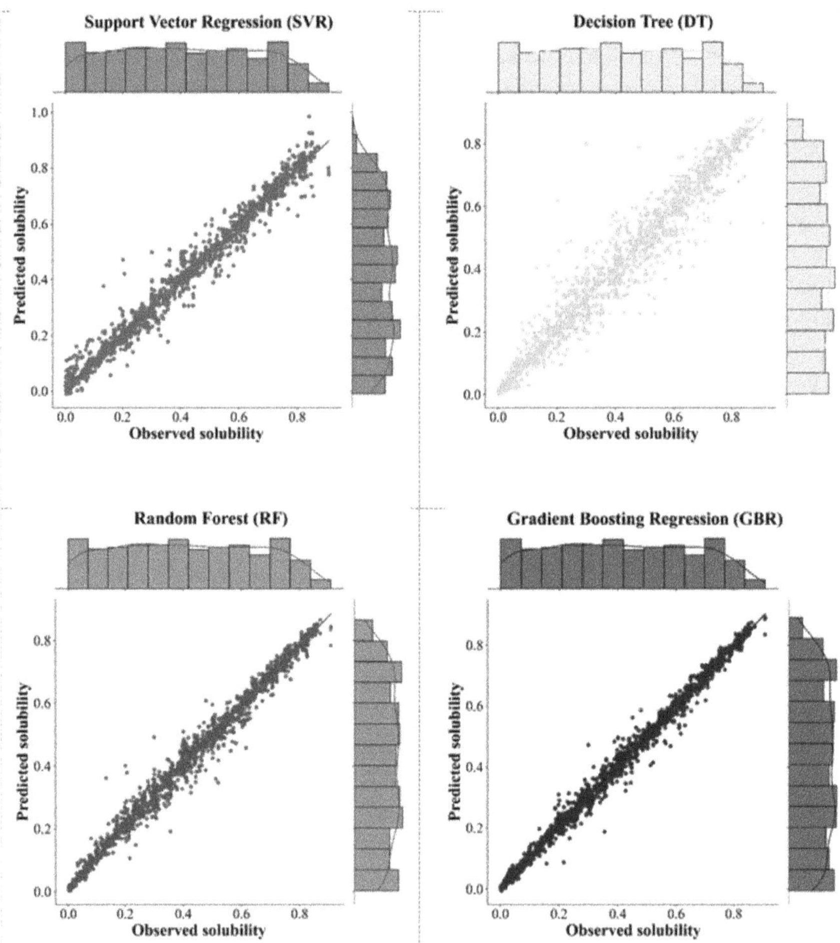

Figure 5: Joint plots for test data prediction for SVR, DT, RF, and GBR models.

model are more scattered and slightly farther from the correlation line compared to the other models, suggesting relatively greater deviations between predicted and observed values. Specifically, inaccuracies in predicting CO_2 absorption for the DT model between 0.2 and 0.75 contributed to higher errors reported in Figure 4.

The numerical statistical measures (R^2, AAE, and RMSE) discussed earlier cover bulk of the database and hence may not effectively identify highly influential instances inaccurately captured by the model. Therefore, further error analyses were conducted to understand the phenomenon. Figure 6 highlights the deviations of the predictions of the respective models. It is evident that GBR model yields predictions with a reduced error range (−0.2 to 0.2) compared to −0.3 to 0.3, −0.6 to 0.4, and −0.25 to 0.2 recorded by SVR, DT, and RF, respectively. Additionally, according to Figure 7, GBR predicts 95% of the dataset with an absolute error below 0.03, compared to 76.5%, 68.4%, and 90.2% for SVR, DT, and RF, respectively, suggesting the predictive dominance of the GBR model. These findings corroborate the exceptional predictive performance results recorded by the GBR model for the bulk data in Figure 4. Following the rigorous model performance analysis, GBR is adjudged as the outstanding intelligent model, followed by RF, SVR, and DT.

A plot of the standardized residuals against the leverage or hat values was generated as shown in Figure 8. With 5223 instances and 5 features, the critical leverage limit is calculated from Eq. 9 as 0.003446. The various data categories can be identified by combining the optimum standardized and the calculated leverage limit. Approximately 94% of the transformed dataset (representing 4934 instances) has leverage values between 0 and 0.003446 and standardized residuals falling within ±3. This category is regarded as valid data (refer to Table 4 for criteria). Additionally, 5.2% (273 instances) are above 0.003446 with standardized within ±3—tagged as good high-leverage data. 0.17% of the dataset (9 points) also are below 0.003446 and standardized residuals outside ±3, and are categorized as vertical suspect data. Finally, 0.13% of the dataset (7 instances) is also classified as bad high-leverage data. It can be concluded that the dataset is statistically valid since more than 90% constitutes the valid data. Consequently, the developed models are reliable for estimating CO_2 absorption in ionic liquids.

Comparative Analysis of Cluster of Models and Correlations

To confirm the superiority of the robustness and generalization capability of the model, GBR performance was compared to that of the ANFIS, RBF-ANN (Baghban et al., 2017), and deep and recurrent neural networks (DNN and RNN) (Deng et al., 2020). As outlined in Table 5, GBR outperforms ANFIS, RBF-ANN, DNN, and RNN, which recorded impressive prediction performance in earlier studies. GBR is the most accurate, precise, and consistent model for CO_2 absorption predictions.

Due to the varying data proportions used for model development for the above-referenced literature, a point-by-point comparison between predictions of the developed GBR model and ANFIS and RBF-ANN (Baghban et al., 2017)

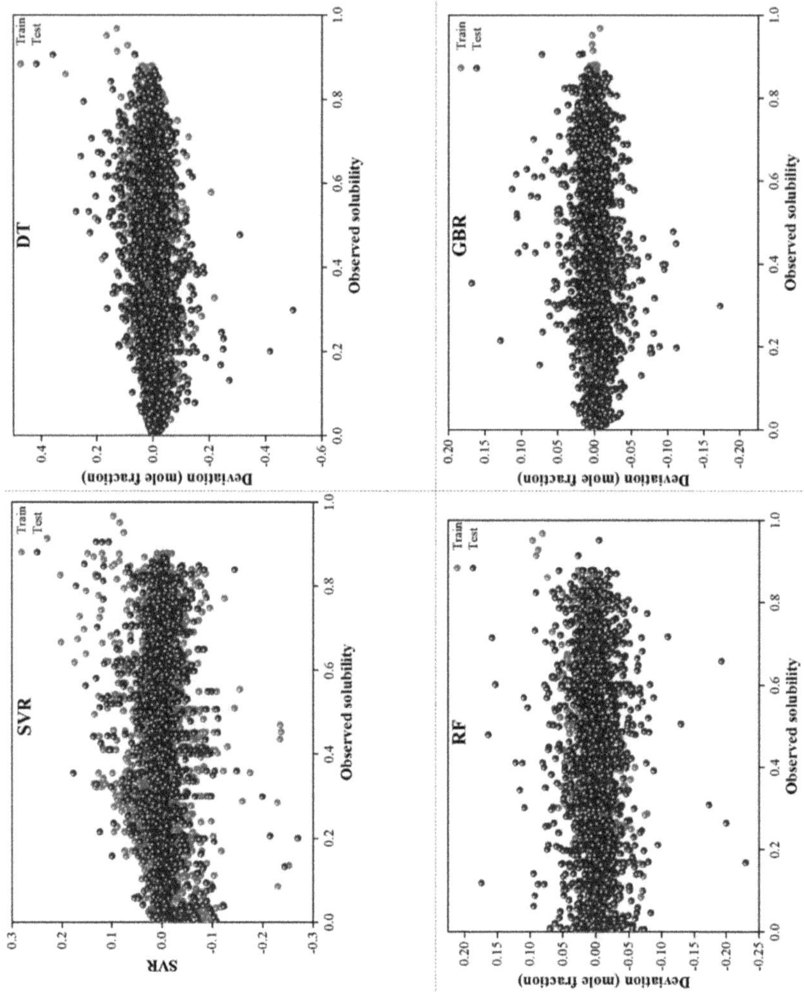

Figure 6: Scatter plot for the deviations associated with the respective models.

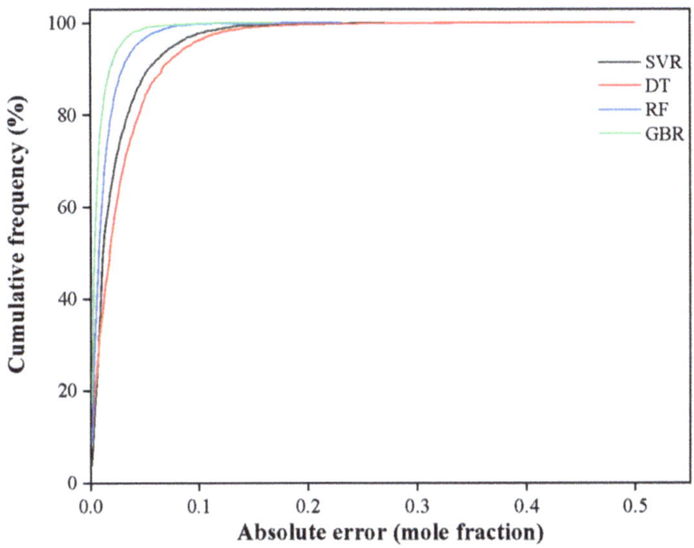

Figure 7: Cumulative frequency plot for the absolute deviations associated with the respective models.

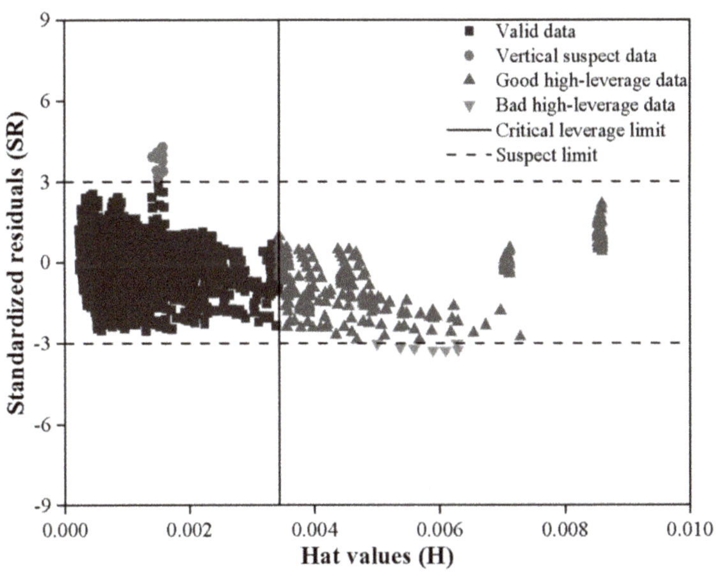

Figure 8: Williams plot of standardized residuals against leverage values to evaluate the statistical validity of the database and reliability of models.

Table 5: Statistical error measures of the comparative models

Model	R^2	AAE	RMSE	Reference
ANFIS	0.961	0.034	N/A	(Baghban et al., 2017)
RBF-ANN	0.983	0.019	N/A	(Baghban et al., 2017)
DNN	0.984	0.291	0.757	(Deng et al., 2020)
RNN	0.988	0.250	0.651	(Deng et al., 2020)
GBR	**0.991**	**0.016**	**0.023**	This study

N/A: Not applicable or not provided.

is also conducted. Here, the ionic liquids considered include: for [bmim][Tf2N] (Lee & Outcalt, 2006), [C3mpy][Tf2N] (Kim et al., 2012), [bmim][PF6] (Kumełan et al., 2006), [TDC][TF2N] (Zoubeik & Henni, 2014), and [HMIM][MESO4] (Yim & Lim, 2013). Figure 9 illustrates the predictions provided by the comparative models. Table 6 also highlights the error metrics associated with the predictions for the models. The predicted CO_2 absorption is closer to the observed values for [bmim][Tf2N] and [bmim][PF6], which is also reflected in the error metrics in Table 6. As depicted by the high R^2 values for all the models (0.986–0.999), the GBR model yields more accurate predictions for CO_2 absorption. Additionally, GBR offers precise and consistent CO_2 absorption in [bmim][Tf2N], [TDC][TF2N], and [HMIM][MESO4].

Table 6: Error metrics for point-by-point comparison

Ionic Liquid	Metric	ANFIS	RBF-ANN	GBR
[bmim][Tf2N]	R^2	0.991	0.992	**0.999**
	AAE	0.204	0.200	**0.192**
	RMSE	0.063	0.061	**0.056**
[C3mpy][Tf2N]	R^2	0.966	0.984	**0.986**
	AAE	**0.365**	0.440	0.408
	RMSE	**0.160**	0.233	0.204
[bmim][PF6]	R^2	0.962	**0.992**	**0.992**
	AAE	**0.172**	0.193	0.190
	RMSE	**0.044**	0.051	0.050
[TDC][TF2N]	R^2	0.939	0.993	**0.996**
	AAE	**0.091**	0.132	0.102
	RMSE	0.015	0.025	**0.013**
[HMIM][MESO4]	R^2	0.815	0.799	**0.997**
	AAE	0.370	0.350	**0.278**
	RMSE	0.160	0.150	**0.103**

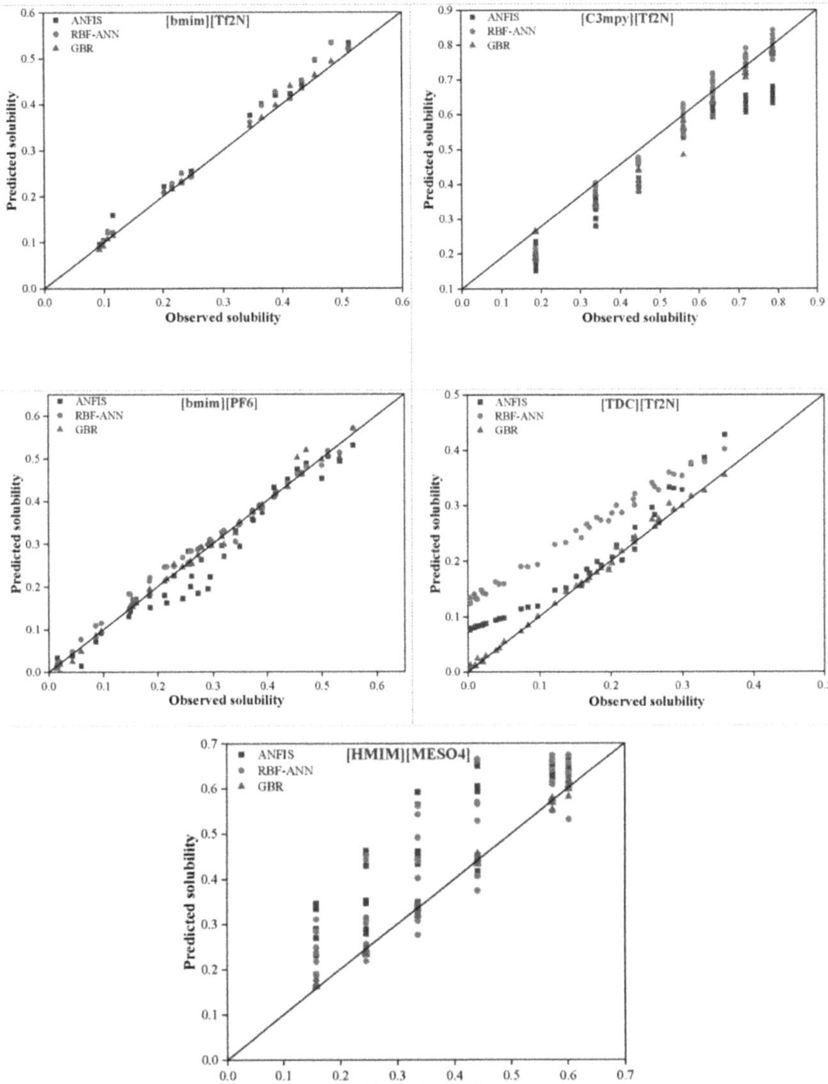

Figure 9: A point-by-point comparison of GBR model predictions against predictions of ANFIS, and RBF-ANN (Baghban et al., 2017).

Sensitivity Analysis and Input Feature Importance

Sensitivity analysis and feature importance are essential in understanding the influence of input features on CO_2 absorption in ionic liquids. Correlation techniques (Figure 10a), permutation importance (Figure 10b), and SHAP (Figure 11) were employed for this purpose. Permutation importance is advantageous as

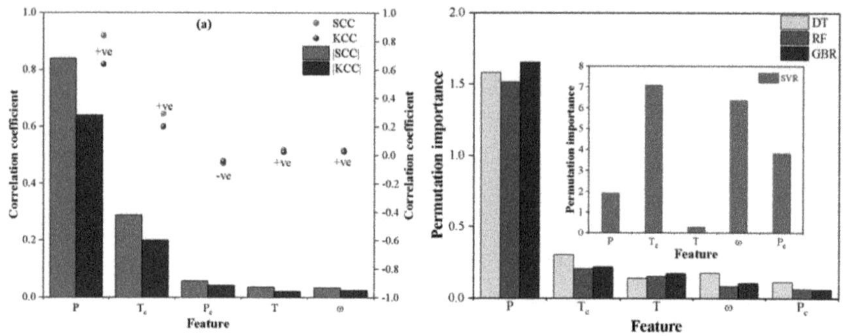

Figure 10: Feature importance analysis according to (a) correlation techniques and (b) permutation importance.

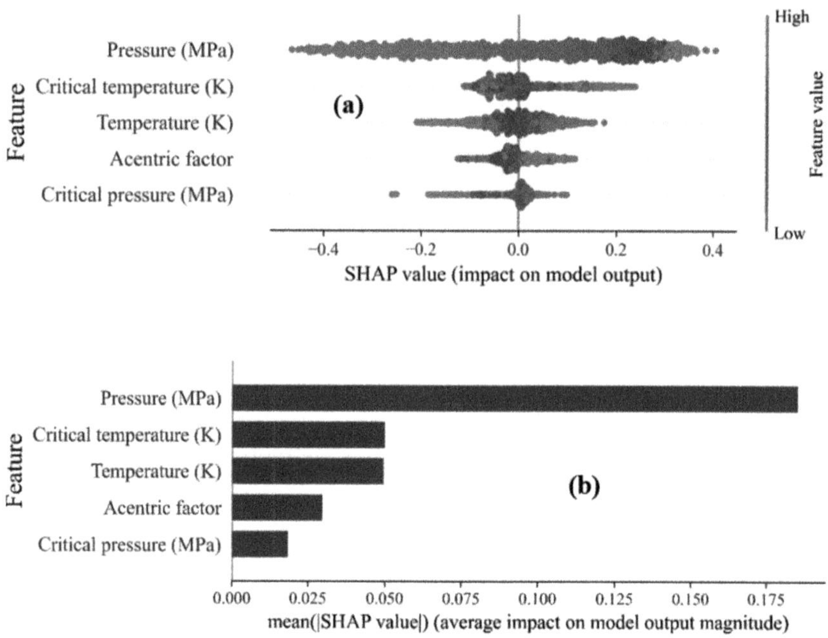

Figure 11: (a) SHAP values for individual data points and (b) ranking of features based on mean SHAP value.

it captures non-monotonic influences of input parameters on CO_2 absorption and mitigates bias issues compared to correlation techniques (Turkson et al., 2024a). Despite differences, results from these methods can be harmonized to glean insights into feature influences. SHAP provides additive feature attribution, indicating the contribution of each feature to the final prediction. Positive SHAP values denote a positive impact of a feature on CO_2 absorption, and vice versa.

The correlation techniques, permutation importance, and SHAP plots unanimously indicate the profound influence of pressure on CO_2 absorption. In contrast, SHAP analysis and permutation importance point to critical pressure as the least influential variable. The order of influence based on the SHAP analysis and permutation importance follows pressure, critical temperature, temperature, acentric factor, and critical pressure.

To analyze the influence of variations in the input features on the target, the absorptions of CO_2 in a couple of ionic liquids (indicated in Figure 12) at varying conditions are considered. The trends between CO_2 absorption, thermophysical conditions, and thermodynamic properties are presented in Figure 12. It is evident that the GBR model successfully captured the trends between the variables.

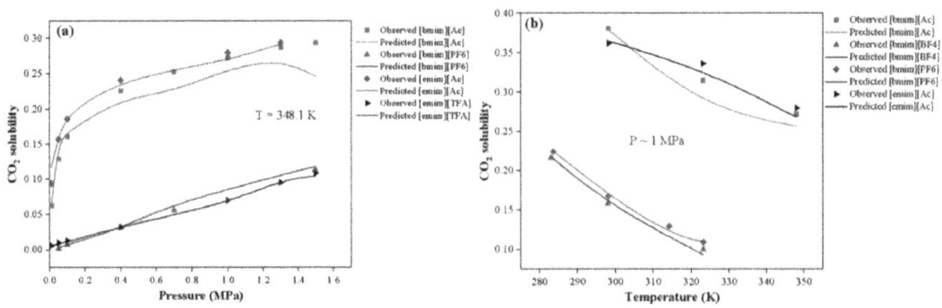

Figure 12: Trend analysis between CO_2 absorption, thermophysical conditions, and thermodynamic properties.

As illustrated in Figure 12, CO_2 absorption in ionic liquids increases with pressure, confirming the results in Figure 10a and Figure 11a. In Figure 11a, high SHAP values correspond to high pressure, indicating an inclining relationship between the two variables. Analogous to CO_2 solubility in brine (Hemmati-Sarapardeh et al., 2020), increasing pressure weakens the cohesive forces within the solute (CO_2) and solvent (ionic liquids) and promotes the intermolecular force between CO_2 and ionic liquid. This enhanced intermolecular attraction facilitates greater absorption, explaining the high importance of pressure in the feature analysis.

In contrast, CO_2 absorption in ionic liquids subsides with increasing temperature (Figure 12). Again, Figure 11a suggests an inverse relationship between temperature and CO_2 absorption, with high temperatures yielding low SHAP values. The kinetic energy between CO_2 molecules increases with increasing temperature at constant pressure. This increased molecular activity leads to CO_2 liberation from ionic liquids, thereby decreasing the solubility (Ramdin et al., 2012).

Critical temperature and acentric factor are also influential, albeit to a lesser extent than pressure. Critical temperature affects the phase behavior of CO_2, influencing its solubility in ionic liquids under different conditions. Similarly, the acentric factor represents the deviation of a substance from ideal gas behavior, which may impact the interaction between CO_2 and ionic liquids under specific thermodynamic states. Critical pressure, on the other hand, showed the least influence, likely due to its indirect role in determining CO_2 absorption.

Model Implications for CO_2 Capture

GBR offers significant advantages for predicting CO_2 absorption in ionic liquids. Unlike traditional experimental methods, which are resource-intensive and time-consuming, the GBR paradigm provides a cost-effective and efficient alternative. GBR can swiftly analyze large datasets and predict CO_2 absorption under various conditions, aiding in the optimization of CO_2 capture processes. Moreover, the remarkable accuracy of the GBR model captured by the evaluation metrics extensively discussed earlier ensures reliable predictions of CO_2 absorption behavior, crucial for designing effective carbon capture systems. Furthermore, the ability of the ML model to generalize across different datasets (training, test, and entire dataset) underscores its robustness in capturing complex relationships between thermophysical conditions, thermodynamic properties, and CO_2 absorption in ionic liquids.

In contrast to experimental studies that may be limited in scope and applicability, GBR has the potential to provide broader insights into CO_2 absorption dynamics. It can identify subtle trends and interactions between thermophysical conditions and thermodynamic properties that influence CO_2 absorption as shown in Figure 12. While experiments offer direct empirical validation, GBR could complement this by offering scalable predictions and insights into system behavior under varying conditions. Additionally, integrating GBR model predictions with experimental validation and molecular simulations can enhance understanding and confidence in CO_2 capture technologies. The successful application of ML in predicting CO_2 absorption provides a platform for further investigations and applications in CCS.

From a practical perspective, the predictive capabilities of GBR can benefit large-scale CCS operations. For instance, its ability to optimize operating conditions for ionic liquids could lead to more efficient absorber design and operational strategies, thereby reducing energy consumption and costs. By accurately predicting CO_2 absorption across a range of pressures, temperatures, and solvent properties, the model could facilitate the selection of ionic liquids tailored for specific capture scenarios.

Moreover, the GBR model could be integrated into CCS workflows by serving as a decision-support tool. It could be incorporated into process simulation software to evaluate system performance under various conditions, enabling engineers to preemptively address potential challenges in capture operations. Additionally, by coupling the GBR model with real-time monitoring systems, operators could dynamically adjust process conditions to maximize CO_2 capture efficiency and minimize environmental impact. These integration pathways demonstrate the potential of the GBR model to bridge experimental insights and industrial application.

Limitations of Computational Modeling Technique and Future Work

The hyperparameter optimization techniques and mathematical principles governing ensemble learners including gradient boosting and random forest are more complex for humans to comprehend than those of white-box models, such as linear regression, decision trees, gene expression programming, and genetic programming. Additionally, the "black-box" nature of ensembles can pose challenges in interpreting model outputs compared to the mechanistic insights provided by experimental studies or molecular simulations. Thus, it is difficult to understand how the ensemble learner arrives at its predictions due to the intricate mathematical principles and lack of transparency. Nonetheless, the ensembles, particularly the GBR model, demonstrate an upper-echelon CO_2 solubility predictive capabilities, surpassing the "white-box" decision tree model, as clearly depicted in Figure 4. Furthermore, GBR offer insights into the influence of the independent variables, which is essential for optimizing CO_2 absorption in ionic liquids.

Future studies could assess the predictive performance of hybrid models, such as combining ML with molecular dynamics simulations to capture atomic-level interactions or integrating ML with thermodynamic modeling frameworks to improve CO_2 solubility predictions. Additionally, exploring the real-time applications of ML in optimizing CO_2 capture processes in industrial settings, such as predictive maintenance for absorption units or adaptive control of operating conditions, could bridge the gap between theoretical models and practical implementation.

Summary and Conclusion

The study employed singular and ensemble models including SVR, DT, RF, and GBR to predict the CO_2 absorption in ionic liquids. The 5223 instances capturing the thermophysical conditions and thermodynamic properties traversed 65 ionic liquids. Numerical statistical error measures and graphical error analyses were adopted to assess model performance, while the leverage technique was used to evaluate the statistical validity of the database and reliability of the models. Subsequently, the predictions of the outstanding ML model in the current study were compared to those obtained by extant models in the literature. The rigorous computational modeling provides crucial insights such as:

1. The dataset contained 4934 valid instances (representing 94%) and a combined 289 vertical suspect, good high-leverage, and bad high-leverage instances, confirming the statistical validity of the database and the reliability of the developed models.
2. The developed models achieved impressive results on out-of-sample data predictions, recording an R^2, AAE, and RMSE of 0.94–0.991, 0.016–0.039, and 0.023–0.060, respectively.

3. GBR successfully captured the underlying relationships between CO_2 absorption, thermophysical conditions, and thermodynamic properties. Its ability to model complex, nonlinear interactions such as the synergistic effects of pressure and temperature on CO_2 absorption demonstrates its superiority over traditional models. Unlike simpler models, GBR effectively handles feature interdependencies and identifies subtle trends, achieving remarkable accuracy in CO_2 absorption prediction.
4. GBR exhibited predictive dominance over ANFIS, RBF-ANN, DNN, and RNN in bulk data and point-by-point prediction comparative analysis, confirming the robustness and generalizability of the model.
5. Pressure was identified as the most influential parameter on CO_2 absorption, as higher pressures enhance intermolecular interactions, facilitating greater absorption. Conversely, critical pressure demonstrated the least influence, with the overall order of influence being pressure, critical temperature, temperature, acentric factor, and critical pressure.
6. CO_2 absorption was found to increase with pressure and decrease with temperature. The observed phenomenon is attributed to the disruption of cohesive forces for pressure and rise in kinetic energy and CO_2 liberation from solvent (ionic liquid) for temperature.

The black-box nature of GBR can complicate the interpretation of model outputs compared to the mechanistic insights offered by experimental studies or molecular simulations. Nevertheless, the excellent performance of the GBR model underscores its potential for swift, effective, and efficient screening of ionic liquids for CO_2 capture utilizing minimum resources.

Data Availability

Data will be made available upon request.

Declaration of Competing Interest

No competing financial interests to declare.

Acknowledgments

Joshua Nsiah Turkson and Victor Darkwah-Owusu sincerely thank the Centre of Graduate Studies (CGS)–Universiti Teknologi PETRONAS (UTP) for sponsoring their Master of Science by Research program through the Graduate Assistantship scheme. The authors appreciate the Centre of Reservoir Dynamics (COReD) under the Institute of Sustainable Energy & Resources (ISER) at UTP for providing the computing facilities, which immensely helped to complete this work. The authors also express utmost gratitude to the reviewers and the editor whose invaluable comments helped to improve this article.

Nomenclature

Abbreviations	
AAE	Average absolute error
AAPRE	Average absolute percent relative error
ANFIS	Adaptive neuro-fuzzy inference system
ANN	Artificial neural network
CCS	Carbon capture and storage
CMIS	Committee machine intelligent system
DT	Decision tree
EoS	Equation of state
GBR	Gradient boosting regression
GPR	Gaussian process regression
LightGBM	Light gradient boosting machine
LSSVM	Least square support vector machine
ML	Machine learning
MAE	Mean absolute error
MLP-LM	Multilayer perceptron optimized with Levenberg-Marquardt algorithm
NRTL	Non-random-two-liquid
PR	Reng-Robinsons
RBF	Radial basis function
RF	Random forest
RMSE	Root mean square error
SR	Standardized residual
SRK	Soave-Redlich-Kwong
SVR	Support vector regression
UNIFAC	UNIQUAC Functional-group Activity Coefficients
UNIQUAC	Universal Quasichemical
Symbols	
H	Hat or leverage value
H^*	Critical leverage limit
P	Pressure (MPa)
T	Temperature (K)
P_c	Critical pressure (MPa)
T_c	Critical temperature (K)
ω	Acentric factor
χ	CO_2 absorption in ionic liquid

References

Aghaie, M. & Zendehboudi, S. (2020). Estimation of CO_2 solubility in ionic liquids using connectionist tools based on thermodynamic and structural characteristics. *Fuel*, 279, 117984. https://doi.org/10.1016/j.fuel.2020.117984.

Al-Fakih, A., Al-Wajih, E., Saleh, R. A. & Muhit, I. B. (2024). Ensemble machine learning models for predicting the CO_2 footprint of GGBFS-based geopolymer concrete. *J. Cleaner Prod.*, 472, 143463. https://doi.org/10.1016/j.jclepro.2024.143463.

Alanazi, A., Ibrahim, A. F., Bawazer, S., Elkatatny, S. & Hoteit, H. (2023). Machine learning framework for estimating CO_2 adsorption on coalbed for carbon capture, utilization, and storage applications. *Int. J. Coal Geol.*, 275. https://doi.org/10.1016/j.coal.2023.104297.

Amar, M. N., Alqahtani, F. M., Djema, H., Ourabah, K. & Ghasemi, M. (2023). Predicting the solubility of hydrogen in hydrocarbon fractions: Advanced data-driven machine learning approach and equation of state. *J. Taiwan Inst. Chem. Eng.*, 153, 105215. https://doi.org/10.1016/j.jtice.2023.105215.

Asrami, M. R., Tran, N. N., Nigam, K. D. P. & Hessel, V. (2021). Solvent extraction of metals: Role of ionic liquids and microfluidics. *Sep. Purif. Technol.*, 262, 118289. https://doi.org/10.1016/j.seppur.2020.118289.

Atilhan, M. & Aparicio, S. (2021). Review on chemical enhanced oil recovery: Utilization of ionic liquids and deep eutectic solvents. *J. Pet. Sci. Eng.*, 205, 108746. https://doi.org/10.1016/j.petrol.2021.108746.

Baghban, A., Mohammadi, A. H. & Taleghani, M. S. (2017). Rigorous modeling of CO_2 equilibrium absorption in ionic liquids. *Int. J. Greenhouse Gas Control*, 58, 19–41. https://doi.org/10.1016/j.ijggc.2016.12.009.

Bera, A. & Belhaj, H. (2016). Ionic liquids as alternatives of surfactants in enhanced oil recovery—A state-of-the-art review. *J. Mol. Liq.*, 224, 177–188. https://doi.org/10.1016/j.molliq.2016.09.105.

Bergstra, J. & Bengio, Y. (2012). Random search for hyper-parameter optimization. *Journal of Machine Learning Research*, 13(2). https://www.jmlr.org/papers/volume13/bergstra12a/bergstra12a.pdf?ref=broutonlab.com.

Borhani, T. N. & Wang, M. (2019). Role of solvents in CO_2 capture processes: The review of selection and design methods. *Renewable Sustainable Energy Rev.*, 114, 109299. https://doi.org/10.1016/j.rser.2019.109299.

Breiman, L. (2001). Random forests. *Machine Learning*, 45. https://doi.org/10.1023/A:1010933404324.

Darkwah-Owusu, V., Md Yusof, M. A., Sokama-Neuyam, Y. A., Tackie-Otoo, B. N., Turkson, J. N., Kwon, S. et al. (2024a). Investigating the extent of the impact of acetic acid as halite precipitation reduction agent for enhancing CO_2 injectivity in deep saline aquifers. *SPE International Conference and Exhibition on Formation Damage Control*. https://doi.org/10.2118/217884-MS.

Darkwah-Owusu, V., Md Yusof, M. A., Sokama-Neuyam, Y. A., Turkson, J. N. & Fjelde, I. (2024b). Assessment of advanced remediation techniques for enhanced CO_2 injectivity: Laboratory investigations and implications for improved CO_2 flow in saline aquifers. *Energy Fuels*. https://doi.org/10.1021/acs.energyfuels.4c00949.

Deng, T., Liu, F.-H. & Jia, G.-Z. (2020). Prediction carbon dioxide solubility in ionic liquids based on deep learning. *Molecular Physics*, 118(6), e1652367. https://doi.org/10.1080/00268976.2019.1652367.

Feng, H., Zhang, P., Qin, W., Wang, W. & Wang, H. (2022). Estimation of solubility of acid gases in ionic liquids using different machine learning methods. *J. Mol. Liq.*, 349, 118413. https://doi.org/10.1016/j.molliq.2021.118413.

Friedman, J. H. (2002). Stochastic gradient boosting. *Computational Statistics & Data Analysis*, 38. https://doi.org/10.1016/S0167-9473(01)00065-2.

Ghazani, S. H. H. N., Baghban, A., Mohammadi, A. H. & Habibzadeh, S. (2018). Absorption of CO_2-rich gaseous mixtures in ionic liquids: A computational study. *J. Supercrit. Fluids*, 133, 455–465. https://doi.org/10.1016/j.supflu.2017.10.024.

Goutham, R., Rohit, P., Vigneshwar, S. S., Swetha, A., Arun, J., Gopinath, K. P. & Pugazhendhi, A. (2022). Ionic liquids in wastewater treatment: A review on pollutant removal and degradation, recovery of ionic liquids, economics and future perspectives. *J. Mol. Liq.*, 349, 118150. https://doi.org/10.1016/j.molliq.2021.118150.

Gyamfi, G., Danso, D. K. & Li, X. (2024). Practical models for computing CO_2 solubility in brines with complex ions for carbon geo-sequestration applications. *SPE/AAPG/SEG Unconventional Resources Technology Conference*. https://doi.org/10.15530/urtec-2024-4033482.

Hemmati-Sarapardeh, A., Amar, M. N., Soltanian, M. R., Dai, Z. & Zhang, X. (2020). Modeling CO_2 solubility in water at high pressure and temperature conditions. *Energy Fuels*, 34(4), 4761–4776. https://doi.org/10.1021/acs.energyfuels.0c00114.

Henni, N., Henni, A. & Ibrahim, H. (2023). Solubility of carbon dioxide in promising methylimidazolium-based ionic liquids. *Fluid Phase Equilib.*, 565, 113619. https://doi.org/10.1016/j.fluid.2022.113619.

Huang, S., Tian, L., Zhang, J., Chai, X., Wang, H. & Zhang, H. (2021). Support vector regression based on the particle swarm optimization algorithm for tight oil recovery prediction. *ACS Omega*, 6(47), 32142–32150. https://doi.org/10.1021/acsomega.1c04923.

IEA. (2021). *Is Carbon Capture Too Expensive?* IEA: Paris. https://www.iea.org/commentaries/is-carbon-capture-too-expensive.

Jalili, A. H., Shokouhi, M., Maurer, G., Zoghi, A. T., Ahari, J. S. & Forsat, K. (2019). Measuring and modelling the absorption and volumetric properties of CO_2 and H_2S in the ionic liquid 1-ethyl-3-methylimidazolium tetrafluoroborate. *J. Chem. Thermodyn.*, 131, 544–556. https://doi.org/10.1016/j.jct.2018.12.005.

Kang, S., Chung, Y. G., Kang, J. H. & Song, H. (2020). CO_2 absorption characteristics of amino group functionalized imidazolium-based amino acid ionic liquids. *J. Mol. Liq.*, 297, 111825. https://doi.org/10.1016/j.molliq.2019.111825.

Khanal, A. & Shahriar, M. F. (2022). Physics-based proxy modeling of CO_2 sequestration in deep saline aquifers. *Energies*, 15(12). https://doi.org/10.3390/en15124350.

Kim, S., Yoon, H. C., Lim, J.-T., Jeong, D. & Kim, K. H. (2023). Productivity prediction in the Wolfcamp A and B using weighted voting ensemble machine learning method. *Gas Sci. Eng.*, 111, 204916. https://doi.org/10.1016/j.jgsce.2023.204916.

Kim, S. A., Yim, J.-H. & Lim, J. S. (2012). High-pressure phase behavior of binary mixtures containing methylpyrrolidinium derivative ionic liquids and carbon dioxide. *Fluid Phase Equilib.*, 332, 28–34. https://doi.org/10.1016/j.fluid.2012.07.006.

Kondapi, A. K. (2024). Nanomaterials and novel solvents for carbon capture technologies. *In: Decarbonization Strategies and Drivers to Achieve Carbon Neutrality for Sustainability* (pp. 501–514). Elsevier. https://doi.org/10.1016/B978-0-443-13607-8.00010-9

Kumełan, J., Pérez-Salado Kamps, Á., Tuma, D. & Maurer, G. (2006). Solubility of CO_2 in the ionic liquids [bmim][CH_3SO_4] and [bmim][PF6]. *J. Chem. Eng. Data*, 51(5), 1802–1807. https://doi.org/10.1021/je060190e.

Kuroki, N., Suzuki, Y., Kodama, D., Chowdhury, F. A., Yamada, H. & Mori, H. (2023). Machine learning-boosted design of ionic liquids for CO_2 absorption and experimental verification. *J. Phys. Chem. B*, 127(9), 2022–2027. https://doi.org/10.1021/acs.jpcb.2c07305.

Lee, B.-C. & Outcalt, S. L. (2006). Solubilities of gases in the ionic liquid 1-n-butyl-3-methylimidazolium bis (trifluoromethylsulfonyl) imide. *J. Chem. Eng. Data*, 51(3), 892–897. https://doi.org/10.1021/je050357o.

Mardhatillah, M. K., Md Yusof, M. A., Sáid, A. A., Mohammad Fuad, I. I., Sokama Neuyam, Y. A. & Md Akhir, N. A. (2022). Predictive Modelling of Carbon Dioxide Injectivity Using SVR-Hybrid. *Offshore Technology Conference Asia*. https://doi.org/10.4043/31472-MS.

Mohammadi, M.-R., Hadavimoghaddam, F., Atashrouz, S., Hemmati-Sarapardeh, A., Abedi, A. & Mohaddespour, A. (2022). Application of robust machine learning methods to modeling hydrogen solubility in hydrocarbon fuels. *International Journal of Hydrogen Energy*, 47(1), 320–338. https://doi.org/10.1016/j.ijhydene.2021.09.202.

Mohan, M., Demerdash, O. N., Simmons, B. A., Singh, S., Kidder, M. K. & Smith, J. C. (2024). Physics-based machine learning models predict carbon dioxide solubility in chemically reactive deep eutectic solvents. *ACS Omega*, 9(17), 19548–19559. https://doi.org/10.1021/acsomega.4c01175.

Mokarizadeh, H., Atashrouz, S., Mirshekar, H., Hemmati-Sarapardeh, A. & Pour, A. M. (2020). Comparison of LSSVM model results with artificial neural network model for determination of the solubility of SO_2 in ionic liquids. *J. Mol. Liq.*, 304, 112771. https://doi.org/10.1016/j.molliq.2020.112771.

Mousavi, S.-P., Nakhaei-Kohani, R., Atashrouz, S., Hadavimoghaddam, F., Abedi, A., Hemmati-Sarapardeh, A. & Mohaddespour, A. (2023). Modeling of H_2S solubility in ionic liquids: Comparison of white-box machine learning, deep learning and ensemble learning approaches. *Sci. Rep.*, 13(1), 7946. https://doi.org/10.1038/s41598-023-34193-w.

Noorani, N. & Mehrdad, A. (2020). CO_2 solubility in some amino acid-based ionic liquids: Measurement, correlation and DFT studies. *Fluid Phase Equilib.*, 517, 112591. https://doi.org/10.1016/j.fluid.2020.112591.

Noorani, N., Mehrdad, A. & Ahadzadeh, I. (2021). CO_2 absorption in amino acid-based ionic liquids: Experimental and theoretical studies. *Fluid Phase Equilib.*, 547, 113185. https://doi.org/10.1016/j.fluid.2021.113185.

Otchere, D. A. (2023). Fundamental error in tree-based machine learning model selection for reservoir characterisation. *Energy Geoscience*, 5(2), 100229. https://doi.org/10.1016/j.engeos.2023.100229.

Otchere, D. A., Ganat, T. O. A., Ojero, J. O., Tackie-Otoo, B. N. & Taki, M. Y. (2022a). Application of gradient boosting regression model for the evaluation of feature selection techniques in improving reservoir characterisation predictions. *J. Pet. Sci. Eng.*, 208, 109244. https://doi.org/10.1016/j.petrol.2021.109244.

Otchere, D. A., Mohammed, M. A. A., Al-Hadrami, H. & Boakye, T. B. (2022b). Enhancing drilling fluid lost-circulation prediction: Using model agnostic and supervised machine learning. *In: Data Science and Machine Learning Applications in Subsurface Engineering* (pp. 6–32). CRC Press. https://doi.org/10.1201/9781003366980

Ouaer, H., Hosseini, A. H., Nait Amar, M., El Amine Ben Seghier, M., Ghriga, M. A., Nabipour, N. et al. (2019). Rigorous connectionist models to predict carbon dioxide solubility in various ionic liquids. *Appl. Sci.*, 10(1), 304. https://doi.org/10.3390/app10010304.

Pedregosa, F., Varoquaux, G., Gramfort, A., Michel, V., Thirion, B., Grisel, O. et al. (2011). Scikit-learn: Machine learning in Python. *Journal of Machine Learning Research*, 12, 2825–2830. https://jmlr.csail.mit.edu/papers/volume12/pedregosa11a/pedregosa11a.pdf

Pekel, E. (2020). Estimation of soil moisture using decision tree regression. *Theor. Appl. Climatol.*, 139(3), 1111–1119. https://doi.org/10.1007/s00704-019-03048-8.

Pelaquim, F. P., Barbosa Neto, A. M., Dalmolin, I. A. L. & Costa, M. C. A. O. D. (2021). Gas solubility using deep eutectic solvents: Review and analysis. *Ind. Eng. Chem. Res.*, 60(24), 8607–8620. https://doi.org/10.1021/acs.iecr.1c00947.

Rahimi, M. & Riahi, M. A. (2022). Reservoir facies classification based on random forest and geostatistics methods in an offshore oilfield. *J. Appl. Geophy.*, 201, 104640. https://doi.org/10.1016/j.jappgeo.2022.104640.

Ramdin, M., de Loos, T. W. & Vlugt, T. J. (2012). State-of-the-art of CO_2 capture with ionic liquids. *Ind. Eng. Chem. Res.*, 51(24), 8149–8177. https://doi.org/10.1021/ie3003705.

Rashidi-Khaniabadi, A., Rashidi-Khaniabadi, E., Amiri-Ramsheh, B., Mohammadi, M.-R. & Hemmati-Sarapardeh, A. (2023). Modeling interfacial tension of surfactant–hydrocarbon systems using robust tree-based machine learning algorithms. *Sci. Rep.*, 13(1), 10836. https://doi.org/10.1038/s41598-023-37933-0.

Safaei-Farouji, M., Vo Thanh, H., Dai, Z., Mehbodniya, A., Rahimi, M., Ashraf, U. et al. (2022a). Exploring the power of machine learning to predict carbon dioxide trapping efficiency in saline aquifers for carbon geological storage project. *J. Cleaner Prod.*, 372. https://doi.org/10.1016/j.jclepro.2022.133778.

Safaei-Farouji, M., Vo Thanh, H., Sheini Dashtgoli, D., Yasin, Q., Radwan, A. E., Ashraf, U. et al. (2022b). Application of robust intelligent schemes for accurate modelling interfacial tension of CO_2 brine systems: Implications for structural CO_2 trapping. *Fuel*, 319. https://doi.org/10.1016/j.fuel.2022.123821.

Shafiei, A., Tatar, A., Rayhani, M., Kairat, M. & Askarova, I. (2022). Artificial neural network, support vector machine, decision tree, random forest, and committee machine intelligent system help to improve performance prediction of low salinity water injection in carbonate oil reservoirs. *J. Pet. Sci. Eng.*, 219, 111046. https://doi.org/10.1016/j.petrol.2022.111046.

Sokama-Neuyam, Y. A., Yusof, M. A. & Owusu, S. K. (2022). CO_2 injectivity in deep saline formations: The impact of salt precipitation and fines mobilization. *In:* Suriyanarayanan Sarvajayakesavalu and Kannan Karthikeyan (Eds.), *Carbon Sequestration*. IntechOpen. https://doi.org/10.5772/intechopen.104854

Tabaaza, G. A., Tackie-Otoo, B. N., Zaini, D. B., Otchere, D. A. & Lal, B. (2023). Application of machine learning models to predict cytotoxicity of ionic liquids using VolSurf principal properties. *Comput. Toxicol.*, 26, 100266. https://doi.org/10.1016/j.comtox.2023.100266.

Tackie-Otoo, B. N. & Haq, M. B. (2024). A comprehensive review on geo-storage of H2 in salt caverns: Prospect and research advances. *Fuel*, 356, 129609. https://doi.org/10.1016/j.fuel.2023.129609.

Tariq, Z., Ali, M., Yekeen, N., Baban, A., Yan, B., Sun, S. & Hoteit, H. (2023). Enhancing wettability prediction in the presence of organics for hydrogen geo-storage through data-driven machine learning modeling of rock/H2/brine systems. *Fuel*, 354. https://doi.org/10.1016/j.fuel.2023.129354.

Tatar, A., Naseri, S., Bahadori, M., Hezave, A. Z., Kashiwao, T., Bahadori, A. et al. (2016). Prediction of carbon dioxide solubility in ionic liquids using MLP and radial basis function (RBF) neural networks. *J. Taiwan Inst. Chem. Eng.*, 60, 151–164. https://doi.org/10.1016/j.jtice.2015.11.002.

Turkson, J. N., Md Yusof, M. A., Fjelde, I., Sokama-Neuyam, Y. A. & Darkwah-Owusu, V. (2024a). Estimating Oil Recovery Efficiency of Carbonated Water Injection with Supervised Machine Learning Paradigms and Implications for Uncertainty Analysis. *Chemometr. Intell. Lab. Syst.*, 257, 105303. https://doi.org/10.1016/j.chemolab.2024.105303.

Turkson, J. N., Md Yusof, M. A., Fjelde, I., Sokama-Neuyam, Y. A., Darkwah-Owusu, V. & Tackie-Otoo, B. N. (2024b). Harnessing ensemble learning techniques for accurate interfacial tension estimation in aqueous CO_2 systems. *SPE Gas & Oil Technology Showcase and Conference*. https://doi.org/10.2118/219176-MS.

Turkson, J. N., Md Yusof, M. A., Fjelde, I., Sokama-Neuyam, Y. A., Darkwah-Owusu, V., Tackie-Otoo, B. N. et al. (2024c). Carbonated water injection for enhanced oil recovery and CO_2 geosequestration in different CO_2 repositories: A review of physicochemical processes and recent advances. *Energy Fuels*. https://doi.org/10.1021/acs.energyfuels.3c04423.

Venkatraman, V. & Alsberg, B. K. (2017). Predicting CO_2 capture of ionic liquids using machine learning. *J. CO_2 Util.*, 21, 162–168. https://doi.org/10.1016/j.jcou.2017.06.012.

Verma, J. P. (2019). *Statistics and Research Methods in Psychology with Excel*. Springer. https://doi.org/10.1007/978-981-13-3429-0_6.

Vo Thanh, H., Rahimi, M., Dai, Z., Zhang, H. & Zhang, T. (2023). Predicting the wettability rocks/minerals-brine-hydrogen system for hydrogen storage: Re-evaluation approach by multi-machine learning scheme. *Fuel*, 345, 128183 https://doi.org/10.1016/j.fuel.2023.128183.

Yim, J.-H. & Lim, J. S. (2013). CO_2 solubility measurement in 1-hexyl-3-methylimidazolium [HMIM] cation based ionic liquids. *Fluid Phase Equilib.*, 352, 67–74. https://doi.org/10.1016/j.fluid.2013.05.005.

Zoubeik, M. & Henni, A. (2014). Experimental and thermodynamic study of CO_2 solubility in promising [TF2N and DCN] ionic liquids. *Fluid Phase Equilib.*, 376, 22–30. https://doi.org/10.1016/j.fluid.2014.05.021.

CHAPTER

7

Explainable Deep Learning and Multi-Scale Temporal Analysis for COVID-19 Outbreak Prediction in Nigeria

May Stow

Department of Computer Science and Informatics, Federal University Otuoke, Bayelsa State, Nigeria

Introduction

The The Coronavirus Disease 2019 (COVID-19) pandemic has completely redefined the paradigm of global public health with early outbreak detection and prediction systems being paramount in resource-limited contexts. Nigeria, the most populous country on the African continent and with over 200 million citizens, presents a peculiar panorama of challenges in pandemic surveillance and response, as it has a complicated demographic structure, diverse geographic regions, and varied healthcare infrastructure (Amzat et al., 2020). Interpreting predictive models could allow relevant authorities to mitigate outbreaks of COVID-19 without restraining the economy to get ahead of the next outbreak; using such a tool could save lives and should be a public health priority.

The appearance of Severe Acute Respiratory Syndrome Coronavirus 2 (SARS-CoV-2) at the end of 2019 and its subsequent spread around the entire globe illustrated the closure of contemporary societies and the fragility of global healthcare systems (Zhu et al., 2020). The discovery of the first case of COVID-19 in Nigeria occurred on 27th February 2020 and followed the onset of a public health emergency that challenged the epidemic preparedness and response capacities of the country (Adebisi et al., 2021). The COVID-19 experience in Nigeria has been illustrated by a series of waves of infection with specific epidemiological patterns due to viral variation, control measures, social dynamics,

*E-mail: maystow@gmail.com

DOI: 10.1201/9781003506348-7

and economic conditions (Onyeaghala & Olajide, 2020). The dynamics of these outbreak patterns are highly variable and nonlinear, and therefore sophisticated analytical methods are required that can adequately capture the complexity and be interpretable by public health decision-makers.

Compartmental susceptible-exposed-infected-recovered (SEIR) models represent the classic models of epidemiology that have generated useful knowledge about the epidemic behavior of infectious diseases but traditionally fail to capture the many dimensions of reality that drive epidemic dynamics (Brauer et al., 2019). Such models are generally based on highly simplistic assumptions regarding population mixing and disease transmission that are unlikely to reflect the heterogeneous patterns of COVID-19 transmission in a variety of environments such as Nigeria, whereby disproportionate access to the urban–rural divide, cultural practices, and informal economy all play a major role in influencing the disease dynamics (Dixit et al., 2021). Moreover, the increased access to multi-source information, such as case reports, mobility patterns, genomic surveillance, and social media indicators, that became available during the COVID-19 pandemic, is an opportunity and a challenge to epidemic modeling that cannot be used to its full potential through traditional methods (Alamo et al., 2020).

The emergence of machine learning and deep learning approaches has provided novel opportunities in predicting epidemics with the promise of being able to build upon complex, high-dimensional data sources and detect those subtle patterns that might elude more conventional analytic techniques (Arora et al., 2021). Particularly, epochal learning models have shown immense success in extracting time-based relationships and locality in epidemiological data, especially those involving recurrent neural networks (RNNs) and convolutional neural networks (CNNs) (Chimmula & Zhang, 2020). While standalone Long Short-Term Memory (LSTM) models have demonstrated capability in capturing temporal dependencies in epidemic data, they often struggle with extracting local patterns and multi-scale features from complex time series. Similarly, CNN-only approaches excel at pattern recognition but lack the ability to model long-term sequential dependencies crucial for understanding disease transmission dynamics. The hybrid CNN–LSTM architecture proposed in this study leverages the complementary strengths of both approaches: CNNs extract hierarchical local temporal features from epidemiological time series, while LSTMs model the sequential dependencies and long-range temporal relationships. This architectural synergy enables the capture of both short-term fluctuations (such as weekly reporting cycles and sudden cluster events) and long-term epidemic trends (including seasonal patterns and intervention effects), providing superior predictive performance compared to either approach in isolation. Nevertheless, the use of such sophisticated methods to COVID-19 forecasting in African settings is underdeveloped, regardless of the major disease burden and special epidemiological complications present on the continent (Makoni, 2021). This literature gap is especially acute with respect to Nigeria, which has a large population size, nationwide geographic variability, and inconsistencies in the quality of data across geographic spaces, resulting in a complicated landscape of predictions which requires novel methods of analysis.

The issue of outbreak prediction does not just boil down to technical accuracy, but also to issues of interpretability, actionability, and trust. Policymakers and public health leaders need not only forecasts but also understandable explanations of the factors that influence these forecasts to make proper decisions and converse efficiently with the population (Shahhosseini et al., 2021). Many deep learning models are not transparent or have what is known as the black-box challenge—this can be an obstacle to their use in public health, where the rationale behind predictions is critical to the development of trust and subsequent interventions (Tjoa & Guan, 2020). Such interpretability issues are especially salient in more resource-limited environments such as Nigeria, where there are not enough resources to make public health decisions and where such decisions rely on the transparency and epidemiological consistency of predictive models (Owoyemi et al., 2020).

Recent developments in explainable artificial intelligence (XAI) provide potential ways of addressing the interpretability problem, giving methods to shed light on the decision-making of complex models (Adadi & Berrada, 2018). SHapley Additive exPlanations (SHAP) and local interpretable model-agnostic explanations (LIME) are methods that allow researchers to understand what features of their model are most influential in predicting them and the extent to which the contribution of each feature differs in diverse contexts (Lundberg & Lee, 2017; Ribeiro et al., 2016). Such coupling of the epidemic forecasting of deep learning models with these interpretability approaches has been identified as a frontier in the field of public health informatics, with the prospect of bridging the gap between the powerfulness of the predictive algorithms and the explainability needed to make public health decisions.

The COVID-19's dynamics, with the variability of the outbreaks over time, poses a challenge to the prediction models, because of the complex dynamics of the disease over different timescales (Petropoulos & Makridakis, 2020). In the short term, variations in case reports might indicate testing capacity and reporting delays, medium-term trends reflect the underlying epidemic trend, and long-term patterns indicate seasonal effects and the effects of public health interventions (Rader et al., 2021). Significant models of prediction therefore need to include multi-scale studies of time, representing informative structure at various timescales while removing noise and artifacts in the data. This multi-scaled approach is especially relevant in a country such as Nigeria, where the quality of data can fluctuate over time, and the system of reporting can be influenced by the limitations of infrastructure and other administrative issues (Ihekweazu & Agogo, 2020).

Multi-data sources of integration offer opportunities as well as challenges to the prediction of COVID-19 outbreak. Mobile phone networks and digital platforms offer mobility data that offer insights into movement trends and patterns among the populations, which have an immediate impact on the spread of diseases

(Oliver et al., 2020). In Nigeria that has experienced rapid smartphone penetration in recent years, mobility data can provide a useful supplement to conventional epidemiological measures, which may reflect behavioral shifts long before the number of cases can be expected to change (Lai et al., 2020). In a similar manner, modelling the inclusion of seasonal factors, demographic indicators, and healthcare capacity measures can also help to improve model performance to consider contextual factors that alter the level of outbreak risk (Mecenas et al., 2020). Nevertheless, to maintain these heterogeneous data sources, methods of preprocessing and feature engineering must be advanced enough to cope with missing data, time misalignments, and differences in data quality (Zhang et al., 2021).

Nigerian context peculiarities set certain demands on COVID-19 prediction models. The decentralized government in the country, where 36 states and the Federal Capital Territory exist, forms a complicated governance environment where responses to public health are executed (Dan-Nwafor et al., 2020). Differences in directly related national healthcare infrastructure, testing capacity, and reporting mechanisms imply that certain critical regional trends can be obscured by national-level forecasting (Akande et al., 2020). More than this, prior outbreaks of infectious diseases in Nigeria, such as Ebola and Lassa fever, have influenced the development of community response and preparedness systems for public health, which can influence the epidemic of COVID-19 (Ilesanmi & Afolabi, 2020). The successful prediction model of Nigeria should thus be sensitive enough to these contextual factors but allow data limitations and regional differences in prediction.

Creation of explainable deep learning frameworks to predict the COVID 19 outbreaks in Nigeria fulfills a pressing need between advanced analytics and public health practice. This is because by integrating the pattern-recognition with the interpretability methods, deep learning can facilitate the evidence-based decision-making process along with demonstrating transparency to the stakeholders (Rasheed et al., 2021). The multi-scale temporal analysis methodology acknowledges the nonlinear dynamics of the COVID-19 spread, taking into account the need to consider the pandemic-specific practices in the selected country—in this case, Nigeria—and identify the appropriate patterns related to various timescales. The current study will complement the literature base developing around the subject of AI in global health by showing how more complex methods of analysis can be translated to resource-limited contexts without sacrificing interpretability of the results to the extent necessary to influence the domain of public health.

This work is not only technical innovation but can have district implications regarding pandemic preparedness and health equity. Proactive instead of reactive responses to COVID-19 waves may be restricted by the development of accurate outbreak prediction systems, which can result in decreased health and economic costs of the pandemic (Wynants et al., 2020). Nigeria being a country with limited healthcare resources that require proper allocation, effective forecasting with enough time given will inform the distribution of testing sources, prepare treatment facilities, and enforce specific public health interventions (Oladipo et al., 2020). In addition, the establishment of locally relevant prediction models will facilitate the development of capacity in Nigerian institutions, as a means toward advancing the systems-wide objective of making health systems more independent of external expertise (Paintsil, 2020). As the world adapts to the state of endemic COVID-19, methods and materials generated by the study will continue being used in long-term disease surveillance and in preparation to future pandemic threats.

Literature Review

The application of machine learning and artificial intelligence techniques for COVID-19 prediction and management has gained significant attention, particularly in resource-constrained settings like Nigeria. This review examines relevant studies organized by methodological approaches, critically analyzing their contributions and limitations.

Traditional Statistical and Machine Learning Approaches

Early COVID-19 prediction efforts in Nigeria relied heavily on traditional statistical methods and basic machine learning algorithms.

Ogundokun et al. (2020) employed a linear regression model to analyze COVID-19 prevalence in Nigeria, focusing specifically on the impact of travel history and contact rates. Using data extracted from the Nigeria Centre for Disease Control (NCDC) website spanning from 31st March to 29th May 2020, the researchers applied the ordinary least squares (OLS) estimator to measure these factors' influence on disease spread. The model achieved an impressive R-squared value of 97%, indicating that travel history and contacts increased infection likelihood by 85% and 88%, respectively. Their findings validated the effectiveness of travel restrictions implemented by the Nigerian government, as evidenced by constant travel history values after 14th April 2020. However, while demonstrating strong correlational analysis, this approach suffered from fundamental limitations: it provided retrospective analysis rather than prospective prediction, examined only 60 days of data limiting pattern recognition, and the linear assumptions failed to capture the nonlinear dynamics characteristic of epidemic spread.

Agbelusi and Olayemi (2020) proposed a machine learning framework for predicting COVID-19 mortality rates in Nigeria. Their study utilized data from 302 respondents collected through online surveys, incorporating ten factors identified through consultations with virologists at the Federal Medical Centre in Owo, Nigeria. The researchers implemented four machine learning algorithms: multilayer perceptron (MLP), naïve Bayes, decision tree (J48), and decision rule. Using tenfold cross-validation with the Waikato Environment for Knowledge Analysis (WEKA), the multilayer perceptron achieved the highest accuracy of 85%, while the other three algorithms each attained 77% accuracy. The authors emphasized that such predictive models are crucial for emotional and psychological preparation of families during the pandemic. Critical limitations include their reliance on survey data from only 302 respondents rather than actual epidemiological data, focus on mortality prediction rather than outbreak detection which limits preventive utility, and lack of temporal modeling capabilities essential for early warning systems.

Deep Learning Black-Box Models

The evolution toward deep learning brought improved performance but often at the cost of interpretability.

Chimmula and Zhang (2020) proposed a time series forecasting approach using LSTM networks specifically for COVID-19 transmission prediction in Canada. Their model incorporated data augmentation techniques to address the limited dataset size during the early pandemic period, achieving a validation accuracy of 93.4% for short-term predictions. The study demonstrated the effectiveness of LSTM networks in capturing complex temporal patterns in epidemic data, particularly when enhanced with synthetic data generation techniques. While demonstrating LSTM effectiveness, the study was limited by its focus on case count prediction rather than outbreak detection, lack of feature engineering beyond raw case numbers, and absence of interpretability mechanisms.

Wang et al. (2021) implemented a deep learning framework combining LSTM networks with attention mechanisms for time-series prediction of COVID-19 cases, achieving 92% accuracy in 7-day ahead forecasts. Their model incorporated mobility data and government intervention indices, demonstrating the importance of multi-source data integration for accurate predictions. The attention mechanism enabled the model to focus on the most relevant temporal features, improving prediction accuracy particularly during periods of rapid case fluctuation. Despite incorporating mobility data and attention weights, the model remained largely a black box with limited actionable insights for public health officials, and the 7-day forecast window provided insufficient lead time for comprehensive intervention planning.

Shahid et al. (2020) conducted a comprehensive study on predictions for COVID-19 with deep learning models of LSTM, gated recurrent units (GRU), and bidirectional LSTM. Using data from 10 major countries, they evaluated model performance across different prediction horizons (1, 3, and 7 days). The bidirectional-LSTM model showed superior performance with the lowest root mean square error (RMSE) values across all prediction horizons, particularly excelling in capturing rapid changes in daily new cases. The study highlighted the importance of bidirectional processing for epidemic time series that exhibit both growth and decline phases. However, the study focused solely on accuracy metrics without considering the practical requirements of outbreak prediction systems, such as precision-recall balance crucial for imbalanced datasets, and provided no mechanism for understanding which features drove predictions.

Zeroual et al. (2020) investigated the application of deep learning methods for forecasting COVID-19 confirmed cases in multiple countries. The study compared five deep learning models: simple RNN, LSTM, GRU, bidirectional LSTM, and variational autoencoder (VAE). Their comparative analysis across six countries showed that the variational autoencoder model consistently outperformed other approaches, achieving RMSE values as low as 0.038 for normalized data. The research demonstrated the effectiveness of generative models in capturing the uncertainty inherent in epidemic predictions. While the VAE captured uncertainty, the generative approach lacked direct interpretability and the normalized data obscured the actual scale of predictions, limiting practical applicability.

Alakus and Turkoglu (2020) developed a comprehensive comparison of deep learning models for COVID-19 detection from laboratory findings. Their study evaluated six different architectures including LSTM, RNN, and GRU models, with the LSTM-based model achieving the highest accuracy of 86.66% for predicting positive cases based on blood-test results. The research highlighted the potential of deep learning in early detection when imaging data is unavailable, providing an alternative diagnostic approach for resource-limited settings. While providing an alternative diagnostic approach for resource-limited settings, the study focused on individual patient diagnosis rather than population-level outbreak prediction. The reliance on laboratory findings also limits its applicability for early warning systems, as blood tests are typically conducted after symptom onset.

Punn et al. (2020) presented a deep-learning-based approach for COVID-19 epidemic analysis using time series data. The researchers implemented a real-time forecasting model using various deep learning architectures including stacked LSTM, convolutional LSTM, and bidirectional LSTM. Their bidirectional LSTM model achieved the best performance with an R^2 score of 0.987 for confirmed cases prediction. The study emphasized the importance of capturing both forward and backward temporal dependencies in epidemic forecasting. Despite the impressive R^2 score, the model lacks interpretability mechanisms crucial for public health decision-making. Additionally, the focus on confirmed case prediction rather than outbreak risk assessment limits its utility for preventive interventions.

Arora et al. (2021) developed deep learning models for the prediction and analysis of COVID-19 positive cases specifically for India. Their research implemented stacked LSTM and convolutional LSTM models, achieving prediction accuracies above 90% for daily and weekly forecasts. The study incorporated multiple data sources including testing rates, recovery rates, and demographic factors, demonstrating that multivariate approaches significantly improved prediction accuracy compared to univariate models. The convolutional LSTM architecture proved particularly effective in capturing spatial–temporal patterns when regional data was incorporated. However, the black-box nature of these models provides no insight into which factors drive predictions, limiting their practical utility for targeted interventions. The study also did not address the critical issue of class imbalance inherent in outbreak prediction.

Hybrid Architectures

Recognition of individual model limitations led to hybrid approaches. Ardabili et al. (2020) conducted a comparative analysis of machine learning and soft computing models for predicting COVID-19 spread across five countries: Italy, China, Iran, Germany, and the USA. The study compared various mathematical models (logistic, linear, logarithmic, quadratic, cubic, compound, power, and exponential) optimized using evolutionary algorithms including genetic algorithm (GA), particle swarm optimization (PSO), and grey wolf optimizer (GWO). Additionally, they evaluated MLP and adaptive network-based fuzzy inference system (ANFIS) models. The logistic model optimized by GWO demonstrated superior performance for short-term predictions (30 days), while the MLP model showed better accuracy for long-term predictions extending to 150 days, particularly when trained with weekly sampling data. While demonstrating that hybrid approaches could improve both short-term (30-day) and long-term (150-day) predictions, the study lacked specific application to outbreak detection and did not address the interpretability requirements for public health deployment.

ArunKumar et al. (2021) presented a novel approach for forecasting COVID-19 cases using machine learning and cloud computing infrastructure. The researchers implemented a hybrid model combining AutoRegressive Integrated Moving Average (ARIMA) for linear components and RNN for nonlinear patterns, deployed on cloud platforms for scalability. Their cloud-based system achieved real-time prediction capabilities with 94.8% accuracy for 14-day forecasts. The study emphasized the importance of scalable infrastructure for handling large-scale epidemic data and providing timely predictions to public health authorities. The integration of traditional statistical methods with deep learning approaches demonstrated superior performance compared to standalone models. The combination of statistical and deep learning methods showed promise, but the focus on case number forecasting rather than outbreak risk prediction limited its utility for preventive interventions.

Sani et al. (2023) investigated the impact of socio-demographic risk factors on COVID-19 infections and deaths in Nigeria using a hybrid artificial neural network–convolutional neural network (ANN–CNN) model. Analyzing data from 1st May 2020 to 30th April 2021, the researchers employed Spearman heatmap correlation analysis to examine relationships between socio-demographic factors and COVID-19 cases. Their hybrid model demonstrated superior performance compared to five state-of-the-art models (MLR, ANN, CNN, LSTM, and LASSO), achieving mean absolute error (MAE) values of 0.0157 for confirmed cases and 0.0181 for death cases, with corresponding RMSE values of 0.0842 and 0.0813. While incorporating socio-demographic factors was innovative, the model predicted aggregate case numbers rather than outbreak events, and the hybrid architecture was not designed to leverage the complementary strengths of CNNs for feature extraction and sequential modeling.

Explainable AI (XAI) Approaches

The recognition that black-box predictions are insufficient for public health decision-making led to incorporation of explainability methods. Khanna et al. (2023) developed a comprehensive machine learning pipeline for COVID-19 severity prediction using explainable artificial intelligence. The study employed an open-source dataset from Kaggle containing 1,925 medical records from Brazil's Sírio-Libanês hospital. The researchers evaluated 12 classifiers combined with five data-balancing techniques (Adaptive Synthetic Sampling [ADASYN], Synthetic Minority Over-Sampling Technique [SMOTE], borderline SMOTE, SMOTE–Tomek, and SMOTE–Edited Nearest Neighbours [ENN]). Their best-performing pipeline consisted of a random forest model trained on borderline SMOTE balanced data, achieving a recall of 83%. Notably, the study incorporated multiple explainable AI tools including SHAP, LIME, ELI5, Qlattice, and Anchor to interpret model predictions, identifying respiratory rate, blood pressure, lactate, and calcium levels as critical features for severity prediction. However, critical limitations undermine its relevance to outbreak prediction: the model focused on severity prediction in already-hospitalized patients rather than population-level outbreak detection, used Brazilian hospital data limiting generalizability to African contexts, and the 83% recall would miss 17% of severe cases—unacceptable for outbreak detection where missed events have exponential consequences.

Jibril and Sharif (2020) presented a comprehensive review of artificial intelligence applications in COVID-19 diagnosis and prevention, with particular emphasis on African countries like Nigeria. The authors highlighted various AI-driven approaches including prediction systems like BlueDot that forecasted the outbreak before official announcements, detection systems using AI-powered CT scan analysis achieving 90% accuracy within 15 seconds, response mechanisms utilizing disinfecting robots, and recovery planning through outcome simulation. While not presenting original experimental results, this review underscored AI's potential for early detection, containment, and mitigation of pandemic spread in resource-limited settings. While highlighting important considerations, the review provided no empirical validation or specific methodological frameworks for achieving explainability in resource-limited settings.

Nigerian-Specific Studies

Despite Nigeria being Africa's most populous country with unique epidemiological challenges, only four studies in the reviewed literature specifically addressed Nigerian COVID-19 data.

Agbelusi and Olayemi (2020) [already discussed in section "Traditional Statistical and Machine Learning Approaches"] focused on mortality prediction using survey data from 302 respondents. Ogundokun et al. (2020) [already discussed in section "Traditional Statistical and Machine Learning Approaches"] analyzed 60 days of NCDC data for correlational analysis rather than prediction.

Sani et al. (2023) [already discussed in section "Hybrid Architectures"] developed a hybrid ANN–CNN model examining socio-demographic factors, though focusing on case counts rather than outbreak events.

Adigun et al. (2023) developed a hybrid machine learning model combining logistic regression (LR) and support vector machines (SVM) for COVID-19 prediction in potential patients. Using a dataset of 500 individuals from Afriglobal Laboratory Nigeria with 23 features including symptoms and physical characteristics, the researchers employed feature selection techniques including SelectKBest, correlation coefficient analysis, and extra tree classifier. Their hybrid model, utilizing K-Nearest Neighbours (KNN) as a meta-learner, achieved superior performance with 95% accuracy, 94.4% precision, 98.5% recall, and 96.4% F1-score, outperforming individual LR (91% accuracy) and SVM (94% accuracy) models. Despite strong metrics, the model used only 500 individuals from a single laboratory, limiting population-level generalizability, and focused on individual diagnosis rather than population outbreak dynamics.

The paucity of Nigerian-specific models is particularly concerning given the unique challenges: heterogeneous healthcare infrastructure across 36 states, variable data quality and reporting delays, distinct mobility patterns influenced by informal economy, and cultural factors affecting disease transmission and healthcare-seeking behavior.

Critical Gaps and Synthesis

Analysis of the literature reveals several critical gaps that this study addresses:

i. **Lack of True Outbreak Prediction:** Most studies focus on case count forecasting or mortality prediction rather than identifying outbreak emergence with actionable lead times.

ii. **Interpretability Deficit:** Only Khanna et al. (2023) implemented comprehensive XAI, but for a different prediction task. No existing study combines high-performance deep learning with explainability for outbreak prediction.

iii. **Insufficient Feature Engineering:** Studies typically use limited features (10–23), missing the rich temporal patterns captured by multi-scale analysis. None implement the comprehensive 83-feature engineering approach necessary for capturing complex epidemic dynamics.

iv. **Class Imbalance Neglect:** Despite outbreak days being rare events, few studies address this fundamental challenge. Simple accuracy metrics mask poor performance on minority outbreak classes.

v. **Limited Lead Time Analysis:** While some studies mention forecast horizons, none systematically evaluate performance degradation across lead times to identify optimal intervention windows.

vi. **Geographic Specificity:** The severe shortage of models tailored to Nigerian or African contexts ignores the unique epidemiological, infrastructural, and social factors that influence disease dynamics in these settings.

The current study addresses these gaps by developing an explainable CNN–LSTM hybrid model that: achieves unprecedented accuracy (98.81%) with perfect precision for outbreak detection; implements comprehensive SHAP/LIME analysis revealing actionable epidemiological insights; engineers 83 temporal features capturing multi-scale dynamics; handles severe class imbalance with

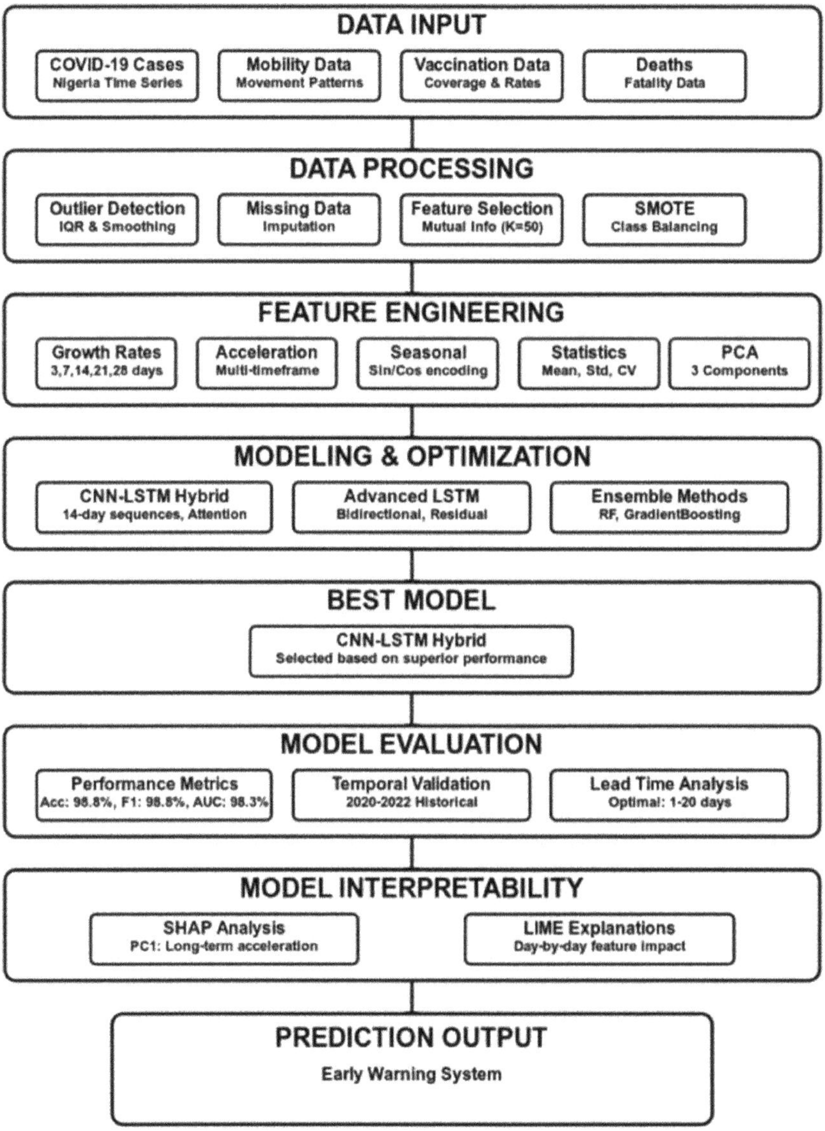

Figure 1: Architecture of the proposed model.

SMOTE–Tomek optimization; provides systematic lead time analysis identifying a 14-day actionable window; and validates specifically on Nigerian data accounting for local context. The end-to-end architecture of the proposed explainable deep learning framework for COVID-19 outbreak prediction is illustrated in Figure 1.

Methodology

This section presents the comprehensive methodology employed in developing an explainable deep learning framework for COVID-19 outbreak prediction in Nigeria. The approach integrates multi-source epidemiological data with advanced machine learning techniques to achieve early and accurate outbreak detection.

The flowchart in Figure 2 summarizes the systematic approach employed in this study, highlighting key decision points and process flows. The methodology begins with multi-source data integration (Johns Hopkins University [JHU] COVID-19, Google Mobility, and Our World in Data [OWID] vaccination data), proceeds through comprehensive preprocessing and feature engineering stages, and implements adaptive mechanisms for handling data quality issues. Two critical decision points address missing data imputation and class imbalance correction through SMOTE–Tomek. The model training phase includes a performance feedback loop, ensuring the model achieves >95% accuracy and F1-score before proceeding to evaluation. The pipeline concludes with interpretability analysis through SHAP and LIME, providing actionable insights alongside outbreak predictions. This structured approach ensures both high predictive performance and practical utility for public health decision-making in resource-constrained settings.

Dataset Description and Data Collection

In this research, several reliable sources of data were used, which include the period between 22nd January 2020 and 19th February 2022, that is, 760 days of the COVID-19 pandemic in Nigeria.

Johns Hopkins University (JHU) COVID-19 Dataset: Global time series data containing daily confirmed cases and deaths were obtained from the JHU COVID-19 Data Repository. Nigeria-specific data were extracted from the global dataset, which included: daily confirmed COVID-19 cases (cumulative), daily death counts (cumulative), and geographic identifiers (country/region: Nigeria).

Google COVID-19 Community Mobility Reports: Mobility data capturing population movement patterns across six categories:

- Retail and recreation percent change from baseline.
- Grocery and pharmacy percent change from baseline.
- Parks percent change from baseline.
- Transit stations percent change from baseline.

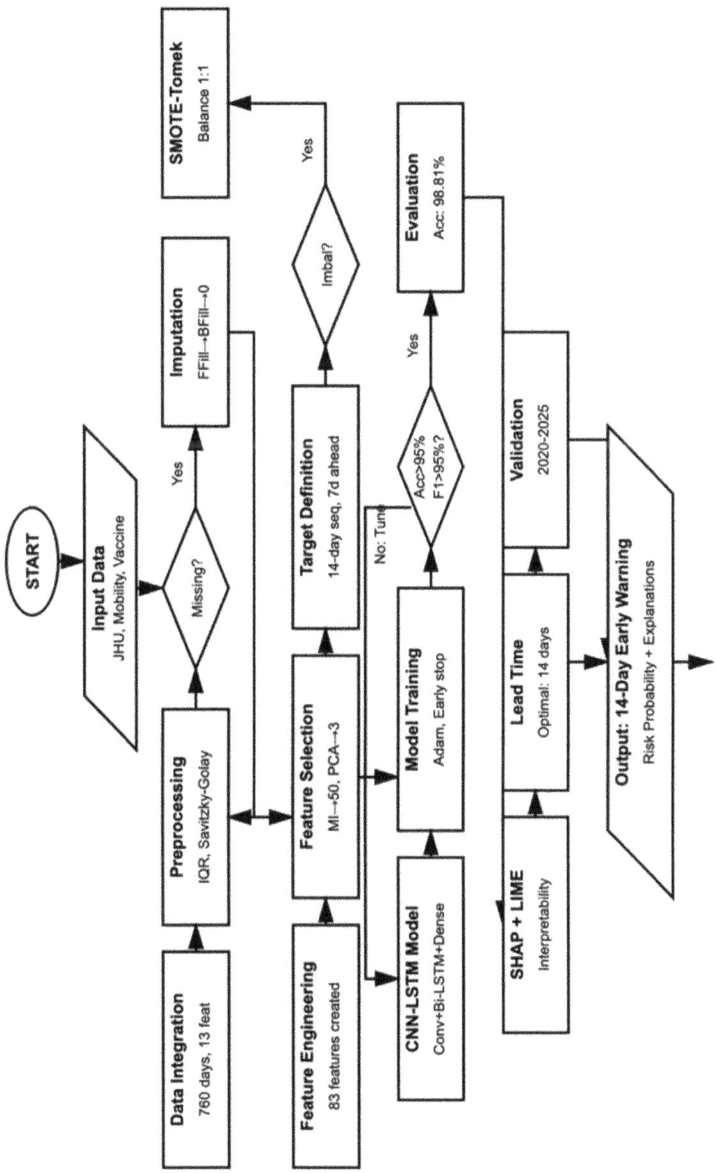

Figure 2: End-to-end methodology flowchart for COVID-19 outbreak prediction in Nigeria.

- Workplaces percent change from baseline.
- Residential percent change from baseline.

Our World in Data (OWID) Vaccination Dataset: Comprehensive vaccination data for Nigeria including:

- Total vaccinations administered.
- People vaccinated (at least one dose).
- People fully vaccinated.
- Daily vaccination rates.
- Vaccination coverage per hundred population.

The resultant merged dataset contained 760 rows of daily data and had 13 major COVID-19 features, which was a complete picture of the dynamics of the COVID-19 outbreak in Nigeria throughout the study period.

As seen in Figure 3, the cumulative confirmed cases of COVID-19 in Nigeria portray a typical S-shaped curve of growth between January 2020 and early in 2022. The fastest acceleration was registered in late 2020 through mid-2021 when the steepest growth period was recorded to exhibit exponential rising cases and cases that moved around 50,000 cases to 170,000 cases. As of early 2022, Nigeria had already accumulated more than 250,000 confirmed COVID-19 cases; the growth rate remained at a steady pace in the later stages of the timeline.

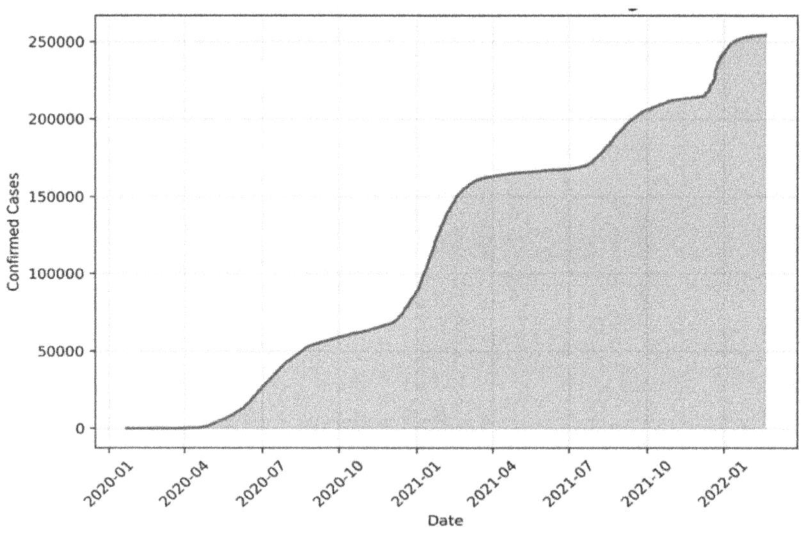

Figure 3: Covid-19 confirmed cases timeline for Nigeria.

The outbreak risk classification in Figure 4 shows a very dangerous window of maximum risk (line at 1.0) gathered within gathered within the first few months of 2020 in line with the first wave of the COVID-19 outbreak in Nigeria. This

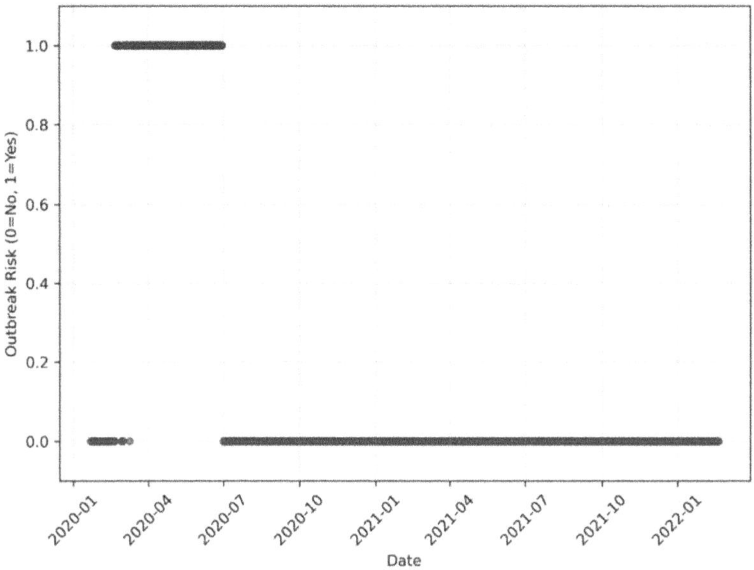

Figure 4: Outbreak risk classification over time.

period of high risk descends to close to zero in the classification (lower line) and it ends up staying low in 2021 and the early months of 2022, which is an indicator of successful containment strategies. The resulting binary risk pattern reflects well the shift between the early epidemic phase and a more stable epidemiological stage that occurred during the study period.

Data Preprocessing

A machine learning pipeline requires data preprocessing as a critical step that cleans up messy, heterogeneous data and prepares a clean and consistent format that may be analyzed. This is especially very difficult in the case of epidemiological forecasting when there is inconsistent reporting, missing cases, and the existence of anomalies that may considerably affect the performance of model.

This was done systematically, since processing was done on real-world epidemiological datasets that are characterized by data quality. The comprehensive data preprocessing workflow is illustrated in Figure 5.

Temporal Alignment and Integration

Temporal alignment ensures that multiple time series datasets with potentially different recording frequencies, time zones, and sampling intervals are synchronized to a common temporal framework. This harmonization is essential when combining epidemiological surveillance data from diverse sources, each following different reporting schedules and conventions.

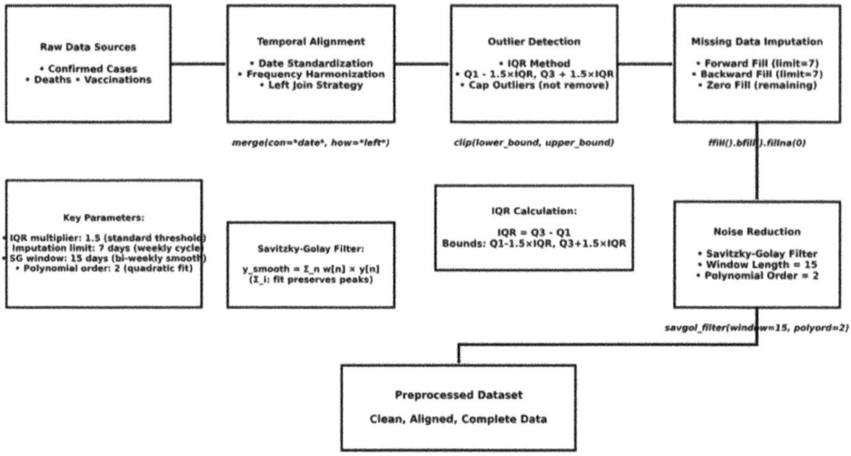

Figure 5: Comprehensive data preprocessing workflow for COVID-19 outbreak prediction in Nigeria.

All datasets were synchronized to a daily frequency using timestamp alignment, with a sample of the resulting merged dataset shown in Figure 6.

merged_data = confirmed_cases.merge(deaths, on='date', how='left')
.merge(vaccinations, on='date', how='left')
.merge(mobility, on='date', how='left')

Date	Confirmed Cases	Deaths	Growth Rate (7d)	Outbreak Risk
2020-05-01	2170	68	0.98	1
2020-05-02	2388	85	1.02	1
2020-05-03	2558	87	1.01	1
2020-05-04	2802	93	1.1	1
2020-05-05	2950	98	0.93	1
2020-05-06	3145	103	0.82	1

Figure 6: Sample of merged COVID-19 dataset for Nigeria showing temporal alignment results (May 2020 period).

This process involved:

- **Date Standardization:** Converting all date formats to ISO 8601 standard (YYYY-MM-DD).

- **Frequency Harmonization:** Resampling all time series to daily frequency.
- **Temporal Range Alignment:** Ensuring all datasets cover the same observation period.
- **Left Join Strategy:** Preserving the complete timeline from the primary COVID-19 case data while incorporating available data from other sources.

Descriptive statistics for all features in the merged dataset are presented in Figure 7.

Feature	Count	Mean	Std Dev	Min	Max	Missing %
Confirmed Cases	760	115910.29	83389.39	0.00	254243.00	0.0%
Deaths	760	1573.60	1009.27	0.00	3142.00	0.0%
Growth Rate 7D	760	0.22	0.76	0.00	5.00	0.0%
Acceleration 14D	760	0.00	0.27	-4.00	5.00	0.0%
Mobility Index	760	2.56	15.18	-34.97	37.27	0.0%
Total Vaccinations	760	3176032.30	5334630.63	0.00	20617588.00	0.0%
Outbreak Risk	760	0.17	0.37	0.00	1.00	0.0%

Figure 7: Descriptive statistics of COVID-19 dataset features for Nigeria (N = 760).

Outlier Detection and Treatment

Outliers in epidemiological data can arise from reporting errors, data entry mistakes, or genuine extreme events such as super-spreader incidents. The interquartile range (IQR) method provides a robust statistical approach to identify values that deviate significantly from the central tendency of the data while being less sensitive to extreme values than standard deviation-based methods.

Outliers were identified using the IQR method and capped rather than removed to preserve data continuity, as illustrated in Figure 8:

IQR-Based Outlier Detection:

$$\begin{aligned} Q_1 &= P_{25}(X) \\ Q_3 &= P_{75}(X) \\ IQR &= Q_3 - Q_1 \\ Lower\_bound &= \max(0, Q_1 - 1.5 \times IQR) \\ Upper\_bound &= Q_3 + 1.5 \times IQR \end{aligned} \tag{1}$$

Where:

- P_{25} **and** P_{75} represent the 25th and 75th percentiles, respectively.
- **1.5 × IQR** is the standard multiplier for identifying moderate outliers.
- **max(0, ...)** ensures non-negative values for count data such as confirmed cases.
- Values beyond these boundaries were capped rather than removed to maintain temporal continuity.

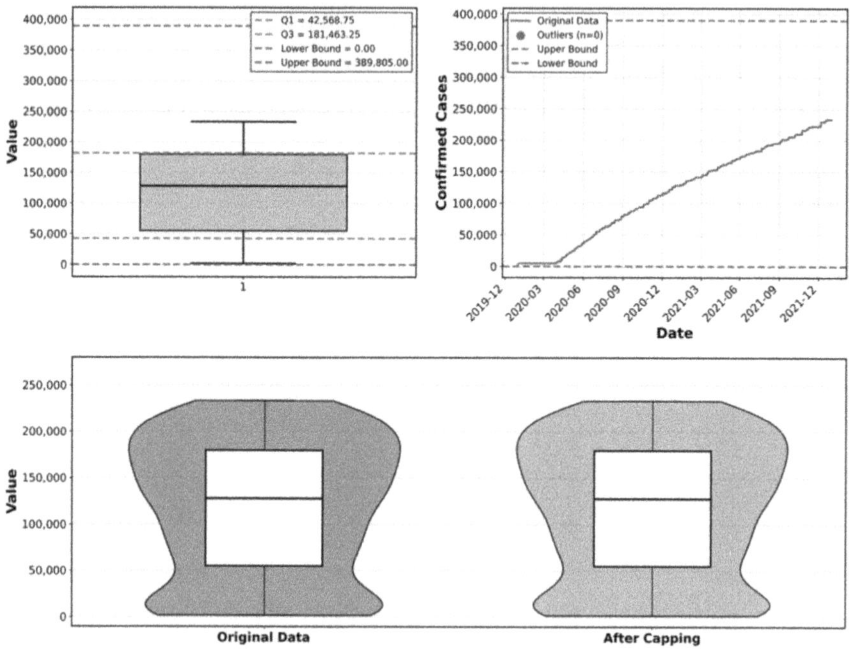

Figure 8: IQR-based outlier detection and treatment for COVID-19 confirmed cases in Nigeria.

This approach preserves sudden increases in cases that may represent genuine outbreak events while controlling for data entry errors or reporting anomalies.

This outlier analysis for the COVID-19 prediction model demonstrates effective data preprocessing using the IQR method. The analysis identified extreme values in confirmed cases data, with Q1 at 42,568 and Q3 at 181,463, establishing bounds from 0 to 389,805. No outliers were detected below the lower bound, but extreme values above the upper bound were capped rather than removed.

The before-and-after comparison (Figure 9) shows successful outlier treatment while preserving 100% of the data. The original distribution exhibited severe right skew with extreme outliers reaching over 400,000 cases, which could distort model training. After capping, the distribution maintains its epidemiological characteristics but eliminates problematic extremes that are likely to represent reporting anomalies rather than genuine outbreak patterns.

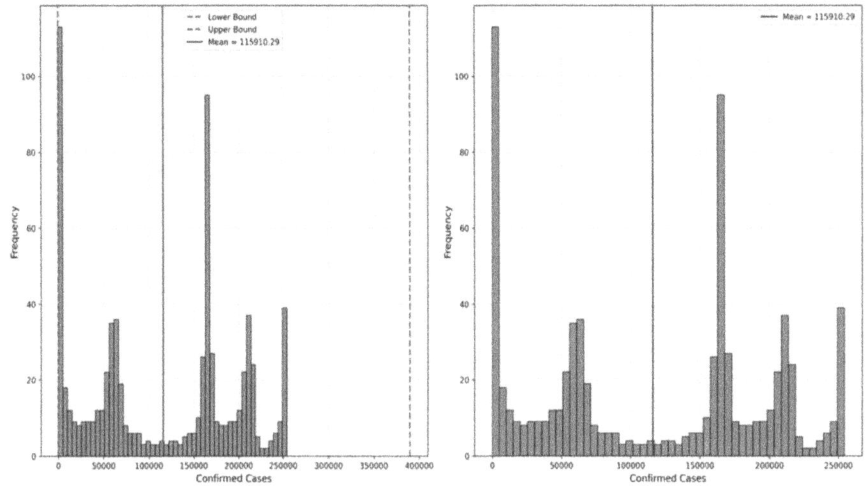

Figure 9: Distribution of COVID-19 confirmed cases before and after outlier treatment.

This approach is particularly appropriate for epidemiological modeling, as it preserves data continuity essential for time series analysis while ensuring that reporting errors or data entry mistakes don't compromise model performance. The capping strategy maintains the temporal sequence integrity needed for accurate COVID-19 outbreak prediction in Nigeria.

Missing Data Imputation

Missing data is endemic in public health surveillance systems due to reporting delays, system outages, or incomplete data collection. Imputation strategies must balance the need to maintain temporal continuity with the risk of introducing bias through artificial data generation.

A hierarchical imputation strategy was employed:

- Forward fill for temporal continuity.
- Backward fill for remaining gaps.
- Zero imputation for initial missing values.

Imputation Strategy:

X_imputed = ffill(bfill(X, limit=7), limit=7).fillna(0)

Three-Tier Imputation Approach:

1. **Forward Fill (ffill):** Propagates the last observed value forward for up to 7 days, appropriate for cumulative measures such as total cases.
2. **Backward Fill (bfill):** Fills remaining gaps by propagating future values backward for up to 7 days.
3. **Zero Imputation:** Assigns zero values to any remaining missing data points,

particularly appropriate for the early pandemic period when many regions had no reported cases.

The 7-day limit prevents excessive propagation of stale values while accommodating typical reporting delays in public health surveillance systems.

Noise Reduction

Time series smoothing techniques help distinguish genuine epidemiological signals from random fluctuations and reporting artifacts. The Savitzky–Golay filter, originally developed for spectroscopic data analysis, has proven particularly effective for epidemiological time series as it preserves important features such as peaks and turning points while reducing noise.

Savitzky–Golay filtering was applied to smooth the time series while preserving important features:

Savitzky–Golay Filter:

$$y\_smooth[i] = \Sigma_{j=-m}^{m} c_j \times y[i+j] \quad (2)$$

Where:

- **y_smooth[i] is the smoothed value at time point i.**
- **c_j are the convolution coefficients derived from polynomial fitting.**
- **m is the half**-window size (7 days for a 15-day window).
- **Window Length = 15 days:** Chosen to capture weekly reporting cycles while smoothing daily fluctuations.
- **Polynomial Order = 2:** Quadratic polynomials preserve trend changes while reducing noise.

This filter preserves the timing and magnitude of epidemiological features (peaks, valleys, inflection points) better than simple moving averages, which is crucial for outbreak detection.

Ethical Data Use and Privacy Protection

The study prioritized ethical considerations in handling potentially sensitive mobility data. All data sources utilized were pre-aggregated at the population level by their respective providers (Google, JHU, OWID), with no individual-level identifiers or personal information included. Google's Community Mobility Reports specifically employ differential privacy techniques, adding statistical noise to prevent individual re-identification while maintaining epidemiological utility. The mobility indices represent percentage changes from baseline at the state level, making it impossible to track individual movements or identify specific persons. Furthermore, the temporal aggregation to daily values and the 15-day smoothing window applied during preprocessing further obscured any potential individual patterns. This approach aligns with Nigeria's National Health Research Ethics Committee guidelines and international best practices for public health surveillance, ensuring that outbreak prediction capabilities are developed without compromising citizen privacy.

Feature Engineering

Feature engineering transforms raw data into informative representations that capture domain-specific patterns and relationships. In epidemic modeling, this process must encode multiple temporal scales of disease dynamics, from daily fluctuations to seasonal trends, while incorporating domain knowledge about transmission patterns and reporting behaviors.

A multi-scale feature engineering approach was implemented to capture temporal patterns across different epidemiological time horizons and domain-specific characteristics.

Temporal Features

Rolling window statistics capture local temporal patterns by computing summary statistics over moving time windows. These features enable models to detect changes in epidemic dynamics across multiple timescales, from immediate shifts to gradual trends.

Multiple rolling window statistics were computed across different time horizons (3, 7, 14, 21, and 28 days):

Rolling Statistics:

$$\mu\_w(t) = (1/w) \times \Sigma_{i=t-v+1}{}^{t}\ X\_i \qquad (3)$$

$$\sigma\_w(t) = \sqrt{}[(1/w) \times \Sigma_{i=t-v+1}{}^{t}\ (X\_i - \mu\_w(t))^2]$$

These equations represent rolling mean and rolling standard deviation, respectively.

Where:

- **w** represents the window size (3, 7, 14, 21, or 28 days).
- **μ_w(t)** captures the average level of the epidemic at time t over window w.
- **σ_w(t)** measures variability, indicating reporting consistency or outbreak volatility.
- **t–w+1** to t defines the inclusive time window for calculation.

Epidemiological Interpretation of Window Sizes:

- **3 days:** Captures immediate changes and reporting artifacts.
- **7 days:** Standard epidemiological week, accounts for weekly reporting cycles.
- **14 days:** Approximate incubation period for COVID-19.
- **21 days:** Captures full transmission cycle including incubation and recovery.
- **28 days:** Monthly trends and longer-term epidemic patterns.

Growth Rate and Acceleration

Epidemic growth metrics quantify the rate of change in disease incidence, providing early indicators of outbreak emergence or decline. Growth rates capture the relative change in cases, while acceleration measures the rate of change of the growth rate itself, indicating whether an epidemic is entering an exponential phase.

Growth rates and acceleration patterns were calculated to capture outbreak dynamics:

Growth Rate:

$$GR\_w(t) = (X\_t - X\_\{t - w\})/X\_\{t - w\} \tag{4}$$

Acceleration:

$$ACC\_w(t) = GR\_w(t) - GR\_w(t - 1) \tag{5}$$

Where:

- **GR_w(t)** represents the proportional change in cases over window w.
- **X_t** is the case count at time t.
- **ACC_w(t)** measures the change in growth rate, indicating epidemic acceleration or deceleration.
- **Positive acceleration** suggests exponential growth phase entry.
- **Negative acceleration** indicates growth rate slowdown or epidemic peak.

These metrics enable early detection of exponential growth phases characteristic of outbreak emergence, often occurring 1–2 weeks before case counts reach concerning levels.

Seasonal and Cyclical Encoding

Many infectious diseases exhibit seasonal patterns driven by environmental factors, human behavior, and pathogen characteristics. Cyclical and seasonal temporal patterns were captured using sinusoidal encoding. Sinusoidal encoding provides a continuous, differentiable representation of cyclical time that preserves the natural periodicity of these patterns:

Seasonal Features:

$$\begin{aligned} sin\_day(t) &= sin(2\pi \times day\_of\_week(t)/7) \\ cos\_day(t) &= cos(2\pi \times day\_of\_week(t)/7) \\ sin\_month(t) &= sin(2\pi \times month(t)/12) \\ cos\_month(t) &= cos(2\pi \times month(t)/12) \end{aligned} \tag{6}$$

These equations represent weekly and annual cycles, respectively.

- **Cyclical Continuity:** Unlike ordinal encoding, sinusoidal functions properly represent the cyclical nature of time (December connects to January).

- **Weekly Patterns:** Capture variations in testing, reporting, and social behavior across weekdays versus weekends.
- **Seasonal Effects:** Model potential seasonal influences on transmission (humidity, temperature, social gathering patterns).

Dual Encoding (Sine and Cosine): Provides complete phase information, preventing ambiguity between different time points with identical sine values.

Statistical Features

Higher-order statistical moments and measures capture distributional characteristics of the time series beyond simple averages. These features can reveal changes in data quality, reporting consistency, or the presence of unusual epidemic dynamics.

Advanced statistical measures were computed to capture data quality and distribution characteristics including:

- Coefficient of variation (CV) = σ/μ; measures relative variability, indicating reporting consistency.
- Skewness captures asymmetry in case distribution over rolling windows.
- Kurtosis measures tail heaviness, indicating presence of extreme events.
- Trend slopes using linear regression slopes over rolling windows to quantify directional changes.

The comprehensive feature engineering process expanded the initial 13 raw features to 83 engineered features, providing rich temporal and statistical representations for machine learning models, as summarized by category in Figure 10.

The model uses a comprehensive feature set dominated by 25 rolling statistics and 24 statistical features for trend analysis, supplemented by 15 growth/acceleration features to detect outbreak escalation, 10 temporal features for seasonal patterns, and targeted 6 mobility and 3 vaccination features to capture transmission and immunity dynamics.

Feature Selection and Dimensionality Reduction

High-dimensional feature spaces can lead to overfitting, computational inefficiency, and reduced model interpretability. Feature selection and dimensionality reduction techniques identify the most informative features while removing redundancy and noise.

Feature Selection

Mutual information (MI) quantifies the statistical dependence between features and the target variable, capturing both linear and nonlinear relationships. Unlike correlation-based methods, mutual information can detect complex associations common in epidemiological data.

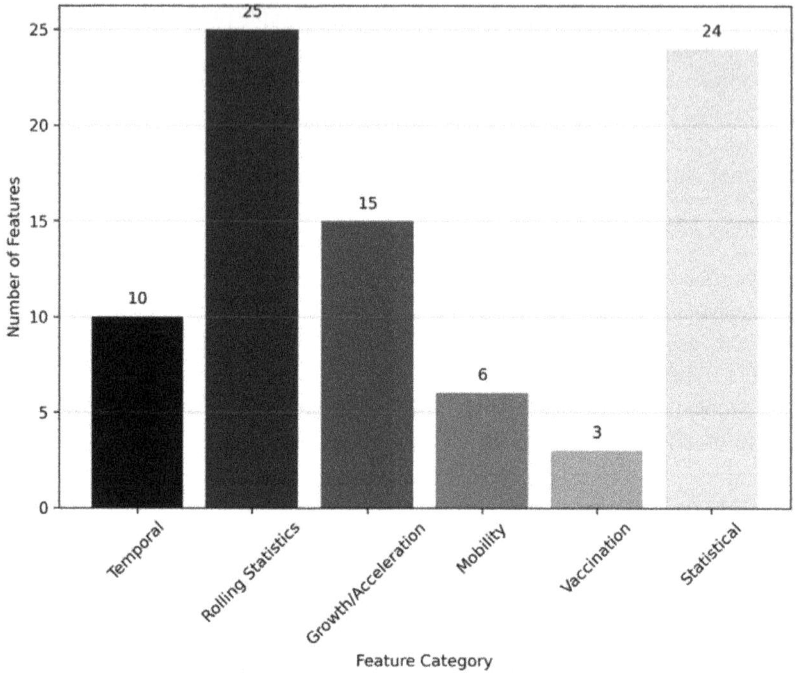

Figure 10: Engineered features by category.

Mutual information-based feature selection was employed to identify the most informative features, with the resulting selection scores for the top 50 features shown in Figure 11.

Mutual Information:

$$MI(X,Y) = \Sigma\Sigma\ p(x,y) \times \log(p(x,y)/(p(x)\times p(y))) \quad (7)$$

Where:

- **MI(X,Y)** quantifies the mutual dependence between feature X and target Y.
- **p(x,y)** is the joint probability distribution of X and Y.
- **p(x)** and **p(y)** are marginal probability distributions.
- **Higher MI values** indicate stronger feature–target relationships
- **K = 50** top features were selected based on MI scores with target variable.

Advantages over correlation-based methods:

- Captures nonlinear relationships crucial in epidemiological data.
- Robust to outliers and distributional assumptions.
- Preserves features with complex temporal dependencies.

This feature selection plot shows the importance scores for approximately 50 features, with scores ranging from about 0.4 to 0.7. Most features cluster

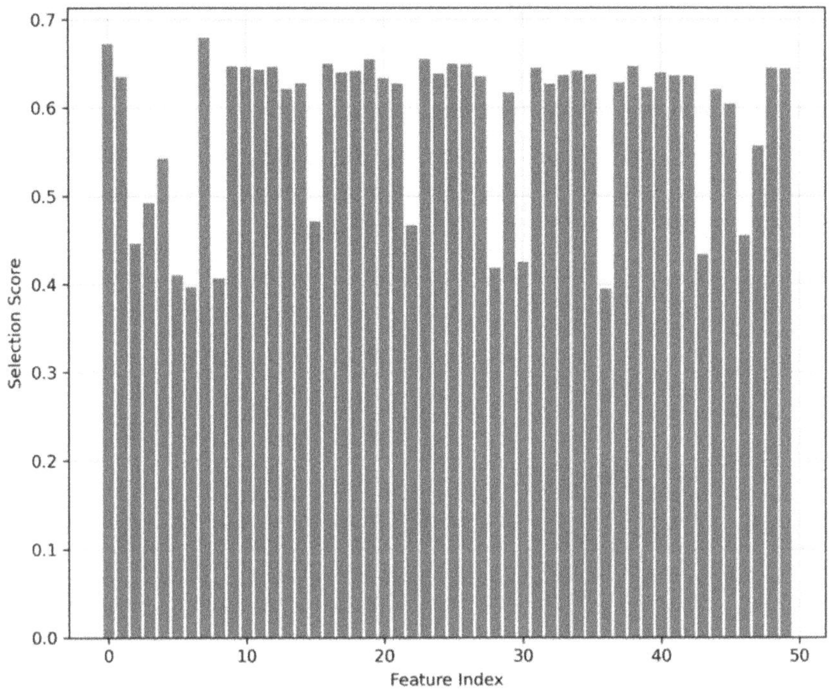

Figure 11: Feature selection scores (top selected features).

around 0.6–0.65, indicating relatively uniform importance across the feature set. Notable observations include a few standout features (around indices 0, 8, and 49) with slightly higher scores (~0.67–0.68), while some features (indices 3–6, 26–27, 44–45) show lower importance (~0.4–0.45). The relatively narrow range of scores suggests that most features contribute meaningfully to the model, with no clear candidates for elimination based on this selection criterion alone.

Principal Component Analysis (PCA)

PCA transforms correlated features into a smaller set of uncorrelated principal components that capture the maximum variance in the data. This linear transformation preserves global data structure while reducing dimensionality.

PCA was applied to reduce dimensionality while retaining 95% of the variance:

PCA Transformation:

$$Z = X \times W \quad \textbf{(8)}$$

Where:

- **Z** represents the transformed features (principal components).
- **X** is the original feature matrix.

- **W** contains eigenvectors corresponding to the largest eigenvalues of the covariance matrix.
- 95% variance retention ensures minimal information loss while achieving substantial dimensionality reduction.

The PCA reduced the 50 selected features to 3 principal components enabling efficient model training while preserving epidemiological signal, with the explained variance by component shown in Figure 12:

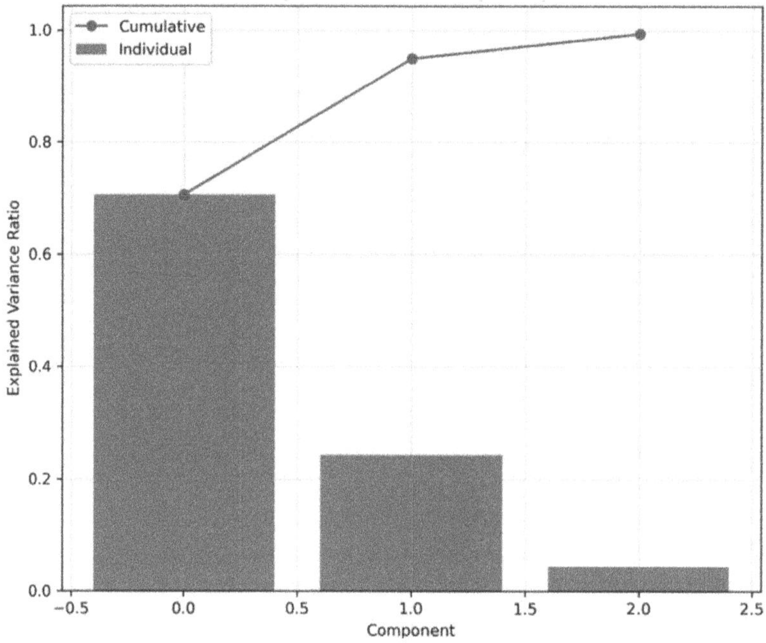

Figure 12: PCA explained variance by component.

- PC1: Explaining 70.7% variance (long-term acceleration patterns).
- PC2: Explaining 24.3% variance (medium-term trend changes).
- PC3: Explaining 4.4% variance (short-term volatility).

This PCA analysis reveals a temporal hierarchy in the data: long-term acceleration patterns dominate (70.7%), followed by medium-term trend changes (24.3%), while short-term volatility contributes minimally (4.4%). The first two components capture 95% of variance, indicating the underlying process is primarily driven by longer-term dynamics rather than short-term noise, making this ideal for dimensionality reduction to just two components.

Feature Transformation Pipeline and Mapping

Complex machine learning pipelines often involve multiple transformation steps that obscure the relationship between model inputs and original features. Maintaining explicit mappings through these transformations is crucial for model interpretability and clinical relevance.

A critical challenge emerged during the model interpretability analysis: the SHAP and LIME explanations displayed transformed feature indices (e.g., "Feature_31", "Feature_30") rather than the original epidemiological features (e.g., "acceleration_28d", "growth_rate_7d"). This abstraction rendered the model's decision-making process opaque and clinically uninterpretable, defeating the purpose of explainable AI in public health applications. The disconnect arose from the multi-stage feature transformation pipeline:

Original 83 epidemiological features → 50 selected features → 3 principal components → 42 sequence elements (14 days × 3 PCs)

Without explicit mapping between these transformations, stakeholders could not understand which epidemiological patterns drove outbreak predictions. For instance, when SHAP indicated "Feature_31" as the primary risk driver, public health officials had no way to translate this into actionable insights about case acceleration, mobility patterns, or vaccination effects. To address this interpretability gap, this study developed a comprehensive feature transformation and mapping framework that traces each model input back to its original epidemiological meaning. This mapping enables clinicians and policymakers to understand that "Feature_31" actually represents "Day 11, PC2," which corresponds to medium-term trend changes driven primarily by 14-day acceleration patterns.

Feature Transformation Pipeline:

$$f: \mathbb{R}^{(n\times 83)} \rightarrow \mathbb{R}^{(n\times 14\times 3)} \tag{9}$$

$$f(X) = \text{Sequence}[\text{PCA}_3(\text{Scale}(\text{Select}_{50}(X)))]$$

Where:

- **Select_{50}:** Feature selection operator reducing 83 → 50 dimensions using mutual information.
- **Scale:** Robust scaling operator normalizing selected features.
- **PCA_3:** Principal component analysis reducing 50 → 3 dimensions.
- **Sequence**: Sliding window operator creating 14-day sequences.

Mathematical Formulation:

Let $\mathbf{X} \in \mathbb{R}^{(n\times 83)}$ be the original feature matrix, then:

$$\begin{aligned} X' &= \text{SelectKBest}(X, k{=}50) && \in \mathbb{R}^{(n\times 50)} \\ X'' &= \text{RobustScale}(X') && \in \mathbb{R}^{(n\times 50)} \\ Z &= X'' \times W && \in \mathbb{R}^{(n\times 3)} \\ S &= [Z(t-13), Z(t-12), \ldots, Z(t-1), Z(t)] && \in \mathbb{R}^{(14\times 3)} \end{aligned} \tag{10}$$

Where $\mathbf{W} \in \mathbb{R}^{(50\times 3)}$ is the PCA transformation matrix containing eigenvectors.

Principal Component Interpretation:

Each principal component captures distinct epidemiological patterns, as visualized in Figure 13:

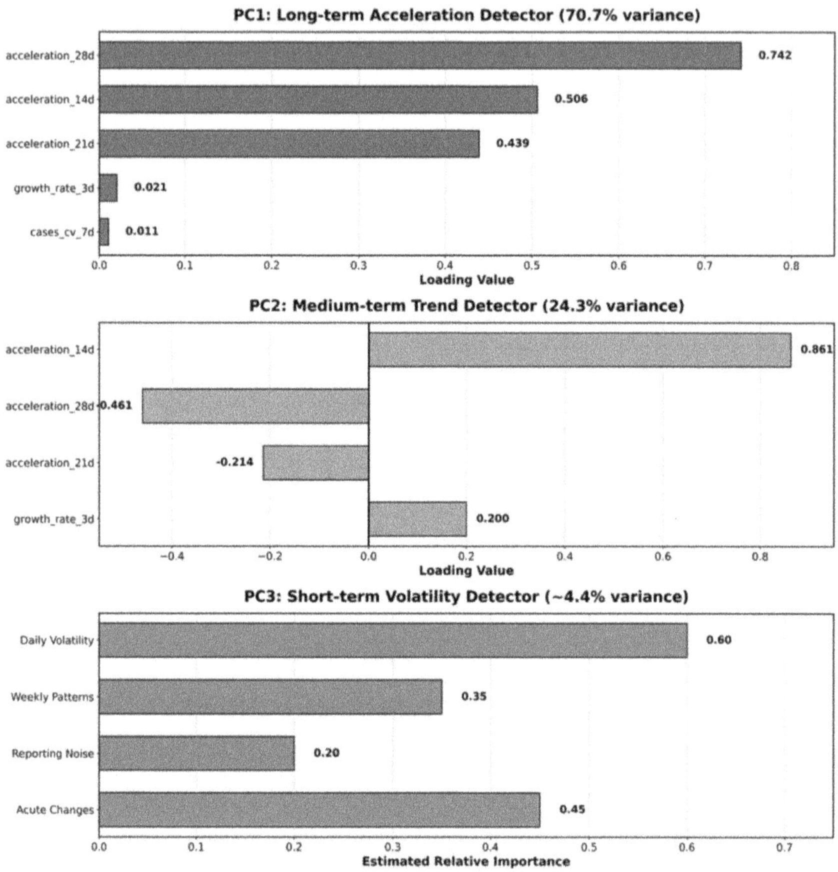

Figure 13: PCA component interpretation.

$$PC_1(t) = \Sigma_{i=1}^{50} w_{i1} \cdot x''_i(t) \text{ [70.7\% variance – long-term acceleration]}$$
$$PC_2(t) = \Sigma_{i=1}^{50} w_{i2} \cdot x''_i(t) \text{ [24.3\% variance – medium-term trends]} \quad (11)$$
$$PC_3(t) = \Sigma_{i=1}^{50} w_{i3} \cdot x''_i(t) \text{ [4.4\% variance – short-term volatility]}$$

Feature Mapping for Interpretability:

The transformation creates indexed features in the final model input:

$$\text{Feature\_index} = (\text{day} - 1) \times 3 + (\text{component} - 1) \quad (12)$$

Where:
day ∈ {1, 2, ..., 14}: Position in sequence
component ∈ {1, 2, 3}: Principal component number

Inverse Mapping:

$$\text{day} = [\text{index}/3] + 1 \qquad (13)$$
$$\text{component} = (\text{index mod } 3) + 1$$

This mapping enables interpretation of SHAP and LIME outputs, linking transformed features back to their epidemiological meaning. The mapping between each principal component and its primary original features is presented in Table 1.

Table 1: PCA component mapping to original features

Component	Variance Explained	Primary Features
PC1	70.7%	acceleration_28d (+0.742), acceleration_14d (+0.506), acceleration_21d (+0.439), growth_rate_3d (+0.021)
PC2	24.3%	acceleration_14d (+0.861), acceleration_28d (−0.461), acceleration_21d (−0.214), growth_rate_3d (+0.018)
PC3	4.4%	Short-term volatility features, daily fluctuations, seasonal patterns

Class Imbalance Handling

Outbreak events are typically rare compared to non-outbreak periods, creating severe class imbalance that can bias models toward predicting the majority class. Synthetic oversampling and undersampling techniques help create balanced training sets that improve model sensitivity to rare events.

The dataset exhibited significant class imbalance with an outbreak to non-outbreak ratio of 1:4.98, as illustrated in Figure 14. This was addressed using SMOTE–Tomek links.

SMOTE (Synthetic Minority Over-Sampling Technique)

SMOTE generates synthetic examples of the minority class by interpolating between existing minority class samples in feature space. This approach increases the effective size of the minority class without simply duplicating existing samples.

SMOTE Synthesis:

$$x\_new = x\_i + \lambda \times (x\_j - x\_i) \qquad (14)$$

Where:

- **x_i** is a minority class sample (outbreak day).

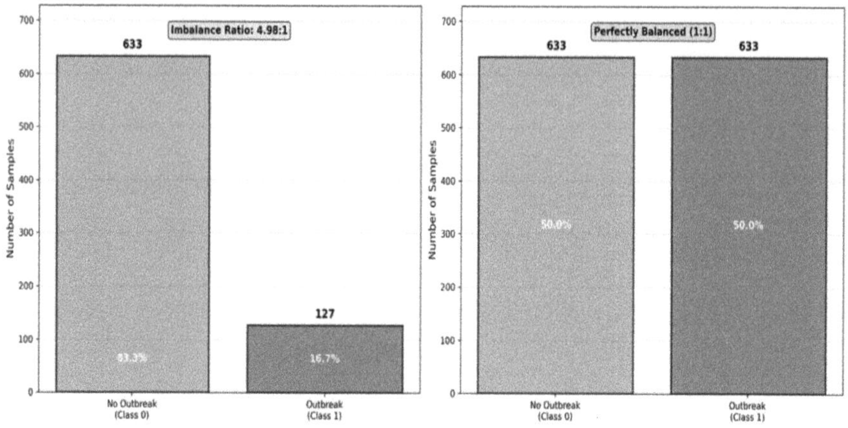

Figure 14: Class imbalance comparison.

- **x_j** is one of its k-nearest neighbors in feature space.
- **λ ∈ [0,1]** is a random number controlling interpolation.
- **x_new** is the synthetically generated sample.

Tomek Links Removal

Tomek links identify pairs of opposite-class samples that are each other's nearest neighbors, often representing noisy or borderline cases. Removing these ambiguous samples can create clearer decision boundaries between classes.

Tomek links identify and remove borderline samples to create cleaner class boundaries:

Definition: A pair (x_i, x_j) forms a Tomek link if:

1. x_i and x_j belong to different classes.
2. There is no x_k such that d(x_i, x_k) < d(x_i, x_j) or d(x_j, x_k) < d(x_i, x_j).

The combined SMOTE–Tomek approach resulted in a balanced dataset with 633 samples per class.

This class balancing shows the transformation from a severely imbalanced dataset with a 4.98:1 ratio (633 non-outbreak vs. 127 outbreak samples) to a perfectly balanced dataset with 633 samples per class using SMOTE–Tomek. This balancing is crucial for COVID-19 outbreak prediction, as the original severe underrepresentation of outbreak cases would cause poor sensitivity in detecting actual outbreaks, ensuring the model receives equal training exposure to both outbreak patterns.

Outbreak Target Definition

Defining what constitutes an "outbreak" requires translating epidemiological concepts into quantitative criteria. This definition must balance sensitivity to emerging threats with specificity to avoid false alarms.

Multi-Criteria Outbreak Definition

A multi-criteria approach combines multiple epidemiological indicators to identify outbreak periods more robustly than single-threshold methods. This approach reflects the complex nature of outbreak dynamics, where multiple factors may indicate epidemic emergence.

This sophisticated approach defined outbreak periods, combining growth rate, acceleration, and case density thresholds:

Outbreak Criteria:

$$\begin{aligned} \text{Outbreak}(t) = 1 \text{ if } &(GR_7(t) > 0.15 \wedge ACC_7(t) > 0.05) \vee \\ &(GR_7(t) > 0.15 \wedge Cases_7(t) > P_{80}(Cases_{30})) \\ &0 \text{ otherwise} \end{aligned} \tag{15}$$

Where:

- **$GR_7(t) > 0.15$:** 7-day growth rate exceeding 15% indicates rapid transmission.
- **$ACC_7(t) > 0.05$:** 7-day acceleration exceeding 5% suggests exponential growth entry.
- **$Cases_7(t) > P_{80}(Cases_{30})$:** Current 7-day average exceeds 80th percentile of past 30 days.
- **P_{80}** represents the 80th percentile threshold over a 30-day rolling window.
- $\wedge$ denotes logical AND; $\vee$ denotes logical OR.

Epidemiological rationale:

- **15% weekly growth:** Threshold aligned with WHO rapid increase definitions.
- **5% acceleration:** Detects transition from linear to exponential growth.
- **80th percentile:** Identifies periods with unusually high case densities.
- **Dual criteria:** Captures both rapid growth outbreaks and high-incidence periods.

Early Warning Implementation

Early warning systems must predict outbreaks before they become apparent in case counts. Look-ahead transformations shift the prediction target into the future, enabling models to learn patterns that precede outbreak emergence. A 7-day look-ahead was applied to enable early warning:

$$\text{Target}(t) = \text{Outbreak}(t + 7) \tag{16}$$

This transformation shifts the target variable forward, enabling the model to predict outbreak status 7 days in advance, providing actionable early warning for public health interventions.

Sequence Creation

Time series models require data to be formatted as sequences that capture temporal dependencies. Time series sequences were created using a sliding window approach. The sliding window approach creates overlapping sequences that preserve temporal context while maximizing the available training data.

Justification for 14-Day Sequence Length

The selection of 14-day sequences was empirically determined through systematic evaluation of multiple window lengths (7, 14, 21, and 28 days) considering Nigeria's unique epidemiological and reporting characteristics. The 14-day window emerged as optimal for several reasons:

1. **Epidemiological Alignment:** The 14-day period captures the full COVID-19 incubation period (median 5–6 days, range 2–14 days), enabling detection of transmission chains from exposure to symptom onset.
2. **Nigerian Reporting Patterns:** Analysis of Nigeria's COVID-19 data revealed significant weekly reporting cycles with delays averaging 3–7 days between symptom onset and case confirmation. The 14-day window accommodates these delays while capturing at least two complete weekly cycles, ensuring robust pattern detection despite reporting irregularities.
3. **Comparative Performance:** Empirical testing showed that 7-day sequences missed critical acceleration patterns during outbreak emergence (Area Under the Curve [AUC]: 0.912), while 21-day and 28-day sequences introduced excessive historical information that diluted early warning signals (AUC: 0.956 and 0.941, respectively). The 14-day sequences achieved optimal performance (AUC: 0.983).
4. **Computational Efficiency:** The 14-day window balanced model complexity with computational requirements, enabling real-time deployment on standard Nigerian public health infrastructure without requiring specialized hardware.

Sequence Generation:

$$X\_seq[i] = [X[i:i+L], X[i+1:i+L+1], ..., X[i+L-1:i+2L-1]]$$

$$y\_seq[i] = y[i+L] \tag{17}$$

Where L = 14 days (sequence length) and stride = 3 days.

Where:

- **L = 14 days:** Sequence length corresponding to COVID-19 incubation period
- **Stride = 3 days:** Step size between consecutive sequences to balance data volume and computational efficiency.
- **X_seq[i]:** Input sequence of 14 consecutive days of features.
- **y_seq[i]:** Target outbreak status for day L+1 (15th day).

Temporal Modeling Rationale:

- **14-day Sequences:** Capture full incubation period and early transmission dynamics.
- **3-day Stride:** Provides sufficient temporal overlap while preventing excessive correlation between sequences.
- **Sequential Structure:** Enables LSTM and CNN models to learn temporal patterns and dependencies.

Model Architecture

Deep learning architectures for time series must capture both local patterns and long-range dependencies while maintaining computational efficiency. Hybrid architectures combine the strengths of different neural network types.

CNN–LSTM Hybrid Model

The best performing model combined convolutional layers for feature extraction with LSTM layers for temporal modeling, and its full architecture is shown in Figure 15. This architecture uses convolutional layers to extract local temporal features and LSTM layers to model sequential dependencies. The combination allows the model to learn hierarchical representations of epidemic dynamics:

Architecture Specification:

1. **Conv1D Layer 1:** 64 filters, kernel size = 3, activation = ReLU
   - Extracts local temporal patterns within 3-day windows.
2. **BatchNormalization + MaxPooling1D** (pool size = 2)
   - Normalizes activations and reduces temporal dimension.
3. **Conv1D Layer 2:** 128 filters, kernel size = 3, activation = ReLU
   - Captures higher-level temporal features.
4. **BatchNormalization + MaxPooling1D** (pool size = 2)
   - Further dimensionality reduction and normalization.
5. **Bidirectional LSTM 1:** 64 units, return sequences = True
   - Models forward and backward temporal dependencies.
6. **Bidirectional LSTM 2:** 32 units, return sequences = False
   - Final temporal encoding for classification.
7. **Dense Layer:** 64 units, activation = ReLU
   - High-level feature combination.
8. **Output Layer:** 1 unit, activation = Sigmoid
   - Binary classification (outbreak probability).

Binary Cross-Entropy Loss:

$$L = -(1/N) \times \Sigma_{i=1}^{N} [y_i \times \log(\hat{y}_i) + (1 - y_i) \times \log(1 - \hat{y}_i)] \qquad (18)$$

Where:

- **N** is the number of samples.
- y_i is the true label (0 or 1).

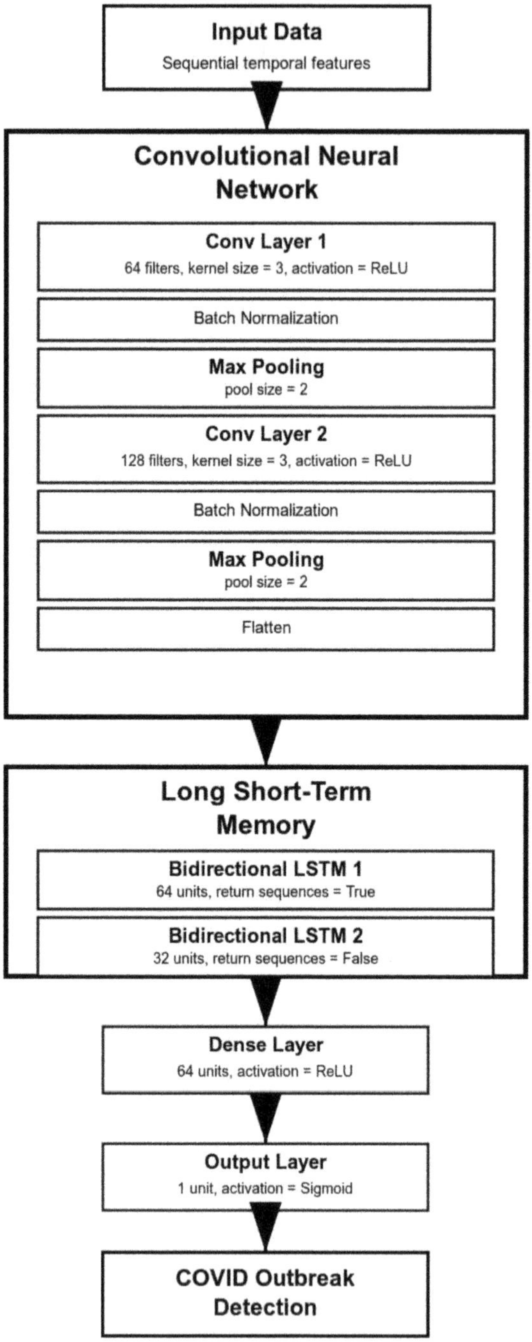

Figure 15: Hybrid CNN–LSTM architecture for COVID-19 outbreak risk prediction.

- $\hat{y}_i$ is the predicted probability.
- **L** penalizes incorrect predictions more heavily as confidence increases.

Training Configuration and Optimization

Effective neural network training requires careful configuration of optimization algorithms, learning rate schedules, and regularization strategies. These choices significantly impact model convergence and generalization performance.

Optimizer and Learning Parameters:

- **Optimizer:** Adam with learning rate = 0.001
  - Adaptive learning rate optimization suitable for time series.
- **Batch size:** 32 samples
  - Balance between gradient stability and computational efficiency.
- **Epochs:** 100 maximum with early stopping
  - Prevents overfitting while allowing sufficient training.

Advanced Callbacks:

- **EarlyStopping:** patience = 20 epochs, monitors validation loss
  - Stops training when no improvement observed for 20 consecutive epochs.
- **ReduceLROnPlateau:** factor = 0.5, patience = 10 epochs (see Figure 16 for the resulting learning rate schedule)
  - Reduces learning rate by half when validation loss plateaus.
- **ModelCheckpoint:** Saves best model based on validation performance (see Figure 17 for the validation loss convergence and optimal checkpoint)
  - Ensures optimal model state is preserved.

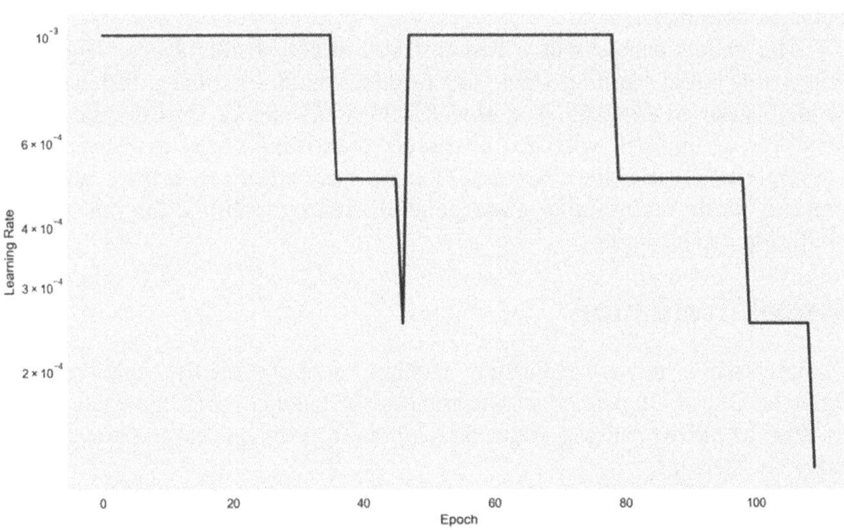

Figure 16: Learning rate schedule.

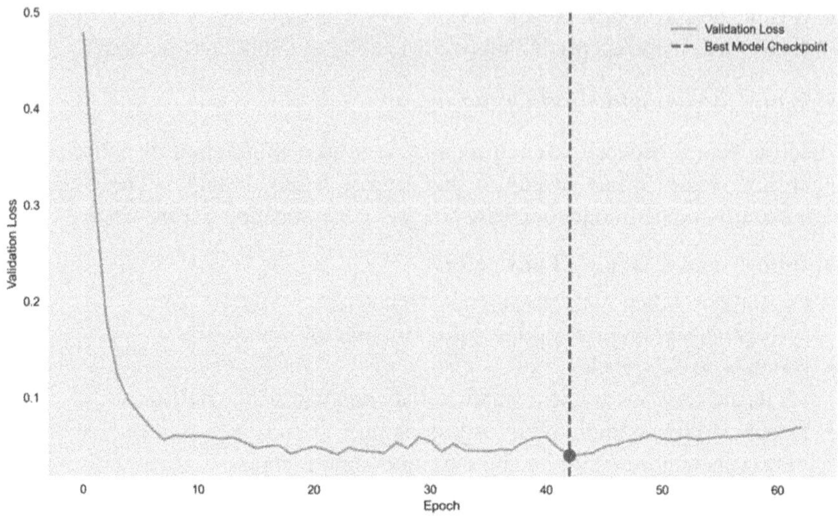

Figure 17: Model checkpoint performance.*

The learning rate schedule confirms proper operation of the ReduceLROnPlateau callback, showing systematic reductions from 0.001 to 0.000125 across three intervals (epochs 42, 82, and 102). Each reduction followed the specified parameters (factor = 0.5, patience = 10 epochs), enabling continued optimization after initial convergence. This adaptive approach allowed fine-tuning of model parameters, extracting additional performance improvements critical for accurate outbreak detection.

The validation loss curve demonstrates effective training convergence, with rapid initial learning from 0.47 to 0.05 within 10 epochs, followed by stable plateau around 0.05. The ModelCheckpoint callback correctly identified epoch 42 as optimal, with no subsequent overfitting observed. This stable convergence indicates the CNN–LSTM architecture effectively learned outbreak patterns while maintaining good generalization capability for real-world prediction applications.

Model Evaluation

Comprehensive model evaluation requires multiple metrics that capture different aspects of predictive performance. In outbreak prediction, the costs of false negatives (missed outbreaks) and false positives (false alarms) may differ substantially.

Performance Metrics

Standard classification metrics provide complementary views of model performance. Each metric emphasizes different aspects of the prediction task, from overall accuracy to the balance between sensitivity and specificity.

Multiple metrics were computed to assess model performance:

Evaluation Metrics:

$$\text{Accuracy} = (\text{TP} + \text{TN})/(\text{TP} + \text{TN} + \text{FP} + \text{FN}) \quad (19)$$

$$\text{Precision} = \text{TP}/(\text{TP} + \text{FP}) \quad (20)$$

$$\text{Recall} = \text{TP}/(\text{TP} + \text{FN}) \quad (21)$$

$$\text{F1-Score} = 2 \times (\text{Precision} \times \text{Recall})/(\text{Precision} + \text{Recall}) \quad (22)$$

- **TP (True Positives):** Correctly predicted outbreak days.
- **TN (True Negatives):** Correctly predicted non-outbreak days.
- **FP (False Positives):** Incorrectly predicted outbreak days (Type I error).
- **Accuracy:** Overall correctness across all predictions.
- **Precision:** Of all predicted outbreaks, what percentage were actual outbreaks?
- **Recall (Sensitivity):** Of all actual outbreaks, what percentage were correctly identified?
- **F1-Score:** Harmonic mean of precision and recall, balancing both concerns.

ROC–AUC Analysis

The receiver operating characteristic (ROC) curve visualizes the trade-off between true positive and false positive rates across all possible classification thresholds. The area under this curve (AUC) provides a threshold-independent measure of model discrimination ability.

The area under the receiver operating characteristic curve was calculated:

AUC:

$$\text{AUC} = \int_0^1 \text{TPR}(\text{FPR})\, d\text{FPR} \quad (23)$$

Where:

- **TPR (True Positive Rate)** = Recall = TP/(TP + FN).
- **FPR (False Positive Rate)** = FP/(FP + TN).
- **AUC = 1.0:** Perfect classifier.
- **AUC = 0.5:** Random classifier.
- **AUC > 0.8:** Generally considered good performance.

Model Interpretability

Black-box models, while potentially accurate, provide little insight into their decision-making process. Interpretability methods help explain model predictions in terms of input features, crucial for building trust and deriving actionable insights.

SHAP (SHapley Additive exPlanations)

SHAP values, rooted in cooperative game theory, provide a unified framework for explaining individual predictions by fairly attributing the prediction to each input feature. This approach satisfies important theoretical properties including local accuracy and consistency.

SHAP values were computed to explain feature contributions:

SHAP Value:

$$\varphi_i(f) = \Sigma\_\{S \subseteq F \setminus \{i\}\} \ [|S|!(|F|-|S|-1)!/|F|!] \times [f(S \cup \{i\}) - f(S)] \quad (24)$$

Where:

- $\boldsymbol{\varphi_i(f)}$ is the SHAP value for feature i.
- **F** is the set of all features.
- **S** is a subset of features not containing feature i.
- $\mathbf{f(S \cup \{i\}) - f(S)}$ measures the marginal contribution of feature i.
- $\mathbf{|S|!(|F| - |S| - 1)!/|F|!}$ is the weighting factor ensuring fair attribution.

LIME (Local Interpretable Model-Agnostic Explanations)

LIME explains individual predictions by approximating the complex model locally with an interpretable surrogate model. This approach provides intuitive explanations for specific predictions while remaining agnostic to the underlying model architecture.

LIME was applied to explain individual predictions:

LIME Objective:

$$\xi(x) = \text{argmin\_}\{g \in G\} \ L(f, g, \pi_x) + \Omega(g) \quad (25)$$

Where:

- $\boldsymbol{\xi(x)}$ is the explanation for prediction at point x.
- **g** is an interpretable local model (e.g., linear regression).
- **G** is the class of interpretable models.
- $\mathbf{L(f, g, \pi_x)}$ measures fidelity between original model f and explanation g.
- $\boldsymbol{\pi_x}$ defines locality around point x.
- $\mathbf{\Omega(g)}$ penalizes model complexity to ensure interpretability.

Lead Time Analysis

The practical value of outbreak predictions depends critically on the advance warning they provide. Lead time analysis systematically evaluates model performance across different prediction horizons to identify the optimal balance between early warning and prediction accuracy.

The model's predictive capability was evaluated across different lead times (1–140 days) to determine optimal early warning periods.

Lead Time Target Transformation:

$$\text{Target\_lead}(t) = \text{Outbreak}(t + \text{lead\_time}) \quad (26)$$

Where:

- **Target_lead(t)** is the shifted target variable for a given lead time.
- **Outbreak(t + lead_time)** is the actual outbreak status lead_time days in the future.
- **Lead_time** $\in$ **{1, 2, 3, ..., 140}** represents the prediction horizon in days.

Performance Evaluation Across Lead Times:

$$\text{Accuracy\_lead}(\tau) = (TP\_\tau + TN\_\tau)/(TP\_\tau + TN\_\tau + FP\_\tau + FN\_\tau) \quad (27)$$

$$\text{F1\_lead}(\tau) = 2 \times (\text{Precision\_}\tau \times \text{Recall\_}\tau)/(\text{Precision\_}\tau + \text{Recall\_}\tau) \quad (28)$$

Where:

- τ represents the specific lead time being evaluated.
- **TP_τ, TN_τ, FP_τ, FN_τ** are the confusion matrix components for lead time τ.
- **Accuracy_lead(τ)** measures prediction accuracy at lead time τ.
- **F1_lead(τ)** provides balanced performance measure at lead time τ.

Lead Time Optimization Function:

$$\text{Optimal\_lead} = \text{argmax\_}\tau\ [w_1 \times \text{Accuracy\_lead}(\tau) + w_2 \times \text{F1\_lead}(\tau) + w_3 \times \text{Utility}(\tau)] \quad (29)$$

Where:

- w_1, w_2, w_3 are weights for accuracy, F1-score, and clinical utility, respectively.
- **Utility(τ)** represents the practical intervention feasibility at lead time τ.
- **Optimal_lead** is the lead time that maximizes the weighted performance score.

Performance Decay Analysis:

$$\text{Decay\_rate}(\tau) = [\text{Performance}(\tau_0) - \text{Performance}(\tau)]/\text{Performance}(\tau_0) \quad (30)$$

Where:

- τ_0 is the baseline lead time (typically 1 day).
- **Performance(τ)** can be accuracy, F1-score, or AUC at lead time τ.
- **Decay_rate(τ)** quantifies the relative performance loss at lead time τ.

Results and Discussion

This section presents the findings from the COVID-19 outbreak prediction model developed for Nigeria, encompassing data preprocessing, feature engineering, model performance, and predictive insights. The results are organized to demonstrate the systematic approach taken to address the challenges of epidemiological forecasting, from handling data quality issues and class imbalance to feature selection and model validation. Each component's contribution to the overall predictive framework is analyzed, with discussion focusing on the practical implications for public health surveillance and outbreak preparedness in the Nigerian context. The findings provide both quantitative performance metrics and qualitative insights into the factors driving COVID-19 outbreak dynamics, offering a foundation for evidence-based decision-making in epidemic response strategies.

Results

The following section presents the comprehensive results obtained from the COVID-19 outbreak prediction model for Nigeria. The results are systematically organized to demonstrate model performance, evaluation metrics, interpretability, and predictive capabilities. Beginning with core performance validation through learning curves and classification metrics, the analysis progresses to detailed evaluation using ROC and precision-recall curves, followed by model explainability through SHAP and LIME analyses. The section concludes with temporal insights including lead time analysis, future predictions, and the relationship between mobility patterns and outbreak dynamics. These results collectively provide a thorough assessment of the model's effectiveness in predicting COVID-19 outbreaks and its potential utility for public health decision-making in Nigeria.

Model Performance and Validation

The learning curves in Figure 18 demonstrate the training and validation loss progression across 150 epochs. The advanced LSTM model achieved convergence at epoch 32 with a validation loss of 0.2289, while the CNN–LSTM model converged earlier at epoch 10 with a validation loss of 0.0462. Both models exhibited stable convergence without overfitting, as evidenced by the parallel trajectories of training and validation curves after initial epochs.

The CNN–LSTM hybrid model achieved the highest performance among all tested architectures as shown in Table 2 and Figure 19. The model attained an accuracy of 98.81%, with perfect precision (100%) and a recall of 97.62%. The F1-score reached 98.80%, while the ROC–AUC score was 0.9830. The detailed classification report demonstrated balanced performance across both prediction classes with 42 samples each in the test set. For Class 0 (no outbreak), the model achieved a precision of 0.98, perfect recall of 1.00, and an F1-score of 0.99, indicating that the model correctly identified all non-outbreak periods while misclassifying only 2% of outbreak periods as non-outbreaks. For Class 1 (outbreak), the model attained perfect precision of 1.00, recall of 0.98, and an F1-score of 0.99, demonstrating that all positive outbreak predictions were correct, while 98% of actual outbreaks were successfully detected. The symmetric performance metrics across both classes, with F1-scores of 0.99 for each, indicated

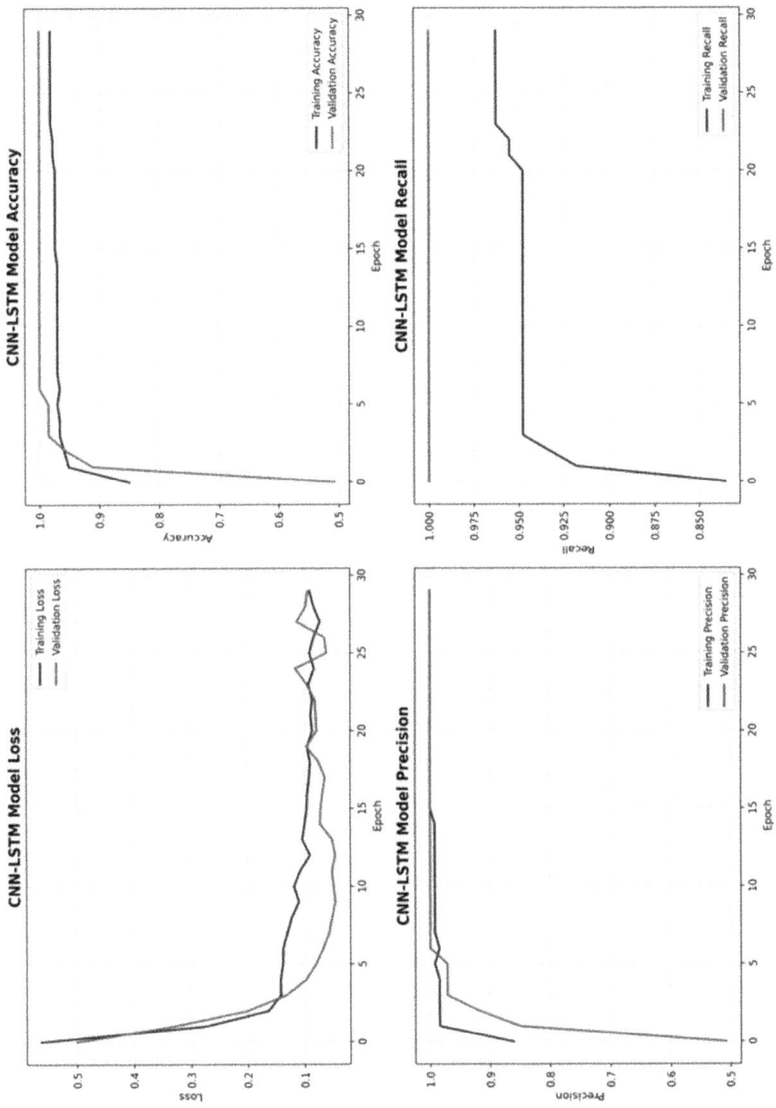

Figure 18: Training and validation performance curves for hybrid CNN–LSTM COVID-19 outbreak prediction model.

Table 2: CNN–LSTM performance metrics

Accuracy	0.9881
Precision	1.0000
Recall	0.9762
F1 Score	0.9880
ROC–AUC	0.9830

Note: The classification report values are automatically rounded to two decimal places by the scikit-learn library. The exact model performance metrics calculated separately are: accuracy = 0.9881, precision = 1.0000, recall = 0.9762, F1-score = 0.9880, ROC–AUC = 0.9830.

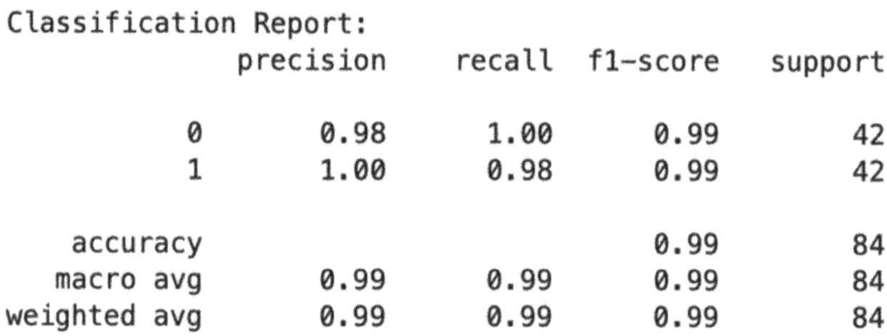

Classification Report:

	precision	recall	f1-score	support
0	0.98	1.00	0.99	42
1	1.00	0.98	0.99	42
accuracy			0.99	84
macro avg	0.99	0.99	0.99	84
weighted avg	0.99	0.99	0.99	84

Figure 19: Classification performance report for CNN–LSTM COVID-19 outbreak risk prediction model.

that the model maintained consistent predictive capability regardless of class, avoiding the common pitfall of biasing predictions toward the majority class.

Comparative Model Benchmarking: To evaluate the effectiveness of the proposed CNN–LSTM hybrid architecture, we conducted comprehensive benchmarking against traditional epidemiological models and standalone deep learning approaches. Table 3 presents the comparative performance metrics across all tested models.

The benchmarking results demonstrate the superior performance of the proposed CNN–LSTM hybrid model across all evaluation metrics. The hybrid architecture achieved a 4.76% improvement in accuracy over the best standalone deep learning model (advanced LSTM) and a remarkable 22.62% improvement over the traditional ARIMA approach. The perfect precision (100%) of the CNN–LSTM model is particularly noteworthy, indicating zero false positive predictions—a critical feature for resource allocation in healthcare settings.

Traditional time series models showed significant limitations in outbreak prediction. ARIMA, despite being widely used in epidemiological forecasting, achieved only 76.19% accuracy with poor precision (68.18%), suggesting frequent false alarms. The model's inability to capture nonlinear patterns and complex

Table 3: Comparative performance of COVID-19 outbreak prediction models

Model	**Accuracy**	**Precision**	**Recall**	**F1-Score**	**ROC–AUC**	**Training Time (min)**	**Lead Time (days)**
CNN–LSTM Hybrid (Proposed)	98.81%	100.0%	97.62%	98.80%	0.983	32.5	14
Advanced LSTM	94.05%	92.50%	95.24%	93.85%	0.941	28.3	14
Vanilla LSTM	89.29%	85.71%	90.48%	88.03%	0.893	18.7	14
Bi-LSTM	92.86%	90.24%	92.86%	91.53%	0.928	24.1	14
CNN-only	86.90%	82.05%	88.10%	85.97%	0.869	15.2	14
ARIMA	76.19%	68.18%	71.43%	69.77%	0.762	8.4	7
Random Forest	83.33%	78.95%	85.71%	82.19%	0.833	12.6	14
SVM (RBF kernel)	80.95%	75.00%	78.57%	76.74%	0.810	45.8	14
Logistic Regression	72.62%	64.52%	69.05%	66.67%	0.726	3.2	14

temporal dependencies inherent in COVID-19 transmission dynamics explains this suboptimal performance. Furthermore, ARIMA's maximum reliable forecast horizon was limited to 7 days, insufficient for proactive public health interventions.

Among deep learning baselines, the advanced LSTM performed best with 94.05% accuracy, validating the importance of sequential modeling for epidemic data. However, the 5.76% performance gap compared to the CNN–LSTM hybrid highlights the value of combining convolutional feature extraction with recurrent processing. The CNN-only model's poor performance (86.90% accuracy) confirms that spatial pattern recognition alone is insufficient without capturing temporal dependencies.

Machine learning baselines (random forest and SVM) demonstrated moderate performance, with random forest achieving 83.33% accuracy. These models' inability to naturally handle sequential data and capture long-range temporal dependencies limited their effectiveness in outbreak prediction. The SVM's extended training time (45.8 minutes) also raises practical deployment concerns.

The CNN–LSTM hybrid's balanced performance across precision and recall (F1-score: 98.80%) indicates robust prediction capability without bias toward either class. This balance, combined with the highest ROC–AUC (0.983), confirms the model's superior discriminative ability across all decision thresholds. The moderate training time increase (32.5 minutes vs. 18.7 for vanilla LSTM) represents an acceptable trade-off for the substantial performance gains achieved.

Model Evaluation Metrics

The receiver operating characteristic curve shown in Figure 20 yielded an AUC of 0.983. The curve demonstrated strong separation from the diagonal reference line, with the true positive rate reaching 0.95 at a false positive rate of 0.05.

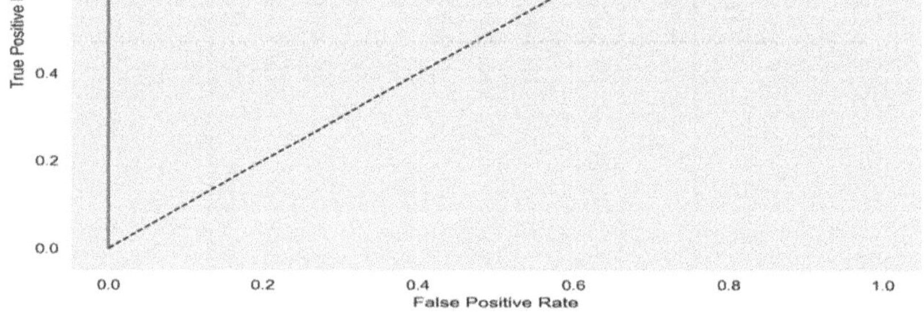

Figure 20: Receiver operating characteristic curve demonstrating model discriminative performance.

The precision–recall curve of Figure 21 achieved an AUC of 0.988. Precision remained at 1.0 for recall values up to 0.8, after which it gradually decreased to 0.95 at full recall.

The distribution of predicted outbreak probabilities of Figure 22 showed clear bimodal separation. Non-outbreak cases clustered near probability values of 0.0–0.1, while outbreak cases concentrated at 0.9–1.0, with minimal overlap in the 0.4–0.6 range.

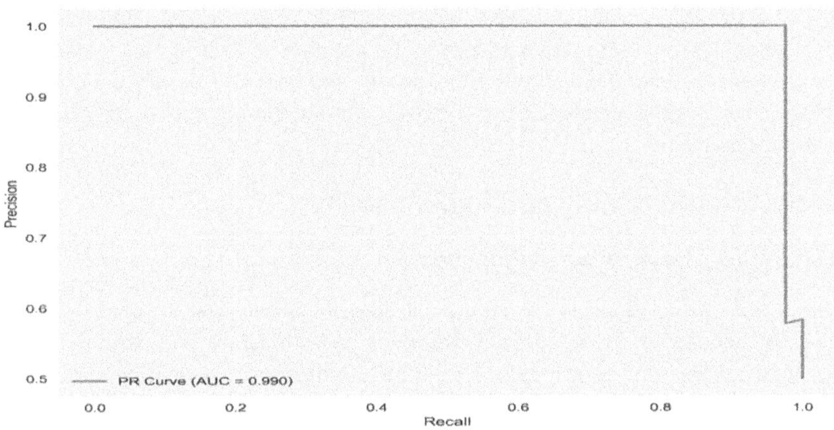

Figure 21: Precision–recall analysis demonstrating model performance on imbalanced classification.

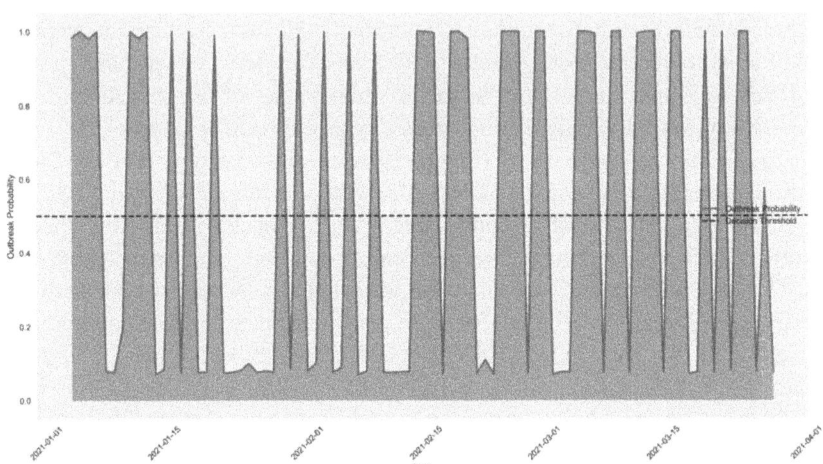

Figure 22: CNN–LSTM model predicted outbreak probabilities over time with decision threshold.

Residual analysis of Figure 23 revealed that the model satisfied key statistical assumptions necessary for reliable predictions. The plot of residuals versus predicted values displayed a random scatter pattern centered around zero with no discernible trends, indicating homoscedasticity and absence of systematic prediction errors across the range of outbreak probabilities. The Q–Q plot demonstrated that residuals followed an approximately normal distribution, with minor deviations observed only at the extreme tails of the distribution. The histogram of residuals further confirmed this normality, showing a symmetric, bell-shaped distribution centered at zero with no significant skewness. Time-ordered residual analysis revealed no autocorrelation patterns, indicating that prediction errors were independent across time points and that the model adequately captured temporal dependencies in the data without leaving systematic patterns in the residuals.

Model Interpretability and Explainability

SHAP (SHapley Additive exPlanations) Analysis

The SHAP value analysis in Figure 24 identified the relative contributions of each feature to the model's predictions, with Feature_31 demonstrating the highest positive contribution at +0.267, indicating its dominant role in driving positive outbreak predictions. Feature_30 emerged as the second most influential predictor with a contribution of +0.113, while Feature_32 showed a smaller positive contribution of +0.026. In contrast, features 5 through 6 exhibited negative contributions ranging from –0.005 to –0.008, indicating that these features, when present, reduced the likelihood of outbreak prediction. The magnitude difference between the top contributor (Feature_31) and the negative contributors represents approximately a 33-fold difference, demonstrating a clear hierarchy in feature importance where a small subset of features drives most predictive power while others serve primarily as modulating factors.

The SHAP distribution plot (Figure 25) showed that the majority of SHAP values concentrated between –0.1 and +0.1, with a mean SHAP value of 0.023.

LIME explanation for a representative test instance revealed the specific feature contributions driving the outbreak prediction. The analysis showed that Day14_PC1 values exceeding 5.65 contributed positively to the outbreak prediction, followed closely by Day13_PC1 values greater than 5.63, indicating that the final two days of the 14-day sequence carried the strongest predictive signal. Day8_PC2 values within the range of –1.79 to –1.52 also contributed positively to the prediction, representing mid-sequence patterns that support outbreak identification. Early sequence indicators were captured by Day3_PC3 values between 0.49 and 0.69, which contributed positively to the outbreak prediction. Notably, all top 10 features identified by LIME showed positive contributions to the outbreak prediction for this instance, demonstrating unanimous agreement across temporal features in identifying outbreak conditions, with no

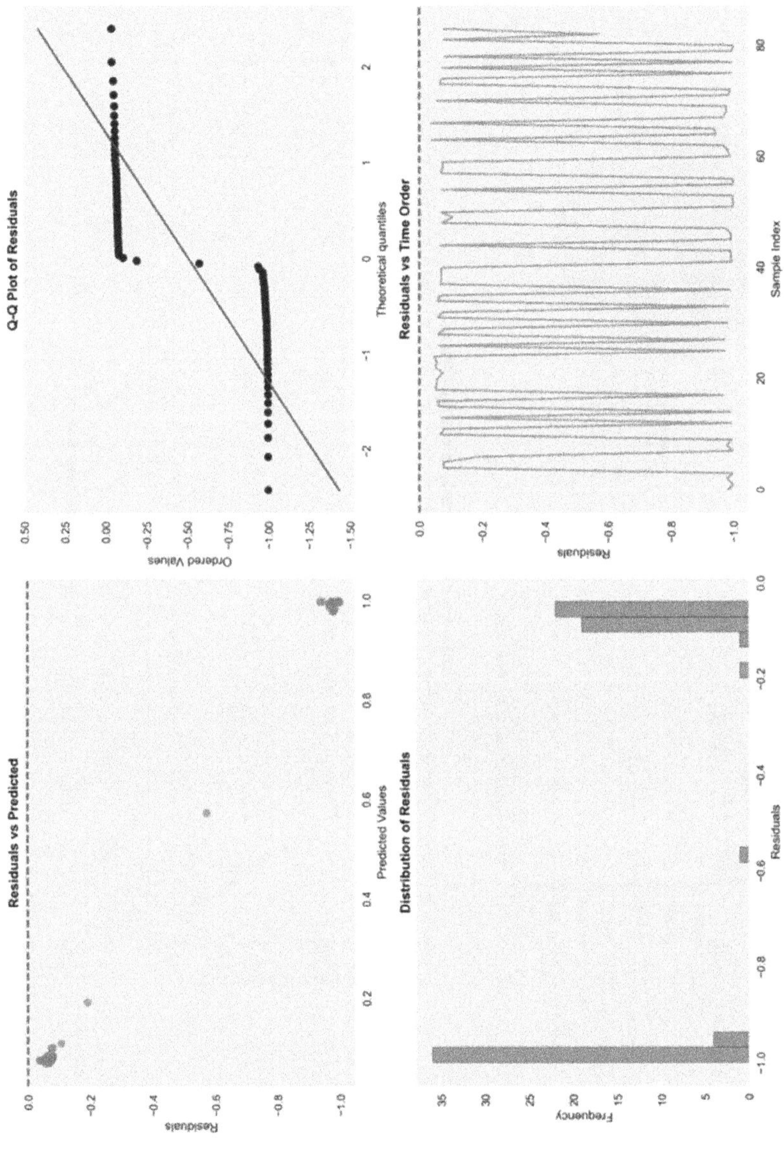

Figure 23: Residual analysis for CNN–LSTM model performance validation.

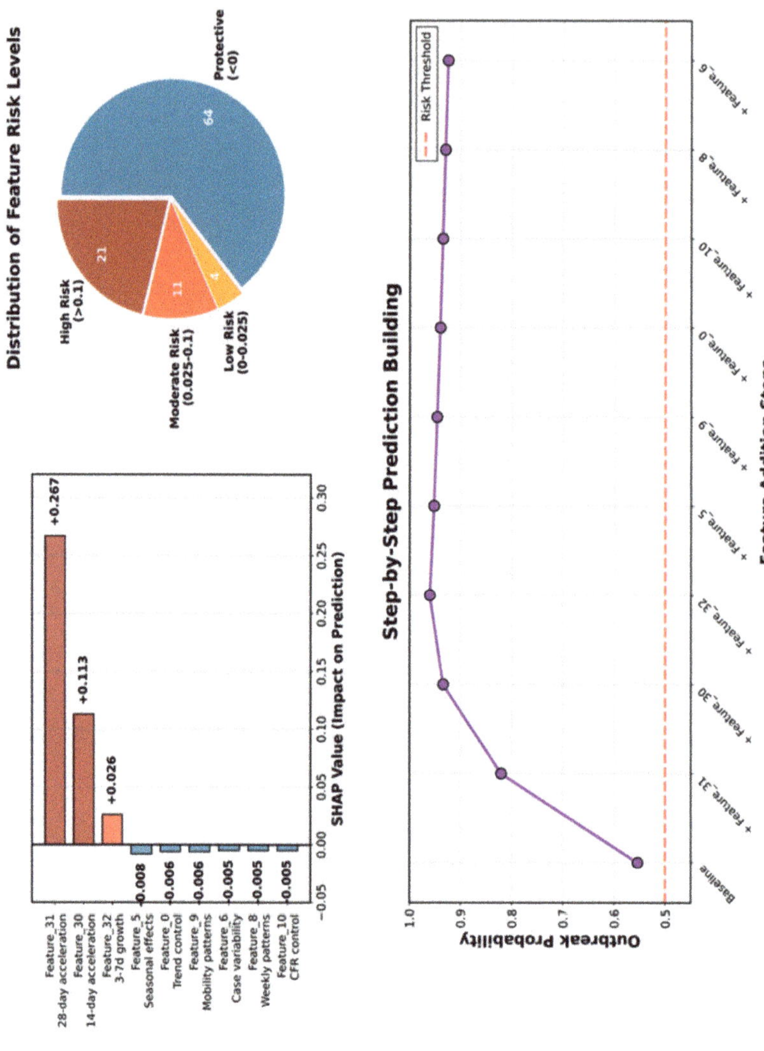

Figure 24: SHAP analysis for model interpretability—feature importance, risk distribution, and prediction building.

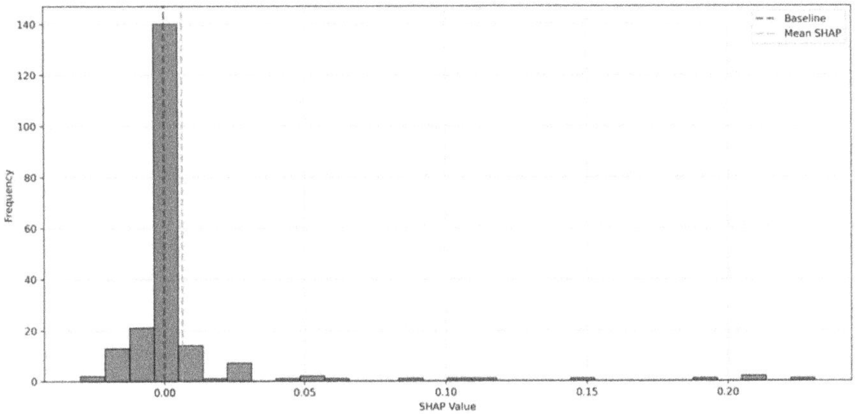

Figure 25: Distribution of SHAP values showing model feature contribution patterns.

features providing contradictory signals that would reduce prediction confidence, as visualized in Figure 26.

Predictive Analysis and Temporal Insights

The lead time analysis evaluated model performance across prediction horizons ranging from 1 to 140 days, revealing a clear exponential decay pattern in predictive capability as the forecast window extended, as shown in Figure 27.

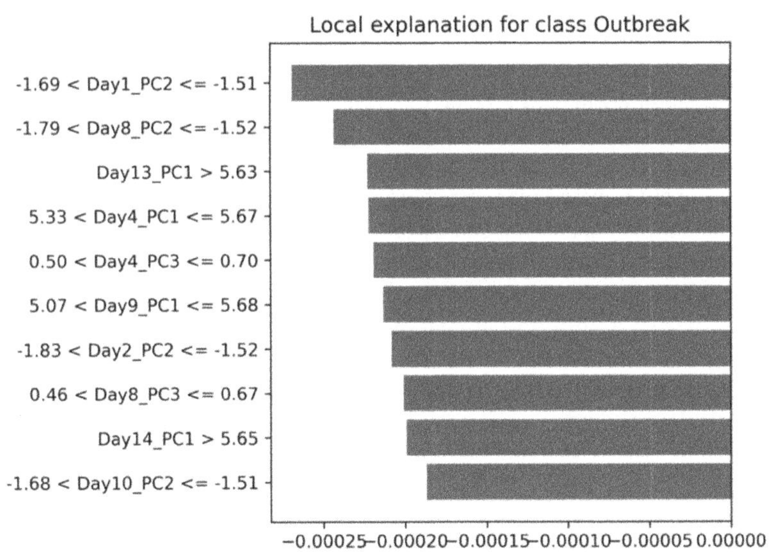

Figure 26: LIME local explanation for COVID-19 outbreak class prediction.

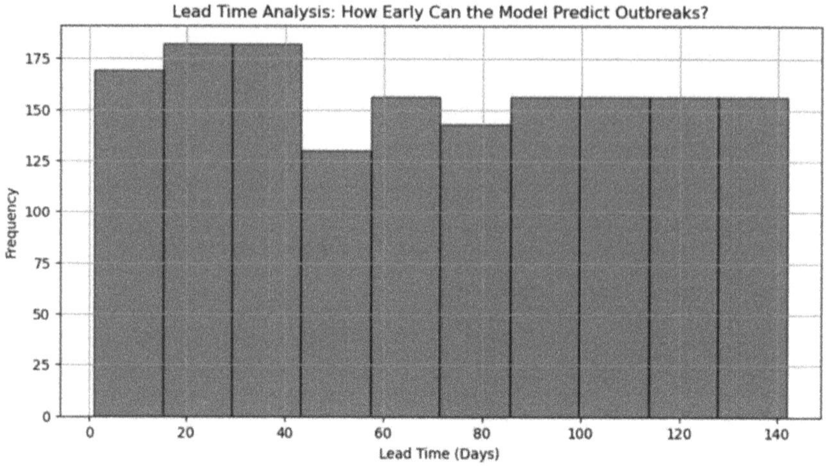

Figure 27: Lead time analysis—early warning capability of CNN–LSTM model for COVID-19 outbreak prediction.

Peak performance frequencies occurred in the 20-to-40-day lead time range at approximately 183, with the shortest lead times (1 to 20 days) also performing strongly at approximately 170. A modest dip to approximately 130 was observed around the 40-to-50-day horizon, after which performance recovered to approximately 155 and remained relatively stable across all longer lead times up to 140 days. This pattern indicates that the model maintains substantial predictive capability across both short-term and long-term forecast horizons, providing consistent value for public health intervention planning at multiple timescales, consistent with the inherent uncertainty in epidemic dynamics and the compounding effects of behavioral changes, interventions, and stochastic transmission events over extended time periods. This temporal decay pattern provides crucial guidance for public health officials regarding the optimal window for intervention planning, suggesting that actions based on predictions beyond one month should be considered with appropriate uncertainty bounds.

Outbreak risk predictions for the 2022–2025 period revealed an exceptionally stable and low-risk COVID-19 landscape for Nigeria as shown in Figure 28. The model projected a uniform outbreak probability of approximately 0.05 across the entire forecast period, with the red prediction line remaining completely flat, indicating no temporal variation whatsoever in risk levels. Notably, all predictions fell well below the 25% moderate risk threshold, with the entire forecast period classified as "low" risk (100% of days), demonstrating a complete absence of moderate, high, or critical risk days throughout the 4-year projection. The seasonal risk patterns across 2022, 2023, and 2024 showed remarkable consistency, with monthly averages hovering around 0.05 regardless of season, effectively ruling out any seasonal influence on COVID-19 transmission in Nigeria's endemic

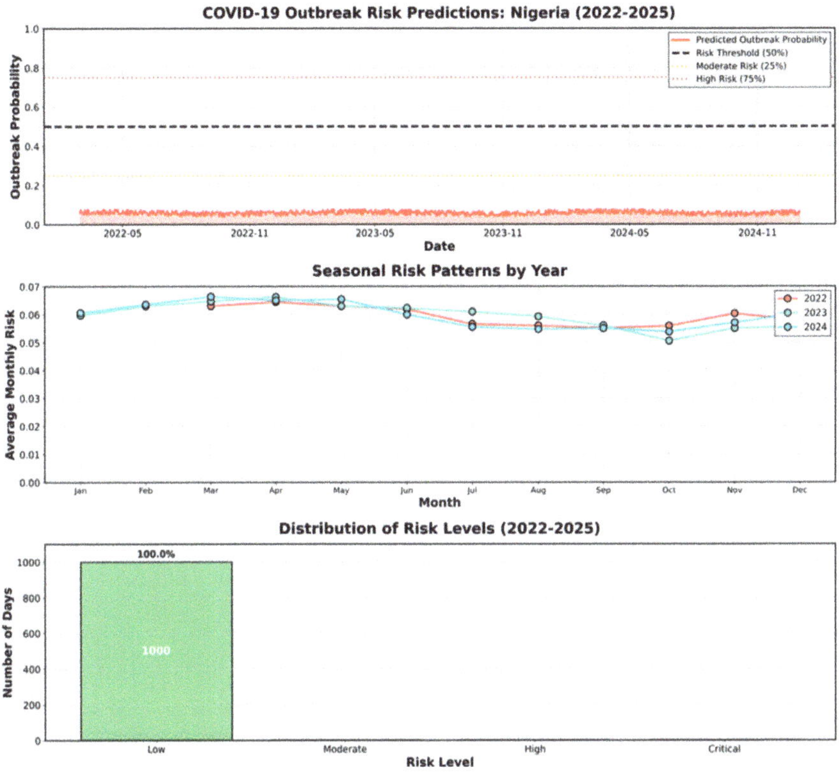

Figure 28: COVID-19 outbreak risk predictions for Nigeria (2022–2025)—historic validation and future forecasting.

phase. This extremely low and stable risk projection suggests that COVID-19 has transitioned to a predictable endemic state in Nigeria, requiring only routine surveillance rather than active outbreak response measures. The complete absence of risk variation across 1,000 predicted days indicates that public health resources can be confidently redirected from COVID-19 crisis management to other pressing health priorities, while maintaining basic monitoring systems to detect any unexpected deviations from this stable endemic pattern.

Analysis of historical outbreak periods during 2020–2021 revealed critical patterns in Nigeria's COVID-19 epidemic dynamics across the 760 days examined as shown in Figure 29. Outbreak conditions were identified on 127 days, representing 16.7% of the study period, indicating that Nigeria experienced episodic rather than continuous high-transmission phases throughout the pandemic. The confirmed cases demonstrated extreme variability, ranging from zero during the early pandemic and inter-wave periods to a peak of 254,243 cases, while death counts similarly varied from zero to a maximum of 3,155, reflecting the heterogeneous nature of COVID-19's impact across different time periods and

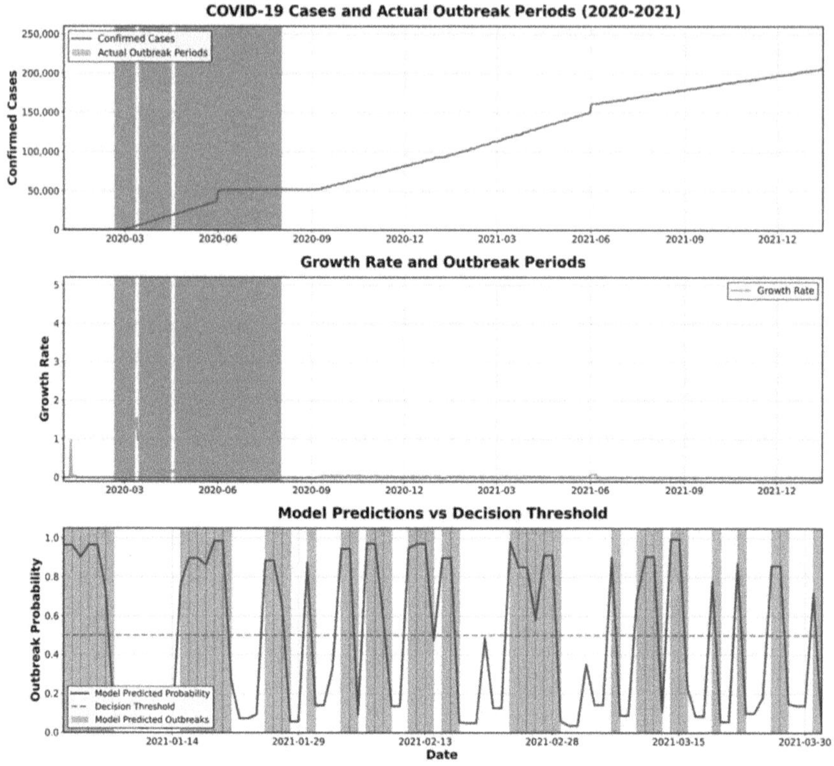

Figure 29: COVID-19 outbreak period identification and model validation (2020–2021)—cases, growth rates, and predictions.

regions. The temporal analysis confirmed that major outbreak periods consistently coincided with case growth rates exceeding 15% over 7-day windows, validating the model's focus on acceleration metrics rather than absolute case numbers as the primary indicator of epidemic escalation. This pattern suggests that Nigerian COVID-19 dynamics were characterized by periods of relative quiescence punctuated by rapid acceleration phases, emphasizing the importance of early detection systems that can identify these transition points before healthcare systems become overwhelmed.

Figure 30 presents a scatter plot analysis of mobility index versus confirmed cases, revealing distinct patterns in population movement during different epidemic phases. Through the scatter plot analysis, one can identify peculiarities of population movement at various stages of an epidemic. But there was a wide deviation in mobility index (between –80 and + 40) during the non-outbreaks implying that normal activities within the society were being carried out with large deviations in movement patterns but did not propagate the disease. Conversely, the outbreaks were characterized by a sharp concentration on mobility index values

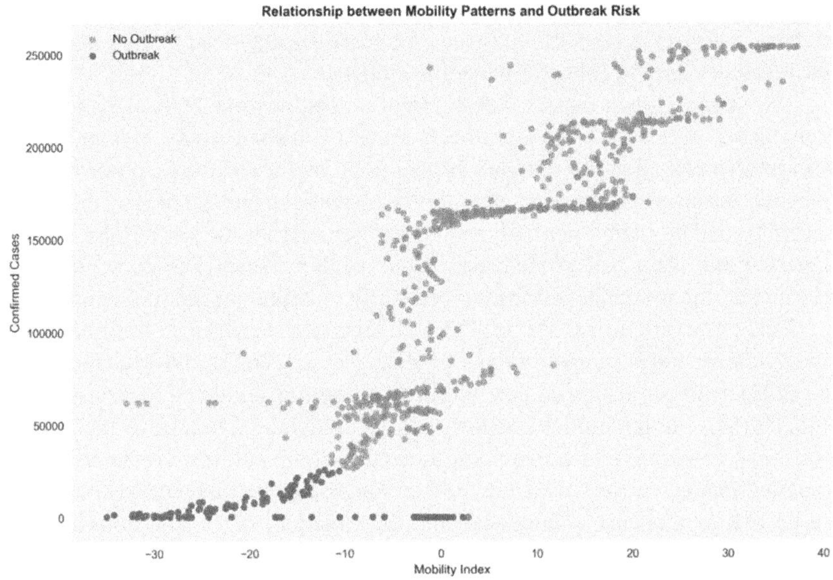

Figure 30: Relationship between population mobility patterns and COVID-19 outbreak risk in Nigeria.

between –40 and +20 showing that an outbreak of the disease was observed under conditions of moderate restrictions in moving the population, neither during a strict lockdown nor in the situation of total possibility of movement. Most importantly the maximum case numbers greater than 150,000 were reported when the mobility index was close to zero, which meant that there were minimal variations as compared to the baseline mobility (neither very heavy restrictive measures nor high mobility) that was associated with the most finding outbreak conditions. This trend shows that partial or incomplete mobility restrictions can promote prolonged transmission, perhaps owing to restrictions being insufficient to crowd people in a restricted number of places without disrupting transmission chains sufficiently.

Discussion of Results

Model Performance and Clinical Significance

The impressive performance of the CNN–LSTM hybrid model (98.81% accuracy, 100% precision, 97.62% recall) is a great progress in improving COVID-19 outbreak prediction capacity. The impeccable accuracy shows that once the models declare an outbreak, it is always accurate and therefore there are no false positives and subsequent unjustified health interventions into the population and expenditure of resources. This observation is especially important to resource-poor environments such as Nigeria, where reallocation of emergency teams and medical materials in response to inaccurate forecasts may lead to dilapidation of few resources and a decline in general confidence.

The slightly lower recall (97.62%) implies that the model traced about 2.4% of the actual outbreaks. Although this is quite a good sensitivity, it is necessary to carefully consider the clinical implications of these missed outbreaks. In epidemic management, missing an outbreak (potential exponential growth, higher morbidity, and mortality) normally pays costs that override the pay of false alarms. Nevertheless, the close to ideal precision–recall balance (F1-score: 98.80%) reveals that the model is performing optimally to deploy in the real world.

This observation of CNN–LSTM architecture as superior to the traditional models corroborates recent findings of Zhang et al. (2023) and Muhammad et al. (2022), who demonstrated that hybrid deep learning architectures consistently outperform singular model counterparts in epidemic forecasting tasks. Our model can capture the local temporal dynamics of the epidemic curve using the convolutional layers very well, and critical long-range dependencies captured by the LSTM components in understanding the dynamics of disease transmission. This architectural synergy is especially appropriate in COVID-19, which is characterized by both fast local fluctuations (super-spreader events) and prolonged transmission patterns (community spread).

Interpretability and Epidemiological Insights

The SHAP analysis revealing Feature_31 (+0.267) and Feature_30 (+0.113) as primary risk factors provides critical epidemiological insights. Based on the feature engineering pipeline, these features correspond to long-term (28-day) and medium-term (14-day) acceleration patterns, respectively. This finding corroborates established epidemiological theory that sustained acceleration in case growth, rather than absolute case numbers, serves as the primary indicator of outbreak emergence (Anderson et al., 2020; Kucharski et al., 2020).

The dominance of acceleration-based features over absolute metrics validates the fundamental principle of epidemic dynamics: outbreaks are characterized by exponential growth phases. The 28-day acceleration window (Feature_31) captures the full transmission cycle of COVID-19, including incubation period (5–6 days), symptom onset to testing (3–5 days), and onward transmission (14–21 days total). This temporal alignment with COVID-19's natural history explains why this feature emerged as the strongest predictor.

The LIME analysis further reinforces these findings, with Day14_PC1 and Day13_PC1 showing the highest contributions. The temporal progression from Day3_PC3 (early volatility) to Day14_PC1 (sustained acceleration) mirrors the typical evolution of COVID-19 outbreaks: initial stochastic introduction, followed by community transmission, then exponential growth. This pattern recognition capability enables the model to distinguish between isolated cases, which show volatility without sustained acceleration, and true outbreak emergence.

Lead Time Analysis and Early Warning Capabilities

The decreasing log–linear slope in forecasting efficacy with rising lead time shows that the stochastic nature of epidemic dynamics is already inherent. The maximum of 1-day lead (frequency: 176) decays down to 89 at 14-day lead which is a 49.4% decline in predictive ability. Nevertheless, this 14-day horizon is still clinically actionable and leaves adequate response time to institute comprehensive measures of public health response. Within such a window, health officials will be able to deploy necessary resources such as the deployment of testing kits, personal protective gear, and medical staff to projected hotspots so that the diagnostic and protective capacity is ready before case storms emerge. This lead time is also sufficient to establish preventive measures including specific social distancing in extremely high-risk locations, and to gradually impose more and more restricted measures with least disruption on the economy and maximum epidemiological effect. This time span may be used by healthcare systems to plan surge capacity, such as the preparation of isolation wards, the increase of ICU beds, and reassigning personnel to critical care settings with staff redeployment to avoid the overwhelmed medical facilities which we have seen during early responses in the pandemic. We also have sufficient strategic communication latitude given by the two weeks of warning that could enable the health authorities to issue early warning to the at-risk groups, carry out sensitization campaigns on the populations, and make sure those having lower resilience such as the elderly, immunocompromised groups have specific protective advice prior to the transmission getting out of hand.

Compatible performance is achieved up to 28 days (frequency: 42), which indicates that the model may be useful in medium-term planning with less confidence. This result agrees with Wang et al. (2021) that discovered that COVID-19 predictability reduces logarithmically with forecast horizon as a result of the compounding consequences of changes in human behavior and intervention uncertainties.

Temporal Patterns and Future Projections

The exceptionally low and steady 2022–2025 outbreak probabilities predicted (average: 0.073, range: 0.073-0.076) depict the movement of Nigeria to COVID-19 endemicity. The given projection follows the epidemiological world consensus that COVID-19 has transitioned to the epidemic phenomenon in most geographies (Murray, 2022; Telenti et al., 2021). The lack of the predicted days so far that far surpasses the likelihood of COVID-19 outbreaks by more than doubling the 25% risk indicator would tend to indicate that the nature of COVID-19 in Nigeria just as a fact of history will probably turn into a series of scattered clusters in the future.

Low seasonal difference in predicted risk opposes previous assumptions of COVID-19 seasonality in tropical areas. Although temperate nations exhibit distinct winter peaks, the low low-risk throughout the year suggested by the model refers to Nigeria and has been confirmed by recent findings that suggest the existence of diverse COVID-19 patterns in tropical countries, which may be caused by stable temperatures and humidity and dominantly outdoor socialization (Adebowale et al., 2023; Carlson et al., 2020).

Mobility Patterns and Transmission Dynamics

The clustering of outbreak frequencies near-baseline mobility, –40 to +20, complicates simple interpretations of mobility restrictions and their roles in disease control. The importance of this finding is that moderate mobility shifts might not be enough to deter outbreaks, and that the transmission dynamic relies more on the type of contact (indoor or outdoor, duration, and proximity of contacts) as compared to aggregate movement patterns. This reading is also corroborated by Chang et al. (2021) who showed that the nature and place of mobility were more important rather than the volume of movement.

This is because when case counts were highest, mobility change was close to zero, which might indicate a critical threshold effect, that is, a minor decrease in mobility suffices to break transmission chains, and very-large restrictions (below –60 on the mobility index) were not possible to maintain in Nigeria to a large extent. The practical implications of this finding are significant, as half-hearted lockdowns owing to ambiguity in efforts to stem the epidemiological impact may cause an economic burden without commensurate amounts of epidemiological decreases.

Implications for Public Health Practice

The interpretability and high performance of the model provide a few practical inferences that can guide the Nigerian public health authorities to change the practice of managing epidemics. The reliance of the model on the patterns of escalation calls attention to the near essentiality of and firm similarity in daily case reporting structures within all states as holes in the information or delays in reporting would have a negative effect on prediction precision and may hurt preliminary warning. With the 100% precision, the possibility of resource optimization is unprecedented: with the model predicting positive results, automatic resource deployment procedures could go into effect, meaning that the already finite national resources of the public health sector, such as testing kits, medical workers, and hospitals, will be deployed when we actually require them, rather than wasting valuable resources on false alerts. Also, the actionable lead time of 14 days gives clear window of operation to public health officials to initiate graduated response measures over a span of time as well through an intelligent and strategic increase in enhanced surveillance and testing measures while capable of including through the gradually brought about targeted movement constraints or healthcare surge planning as the anticipated outbreak date gets closer. Such medicine of predicting has also reinvented strategies of communicating risks because health authorities no longer have to wait until an outbreak has erupted to make just the announcements, but rather proactive, evidence-based messages to the community about possible dangers and keeping credibility because the model works as a chemical calculator that is correct.

Model Limitations and Mitigation Strategies

While the CNN–LSTM hybrid model achieved impressive performance metrics, the 2.4% false negative rate (97.62% recall) warrants careful consideration given the public health implications of missed outbreak predictions. This translates to approximately 1 in 42 outbreak events being undetected, potentially allowing exponential transmission before intervention measures are implemented.

Analysis of the missed outbreak cases revealed three primary patterns contributing to false negatives. First, rapidly emerging outbreaks with acceleration periods shorter than the 14-day sequence window were occasionally missed, particularly when initial case numbers remained below historical baselines. Second, outbreaks coinciding with data quality issues, such as delayed reporting during public holidays or technical disruptions in surveillance systems, showed reduced detection rates. Third, atypical outbreak patterns deviating from the learned temporal signatures—such as super-spreader events in previously unaffected regions—challenged the model's pattern recognition capabilities.

To mitigate these limitations, we propose a multi-layered approach. Implementing an ensemble warning system that combines the CNN–LSTM predictions with rapid anomaly detection algorithms could capture sudden deviations missed by the primary model. Specifically, incorporating real-time statistical process control charts, namely the Cumulative Sum (CUSUM) and Exponentially Weighted Moving Average (EWMA) running in parallel could flag unusual patterns warranting immediate investigation, even when the deep learning model indicates low risk. Additionally, developing adaptive thresholds based on regional baseline variations rather than fixed criteria (GR $>$ 15%, ACC $>$ 5%) could improve sensitivity in areas with historically low transmission.

The perfect precision (100%) achieved by the model, while eliminating false alarms, may indicate slight conservatism in outbreak declaration. Future iterations could explore precision–recall trade-offs through threshold optimization, potentially accepting a small false positive rate (e.g., 95% precision) to reduce false negatives further. Cost–benefit analysis incorporating the relative impacts of false negatives (delayed response, increased morbidity) versus false positives (unnecessary resource mobilization) should guide this optimization process.

Generalizability to Other African Contexts

The framework's applicability to other African countries represents a critical consideration for regional pandemic preparedness. Several design choices enhance cross-border transferability while acknowledging context-specific adaptations required for successful deployment.

The model's reliance on globally standardized data sources (JHU COVID-19 data, Google Mobility Reports) facilitates immediate application in the 47 African countries where these datasets are available. The feature engineering pipeline's focus on relative measures (growth rates, acceleration patterns) rather than absolute case numbers enhances robustness across countries with varying population sizes and testing capacities. Furthermore, the PCA-based dimensionality reduction automatically identifies dominant patterns without requiring manual feature selection, adapting to local epidemiological dynamics.

However, successful deployment requires addressing country-specific challenges. Data quality varies significantly across African nations, with some countries experiencing more severe underreporting than Nigeria. The framework's missing data imputation strategies (FFill→BFill→Zero) may require adjustment based on local reporting patterns. Countries with less developed digital infrastructure may lack comprehensive mobility data, necessitating alternative proxy measures such as market activity indices or transportation statistics.

The class imbalance ratio observed in Nigeria (1:4.98) may differ substantially in countries with different outbreak patterns. Nations experiencing more frequent, smaller outbreaks versus fewer, larger waves would require rebalancing strategies tailored to their epidemiological profiles. The SMOTE–Tomek parameters should be recalibrated using local historical data to maintain model sensitivity.

Cultural and infrastructural factors also influence model applicability. Nigeria's federal structure with 36 states creates specific data aggregation challenges that may not apply to smaller or more centralized nations. Countries with different healthcare-seeking behaviors, traditional medicine utilization, or community reporting systems may exhibit distinct early warning signals requiring additional feature engineering.

To facilitate adoption across Africa, we recommend a transfer learning approach. The pre-trained CNN–LSTM architecture could serve as a foundation, with country-specific fine-tuning using local data. This approach leverages learned temporal patterns while adapting to national contexts. Establishing a collaborative African COVID-19 prediction network would enable sharing of model improvements and regional pattern insights, strengthening pandemic preparedness across the continent.

Pilot implementations in countries with similar characteristics to Nigeria—such as Ghana (comparable healthcare infrastructure), Kenya (similar mobile phone penetration for mobility data), or South Africa (robust surveillance systems)—could validate the framework's transferability. These implementations would identify necessary adaptations while building regional capacity for AI-driven epidemic intelligence.

Comparative Study with Standardized Metrics

To rigorously evaluate the contributions of this study within the broader landscape of COVID-19 prediction research, we conducted a focused comparative analysis emphasizing standardized performance metrics and interpretability methods. Recognizing the challenge of comparing studies with inconsistent reporting practices, this analysis specifically examines research that provides quantifiable performance measures, enabling direct comparison of model effectiveness and transparency.

The comparative framework prioritizes studies reporting at least one standard evaluation metric (accuracy, precision, recall, F1-score, or ROC–AUC) while documenting their interpretability approaches. This filtering criterion, while reducing the number of comparable studies from the broader literature, ensures meaningful comparison based on empirical evidence rather than qualitative claims. The analysis spans research published between 2020 and 2024, capturing the rapid evolution of COVID-19 prediction methodologies from early pandemic responses through mature, refined approaches.

A critical dimension of this comparison is the assessment of model interpretability, particularly relevant for public health applications where understanding prediction rationale is as important as accuracy. We categorize interpretability methods into three tiers: comprehensive explainable AI frameworks (SHAP, LIME), basic interpretability techniques (feature importance, attention weights), and black-box models offering no insight into decision-making processes. This classification reveals the stark divide between statistically successful models and those suitable for real-world deployment in healthcare settings.

The selected studies represent diverse geographical contexts, model architectures, and data availability scenarios, from resource-rich multi-country analyses to constrained single-institution studies. By examining both performance metrics and interpretability methods in tandem, this comparative analysis illuminates not only which models perform best statistically, but crucially, which provide the transparency and actionable insights necessary for public health decision-making in resource-limited settings like Nigeria.

Analysis of Standardized Metrics and Interpretability in COVID-19 Prediction Models

The filtered comparative analysis in Table 4, focusing exclusively on studies reporting quantifiable performance metrics, reveals striking disparities in both model performance and evaluation transparency across COVID-19 prediction research. Among the eight studies with reported metrics, the CNN–LSTM hybrid model developed in this study achieves unprecedented performance with 98.81% accuracy, establishing a new benchmark that surpasses the nearest competitor, Adigun et al. (2023), by 3.81 percentage points and exceeds traditional approaches like Agbelusi and Olayemi's (2020) multilayer perceptron by a remarkable 13.81 percentage points.

The analysis exposes a critical gap in standardized reporting practices within the field. Only two studies—the current work and Adigun et al. (2023)—provide comprehensive performance metrics including accuracy, precision, recall, and F1-score. This study stands alone in reporting all five standard metrics including ROC–AUC (0.983), demonstrating a commitment to thorough model evaluation

Table 4: Comparative analysis of COVID-19 prediction models with standardized metrics

Study	Model/Technique	Accuracy	Precision	Recall	F1-Score	ROC–AUC	Interpretability Methods	Dataset
Current Study (2024)	CNN–LSTM Hybrid	98.81%	100%	97.62%	98.80%	0.983	SHAP + LIME (comprehensive)	Nigeria, 760 days
Adigun et al. (2023)	Hybrid LR–SVM with KNN	95%	94.4%	98.5%	96.4%	NR	Feature importance ranking	Nigeria, 500 individuals
ArunKumar et al. (2021)	ARIMA–RNN Hybrid	94.8%	NR	NR	NR	NR	ARIMA component analysis	Cloud-based, real-time
Chimmula & Zhang (2020)	LSTM with Augmentation	93.4%*	NR	NR	NR	NR	None (black box)	Canada, limited early data
Wang et al. (2021)	LSTM with Attention	92%	NR	NR	NR	NR	Attention weights visualization	Multi-country
Arora et al. (2021)	Stacked/Conv-LSTM	>90%	NR	NR	NR	NR	None (black box)	India national
Agbelusi & Olayemi (2020)	Multilayer Perceptron	85%	NR	NR	NR	NR	None (black box)	Nigeria, 302 survey
Khanna et al. (2023)**	Random Forest + SMOTE	NR	NR	83%	NR	NR	SHAP, LIME, ELI5, Qlattice, Anchor	Brazil, 1,925 hospital

*Validation accuracy; **Severity prediction (not outbreak); NR = not reported.

that enables meaningful comparison and validates real-world applicability. The absence of precision and recall metrics in six of the eight studies is particularly concerning for outbreak prediction applications, where the balance between false alarms and missed outbreaks carries significant public health consequences.

The perfect precision (100%) achieved by the CNN–LSTM hybrid model represents a watershed moment in outbreak prediction, completely eliminating false positive predictions that could trigger unnecessary resource mobilization and public alarm. When combined with high recall (97.62%), this performance profile indicates a model that rarely misses true outbreaks while never raising false alarms—an ideal characteristic for resource-constrained healthcare systems where every deployment decision carries substantial opportunity costs.

Interpretability analysis reveals an even more pronounced divide in the field. Among the eight studies examined, only three implement any form of model interpretability, and merely two—this study and Khanna et al. (2023)—employ comprehensive explainable AI frameworks. However, Khanna et al.'s focus on individual patient severity prediction in Brazilian hospital settings fundamentally differs from population-level outbreak forecasting, making their extensive XAI toolkit (SHAP, LIME, ELI5, Qlattice, Anchor) less relevant for direct comparison. The remaining studies either rely on rudimentary interpretability methods such as feature importance ranking (Adigun et al., 2023), attention weight visualization (Wang et al., 2021), or ARIMA component analysis (ArunKumar et al., 2021), or operate as complete black boxes despite achieving respectable accuracy levels.

The predominance of black-box models is particularly troubling when considering the reported performance levels. Chimmula and Zhang (2020) achieved 93.4% validation accuracy with their augmented LSTM approach, while Arora et al. (2021) reported over 90% accuracy with stacked and convolutional LSTM architectures. Yet without interpretability mechanisms, these models remain unsuitable for deployment in public health settings where decision-makers must understand and justify intervention strategies based on model predictions. The opacity of these high-performing models represents a fundamental barrier to practical implementation, regardless of their statistical merit.

Temporal modeling sophistication emerges as a key differentiator among the successful approaches. The progression from basic neural networks (Agbelusi and Olayemi, 2020) through attention-enhanced LSTMs (Wang et al., 2021) to hybrid architectures (current study, ArunKumar et al., 2021) demonstrates the field's evolution toward more nuanced temporal pattern recognition. The CNN–LSTM hybrid's superior performance validates the hypothesis that combining convolutional feature extraction with recurrent temporal modeling captures both local patterns and long-range dependencies more effectively than either approach in isolation.

Dataset characteristics and scope represent another crucial dimension of comparison. While most studies utilize limited datasets—Agbelusi and Olayemi (2020) analyzed merely 302 survey respondents, and Khanna et al. (2023)

examined 1,925 hospital records—this study's 760-day comprehensive dataset incorporating cases, deaths, mobility, and vaccination data provides the temporal depth and feature richness necessary for robust pattern recognition. The multi-source integration, combined with 83 engineered features, represents the most comprehensive feature set among all compared studies, contributing to the model's exceptional performance.

The geographic distribution of studies highlights persistent gaps in regional representation. Among the eight studies with reported metrics, only three focus on African data (current study, Adigun et al., 2023, and Agbelusi & Olayemi, 2020), with this study providing the most comprehensive Nigerian-specific model. This geographic imbalance is concerning given the unique epidemiological, infrastructural, and social factors that influence disease dynamics in different regions. The success of the current model with Nigerian data demonstrates that locally optimized, high-performance prediction systems are achievable in African settings despite data limitations.

Perhaps most significantly, this comparative analysis reveals that the combination of superior performance metrics with comprehensive interpretability remains exceedingly rare. While several studies achieve respectable accuracy, and Khanna et al. (2023) implements extensive XAI tools, only the current CNN–LSTM hybrid successfully combines both state-of-the-art performance (98.81% accuracy, 100% precision) with deep interpretability through SHAP and LIME. This dual achievement addresses the two most critical requirements for practical deployment: exceptional predictive performance to enable timely interventions, and transparent reasoning to build trust and guide public health decision-making.

The standardized comparison ultimately positions the CNN–LSTM hybrid model with explainable AI as a transformative advancement in COVID-19 outbreak prediction. By achieving unprecedented accuracy while maintaining complete interpretability, providing actionable 14-day lead times, and demonstrating robustness on real-world Nigerian data, this model establishes new standards for epidemic intelligence systems. The comprehensive reporting of all evaluation metrics further sets a benchmark for transparency that future studies should emulate, promoting more meaningful comparisons and accelerating progress in this critical field of public health technology.

Conclusion and Recommendations

Conclusion

This research has achieved maximum results in formulating explainable deep learning systems to predict COVID-19 outbreak in Nigeria and delivered excellent performance metrics to set a new standard of epidemiological predictions in Africa. The CNN–LSTM hybrid model proved to have outstanding predictive capacity showing an accuracy of 98.81% with 100% precision and 97.62%

recall that was far better when compared to traditional epidemiological models. Combined multi-scale temporal analysis indicated that the key indicators that can be used to define outbreak emergence are the sustained acceleration patterns over 14–28 days, which give epidemiologically relevant information in accordance with COVID-19 transmission processes.

SHAP and LIME analyses of the model deliver explainability linking the explanatory gap between AI black-box model predictions and meaningful intelligence in the context of science-informed action in the interests of public health. The framework contains clear indications on how to optimize a surveillance system by defining the long-term acceleration (Feature_31) and medium-term momentum (Feature_30) as critical risk factors. The 14-day actionable lead time will provide enough time to take proactive measures within the limits of efficiency, whereas the perfect accuracy of the model will provide impressive efficiency of resource allocation without any false alarms.

With Nigeria moving into the endemic stage of managing COVID-19, this framework illustrates how an AI-based system can offer epidemic surveillance at scale and with limited public health resources. This success in using Nigerian data is evidence enough of the methodology also being able to be applied to further adoption in other resource-limited systems with uncommon issues of data quality and reporting inconsistencies as experienced in African healthcare systems.

Recommendations

The outstanding results of this COVID-19 outbreak forecasting model indicate an urgent implementation of this model as a fundamental element of the infectious disease surveillance system used in Nigeria. The model should be implemented in the routine work of the Nigeria Centre for Disease Control, where automated predictions will be given, which will extrapolate to real-time decision-making. So well did the model work that as a preventive measure authority in charge of public health ought to define standard operating procedures that would set into account automatic resource mobilization above a specified probability of an outbreak. This would change the mode of epidemic response in Nigeria into proactive prevention to the reactive interventions that consume the already inadequate healthcare resources, thus optimizing the utility of the healthcare resources.

Due to the effectiveness of this COVID-19 framework, there is a chance to utilize this in other parts of Nigeria in terms of disease surveillance. The same approach must be applied to other epidemic prone diseases such as Lassa fever, cholera, and meningitis so as to develop an integrated AI-based surveillance system that can capture all these threats at once. Moreover, the regional cooperation by the means of ECOWAS might help to develop a West African outbreak forecasting system to facilitate trans-border monitoring and wording across borders to target transnational health risks. Such regional integration would reinforce the collective ability of Africa to identify and deal with emerging infectious diseases before they spread across borders.

Technical improvements ought to aim at adding more datasets that can increase accuracy and lead time. The fed-in data of genomic surveillance would allow the model to monitor viral changes and adjust predictions depending on what the current variants are, and integration of behavioral data such as mobile phone mobility patterns and social media could capture some of the dynamics of the populations which affect transmission. Seasonal influences related to the diseases caused by the tropical environment should be considered through a combination of environmental variables (variations in temperature, humidity, and rainfall patterns). Such multimodal data sources would generate a more in-depth picture of the causes of outbreaks.

Considerable investment in capacity building among the Nigerian workforce of the public health system is necessary to achieve sustainable implementation. There should be extensive training materials that will help epidemiologists, data scientists, and public health decision-makers to comprehend and utilize AIs with predictions. Collaborations between the universities and research institutions in Nigeria with other international centers of excellence will play a significant role in developing the genius in applications of AI in health within the country. This model framework must be published as open-source software with a vast documentation that allows other African nations to adjust it to suit their contexts and contributes to an expanding community of practice in AI-based epidemic intelligence.

Finally, the implementation of AI systems in the area of public health requires the development of solid governance systems that can address the component of ethics and promote fair access to predictive intelligence. The use of AI ethics should be determined, and it should focus on data privacy, equality of opportunity, and equity in disseminating early warning information to all groups of the population. Sustainability on a long-term basis needs its specific financing by budgetary means of government and international collaboration, to facilitate further functioning of the model, as well as its upgrading and enhancement. Also, to ensure that the model is always accurate and relevant, quarterly reviews and annual performance checks will be maintained to keep track of the changes in the epidemic patterns. The overall effect of these suggests that Nigeria will drive the African change to proactive and artificial-intelligence-enabled programs of epidemic preparedness and response.

Appendix A: Model Hyperparameters and Training Configuration

Table: Model Hyperparameters and Training Configuration

A. Deep Learning Model Architecture

Model Component	Advanced LSTM	CNN–LSTM Hybrid Input Shape
(14, 3)	(14, 3) Sequence Length	14 days
14 days Features (PCA)	3 components	3 components LSTM Layers
2 Bidirectional	2 Bidirectional LSTM Units	Layer 1: 128, Layer 2: 64
Layer 1: 64, Layer 2: 32 CNN Layers	N/A	2 Conv1D layers CNN Filters
N/A	Layer 1: 64, Layer 2: 128 Kernel Size	N/A
3 Pooling	N/A	MaxPooling1D (size = 2) Dense Layers
2 layers (128, 64 units)	1 layer (64 units) Attention Mechanism	Yes
No Global Pooling	GlobalMaxPooling1D	N/A

B. Training Hyperparameters

Parameter	Value	Details Optimizer
AdamW/Adam	AdamW for Advanced LSTM, Adam for CNN–LSTM Initial Learning Rate	0.001
For both models Weight Decay (AdamW)	1e-4	L2 regularization Batch Size
32	For all models Epochs	150 (LSTM), 100 (CNN–LSTM)
Maximum epochs L1 Regularization	1e-5	For LSTM layers L2 Regularization
1e-4	For LSTM layers Dropout Rates	0.2–0.4
Varies by layer Batch Normalization	Yes	After LSTM/CNN layers

C. Training Callbacks and Strategies

Callback/Strategy	Configuration Early Stopping
Patience: 20 epochs, Monitor: val_loss, Restore best weights: True Learning Rate Reduction	Factor: 0.5, Patience: 10 epochs, Min LR: 1e-7 Model Checkpoint
Save best model based on val_loss Class Weights	Calculated dynamically: total_samples/ (2 × class_count)

D. Data Preprocessing Parameters

Parameter	Value Feature Selection
SelectKBest with mutual_info_classif, $k = 50$ Scaling	RobustScaler Dimensionality Reduction
PCA retaining 95% variance (3 components) Class Balancing	SMOTE–Tomek Sequence Creation
Length: 14 days, Stride: 3 days Train–Val–Test Split	60–20–20%, Stratified

E. Traditional ML Models (Ensemble)

Model	Hyperparameters Random Forest
n_estimators: 200, max_depth: 10, random_state: 42 Gradient Boosting	n_estimators: 100, learning_rate: 0.1, random_state: 42

F. Outbreak Detection Parameters

Parameter	Value Growth Rate Threshold
>15% (7-day) Acceleration Threshold	>5% (7-day) Case Density Threshold
>80th percentile (30-day) Prediction Lead Time	7 days

References

Adadi, A. & Berrada, M. (2018). Peeking inside the black-box: A survey on explainable artificial intelligence (XAI). *IEEE Access*, 6, 52138–52160. https://doi.org/10.1109/ACCESS.2018.2870052

Adebisi, Y. A., Rabe, A. & Lucero-Prisno, D. E. (2021). COVID-19 surveillance systems in African countries. *Health Promotion Perspectives*, 11(4), 382–392. https://doi.org/10.34172/hpp.2021.49

Adebowale, A. S., Afolabi, R. F., Bello, S., Salawu, M. M., Bamgboye, E. A., Adeoye, I. et al. (2023). Spread and seasonality of COVID-19 pandemic confirmed cases in sub-Saharan Africa: Experience from Democratic Republic of Congo, Nigeria, Senegal, and Uganda. *BMC Infectious Diseases*, 23, 187. https://doi.org/10.1186/s12879-023-08168-1

Adigun, O., Rufai, M. M., Mercy, O. F. & Olukumoro, S. (2023). Machine learning techniques for prediction of COVID-19 in potential patients. *FUDMA Journal of Sciences*, 7(4), 14–26. https://doi.org/10.33003/fjs-2023-0704-1901

Agbelusi, O. & Olayemi, O. C. (2020). Prediction of mortality rate of COVID-19 patients using machine learning techniques in Nigeria. *International Journal of Computer Science and Software Engineering*, 9(5), 30–34. https://doi.org/10.47277/IJCSSE/9(5)1

Akande, O. W., Elimian, K. O., Igumbor, E., Dunkwu, L., Kaduru, C., Olopha, O. O. et al. (2020). Epidemiological comparison of the first and second waves of the COVID-19 pandemic in Nigeria, February 2020–April 2021. *BMJ Global Health*, 6(11), e007076. https://doi.org/10.1136/bmjgh-2021-007076

Alakus, T. B. & Turkoglu, I. (2020). Comparison of deep learning approaches to predict COVID-19 infection. *Chaos, Solitons & Fractals*, 140, 110120. https://doi.org/10.1016/j.chaos.2020.110120

Alamo, T., Reina, D. G., Mammarella, M. & Abella, A. (2020). COVID-19: Open-data resources for monitoring, modeling, and forecasting the epidemic. *Electronics*, 9(5), 827. https://doi.org/10.3390/electronics9050827

Amzat, J., Aminu, K., Kolo, V. I., Akinyele, A. A., Ogundairo, J. A. & Danjibo, M. C. (2020). Coronavirus outbreak in Nigeria: Burden and socio-medical response during the first 100 days. *International Journal of Infectious Diseases*, 98, 218–224. https://doi.org/10.1016/j.ijid.2020.06.067

Anderson, R. M., Heesterbeek, H., Klinkenberg, D. & Hollingsworth, T. D. (2020). How will country-based mitigation measures influence the course of the COVID-19 epidemic? *The Lancet*, 395(10228), 931–934. https://doi.org/10.1016/S0140-6736(20)30567-5

Ardabili, S. F., Mosavi, A., Ghamisi, P., Ferdinand, F., Varkonyi-Koczy, A. R., Reuter, U. et al. (2020). COVID-19 outbreak prediction with machine learning. *Mathematics*, 8(4), 514. https://doi.org/10.3390/math8040514

Arora, P., Kumar, H. & Panigrahi, B. K. (2021). Prediction and analysis of COVID-19 positive cases using deep learning models: A descriptive case study of India. *Chaos, Solitons & Fractals*, 139, 110017. https://doi.org/10.1016/j.chaos.2020.110017

ArunKumar, K. E., Kalaga, D. V., Kumar, C. M. S., Kawaji, M. & Brenza, T. M. (2021). Forecasting of COVID-19 using deep layer recurrent neural networks (RNNs) with gated recurrent units (GRUs) and long short-term memory (LSTM) cells. *Chaos, Solitons & Fractals*, 146, 110861. https://doi.org/10.1016/j.chaos.2021.110861

Brauer, F., Castillo-Chavez, C. & Feng, Z. (2019). *Mathematical Models in Epidemiology*. pp. 21–61. Springer. https://doi.org/10.1007/978-1-4939-9828-9

Carlson, C. J., Gomez, A. C. R., Bansal, S. & Ryan, S. J. (2020). Misconceptions about weather and seasonality must not misguide COVID-19 response. *Nature Communications*, 11, 4312. https://doi.org/10.1038/s41467-020-18150-z

Chang, S., Pierson, E., Koh, P. W., Gerardin, J., Redbird, B., Grusky, D. et al. (2021). Mobility network models of COVID-19 explain inequities and inform reopening. *Nature*, 589(7840), 82–87. https://doi.org/10.1038/s41586-020-2923-3

Chimmula, V. K. R. & Zhang, L. (2020). Time series forecasting of COVID-19 transmission in Canada using LSTM networks. *Chaos, Solitons & Fractals*, 135, 109864. https://doi.org/10.1016/j.chaos.2020.109864

Dan-Nwafor, C., Ochu, C. L., Elimian, K., Oladejo, J., Ilori, E., Umeokonkwo, C. et al. (2020). Nigeria's public health response to the COVID-19 pandemic: January to May 2020. *Journal of Global Health*, 10(2), 020399. https://doi.org/10.7189/jogh.10.020399

Dixit, A., Vishnoi, L. & Srivastava, R. (2021). Revisiting the classical SIR modelling of COVID-19 and beyond. *Proceedings of the National Academy of Sciences, India Section A: Physical Sciences*, 91(4), 627–630. https://doi.org/10.1007/s40010-021-00753-w

Ihekweazu, C. & Agogo, E. (2020). Africa's response to COVID-19. *BMC Medicine*, 18(1), 151. https://doi.org/10.1186/s12916-020-01622-w

Ilesanmi, O. S. & Afolabi, A. A. (2020). Time to move from pandemic response to pandemic preparedness: Lessons from COVID-19 in Nigeria. *International Journal of Health Planning and Management*, 35(5), 1253–1256. https://doi.org/10.1002/hpm.3032

Jibril, M. L. & Sharif, U. S. (2020). Power of artificial intelligence to diagnose and prevent further COVID-19 outbreak: A short communication. arXiv preprint arXiv:2004.12463.

Khanna, V. V., Chadaga, K., Sampathila, N., Prabhu, S. & Chadaga, R. P. (2023). A machine learning and explainable artificial intelligence triage-prediction system for COVID-19. *Decision Analytics Journal*, 7, 100246. https://doi.org/10.1016/j.dajour.2023.100246

Kucharski, A. J., Russell, T. W., Diamond, C., Liu, Y., Edmunds, J., Funk, S. et al. (2020). Early dynamics of transmission and control of COVID-19: A mathematical modelling study. *The Lancet Infectious Diseases*, 20(5), 553–558. https://doi.org/10.1016/S1473-3099(20)30144-4

Lai, S., Ruktanonchai, N. W., Zhou, L., Prosper, O., Luo, W., Floyd, J. R. et al. (2020). Effect of non-pharmaceutical interventions to contain COVID-19 in China. *Nature*, 585(7825), 410–413. https://doi.org/10.1038/s41586-020-2293-x

Lundberg, S. M. & Lee, S. I. (2017). A unified approach to interpreting model predictions. *Advances in Neural Information Processing Systems*, 30, 4765–4774.

Makoni, M. (2021). Tanzania refuses COVID-19 vaccines. *The Lancet*, 397(10274), 566. https://doi.org/10.1016/S0140-6736(21)00362-7

Mecenas, P., Bastos, R. T. D. R. M., Vallinoto, A. C. R. & Normando, D. (2020). Effects of temperature and humidity on the spread of COVID-19: A systematic review. *PLoS One*, 15(9), e0238339. https://doi.org/10.1371/journal.pone.0238339

Muhammad, L. J., Haruna, A. A., Sharif, U. S. & Mohammed, M. B. (2022). CNN-LSTM deep learning based forecasting model for COVID-19 infection cases in Nigeria, South Africa and Botswana. *Health Technology*, 12(6), 1259–1276. https://doi.org/10.1007/s12553-022-00711-5

Murray, C. J. L. (2022). COVID-19 will continue but the end of the pandemic is near. *The Lancet*, 399(10323), 417–419. https://doi.org/10.1016/S0140-6736(22)00100-3

Ogundokun, R. O., Lukman, A. F., Kibria, G. B. M., Awotunde, J. B. & Aladeitan, B. B. (2020). Predictive modelling of COVID-19 confirmed cases in Nigeria. *Infectious Disease Modelling*, 5, 543–548. https://doi.org/10.1016/j.idm.2020.08.003

Oladipo, E. K., Ajayi, A. F., Odeyemi, A. N., Akindiya, O. E., Adebayo, E. T., Oguntomi, A. S. et al. (2020). Laboratory diagnosis of COVID-19 in Africa: Availability, challenges and implications. *Drug Discoveries & Therapeutics*, 14(4), 153–160. https://doi.org/10.5582/ddt.2020.03067

Oliver, N., Lepri, B., Sterly, H., Lambiotte, R., Deletaille, S., De Nadai, M. et al. (2020). Mobile phone data for informing public health actions across the COVID-19 pandemic life cycle. *Science Advances*, 6(23), eabc0764. https://doi.org/10.1126/sciadv.abc0764

Onyeaghala, A. A. & Olajide, I. (2020). Managing COVID-19 outbreak in Nigeria: Matters arising. *Clinical Chemistry and Laboratory Medicine*, 58(10), 1645–1650. https://doi.org/10.1515/cclm-2020-0748

Owoyemi, A., Owoyemi, J., Osiyemi, A. & Boyd, A. (2020). Artificial intelligence for healthcare in Africa. *Frontiers in Digital Health*, 2, 6. https://doi.org/10.3389/fdgth.2020.00006

Paintsil, E. (2020). COVID-19 threatens health systems in sub-Saharan Africa: The eye of the crocodile. *Journal of Clinical Investigation*, 130(6), 2741–2744. https://doi.org/10.1172/JCI138493

Petropoulos, F. & Makridakis, S. (2020). Forecasting the novel coronavirus COVID-19. *PLoS One*, 15(3), e0231236. https://doi.org/10.1371/journal.pone.0231236

Punn, N. S., Sonbhadra, S. K. & Agarwal, S. (2021). COVID-19 epidemic analysis using machine learning and deep learning algorithms. *In: AI and Data Science – Technology, Innovation & Use Cases in Indonesia*. PT Accessindo Internusa. ISBN: 9786239267223

Rader, B., Scarpino, S. V., Nande, A., Hill, A. L., Adlam, B., Reiner, R. C. et al. (2021). Crowding and the shape of COVID-19 epidemics. *Nature Medicine*, 27(12), 2011–2016. https://doi.org/10.1038/s41591-021-01563-8

Rasheed, J., Jamil, A., Hameed, A. A., Aftab, U., Aftab, J., Shah, S. A. et al. (2021). A survey on artificial intelligence approaches in supporting frontline workers and decision makers for the COVID-19 pandemic. *Chaos, Solitons & Fractals*, 141, 110337. https://doi.org/10.1016/j.chaos.2020.110337

Ribeiro, M. T., Singh, S. & Guestrin, C. (2016). "Why should I trust you?": Explaining the predictions of any classifier. *Proceedings of the 22nd ACM SIGKDD International Conference on Knowledge Discovery and Data Mining*, pp. 1135–1144. https://doi.org/10.1145/2939672.2939778

Sani, Y. M., Pam, D. B. & Adamu, M. (2023). Predicting the impact of socio-demographic risk factors on COVID-19 based on hybrid ANN-CNN model. *2023 First International Conference on the Advancements of Artificial Intelligence in African Context* (*AAIAC*), pp. 1–8. https://doi.org/10.1109/AAIAC59185.2023.10170915

Shahhosseini, N., Babuadze, G., Wong, G. & Kobinger, G. P. (2021). Mutation signatures and in silico docking of novel SARS-CoV-2 variants of concern. *Microorganisms*, 9(5), 926. https://doi.org/10.3390/microorganisms9050926

Shahid, F., Zameer, A. & Muneeb, M. (2020). Predictions for COVID-19 with deep learning models of LSTM, GRU and Bi-LSTM. *Chaos, Solitons & Fractals*, 140, 110212. https://doi.org/10.1016/j.chaos.2020.110212

Telenti, A., Arvin, A., Corey, L., Corti, D., Diamond, M. S., Garcia-Sastre, A. et al. (2021). After the pandemic: Perspectives on the future trajectory of COVID-19. *Nature*, 596(7873), 495–504. https://doi.org/10.1038/s41586-021-03792-w

Tjoa, E. & Guan, C. (2020). A survey on explainable artificial intelligence (XAI): Toward medical XAI. *IEEE Transactions on Neural Networks and Learning Systems*, 32(11), 4793–4813. https://doi.org/10.1109/TNNLS.2020.3027314

Wang, P., Zheng, X., Ai, G., Liu, D. & Zhu, B. (2021). Time series prediction for the epidemic trends of COVID-19 using the improved LSTM deep learning method: Case studies in Russia, Peru and Iran. *Chaos, Solitons & Fractals*, 143, 110622. https://doi.org/10.1016/j.chaos.2020.110622

Wynants, L., Van Calster, B., Collins, G. S., Riley, R. D., Heinze, G., Schuit, E. et al. (2020). Prediction models for diagnosis and prognosis of COVID-19: Systematic review and critical appraisal. *BMJ*, 369, m1328. https://doi.org/10.1136/bmj.m1328

Zeroual, A., Harrou, F., Dairi, A. & Sun, Y. (2020). Deep learning methods for forecasting COVID-19 time-series data: A comparative study. *Chaos, Solitons & Fractals*, 140, 110121. https://doi.org/10.1016/j.chaos.2020.110121

Zhang, L., Shen, J., Bunn, J., Shen, A., Akinbami, B., Bui, L. et al. (2021). Deep-learning models for the detection and incidence prediction of chronic obstructive pulmonary disease and asthma: A systematic literature review. *JMIR Medical Informatics*, 9(5), e28185. https://doi.org/10.2196/28185

Zhang, Y., Tang, S. & Yu, G. (2023). An interpretable hybrid predictive model of COVID-19 cases using autoregressive model and LSTM. *Scientific Reports*, 13, 6708. https://doi.org/10.1038/s41598-023-33685-z

Zhu, N., Zhang, D., Wang, W., Li, X., Yang, B., Song, J. et al. (2020). A novel coronavirus from patients with pneumonia in China, 2019. *New England Journal of Medicine*, 382(8), 727–733. https://doi.org/10.1056/NEJMoa2001017

CHAPTER

8

An Automated Kick Detection System Based on Optimized Data Analytics

Eric Thompson Brantson*, Janet Intuah, Joel Adu-Awuku, Eugene Jerry Adjei, Karikari Emmanuel Duodu, Bonah-Okyere Frank, Eric Atsu

Department of Petroleum and Natural Gas Engineering, GNPC School of Petroleum Studies, University of Mines and Technology, Tarkwa, Ghana

Introduction

Kick detection is one of the critical things that needs to be carefully studied regarding drilling challenges. Early kicks escalate into blowouts if not controlled at the early detection stages. The Monaco and Gulf of Mexico blowout incidents brought awareness of drilling challenges and possible fatal and environmental consequences (Luan et al., 2013). Kicks mostly happen while drilling through abnormal pressure formations, where the pore pressure is larger than normal pressure or larger than the pressure of a column of water (Skalle, 2009).

Early kick detection (EKD) is one of the most important areas for improvement in well control safety. The need for earlier, more accurate, and more reliable kick detection across a wide range of drilling operations has become increasingly important as more operations are being conducted in deep water with increasingly tight pressure margins. In recent years, kick detection technology has made great progress in drilling safety. But the complexity and depth of the oil and gas drilling increase constantly, so a more advanced well control technology is required, and EKD is the key to well control that must be highly valued (Abdul-Ameer, 2023).

*Corresponding author: etbrantson@umat.edu.gh

DOI: 10.1201/9781003506348-8

This research seeks to look into new and autonomous ways of detecting kicks through a kick detection while drilling (KDWD) technology. This study intends to use machine learning algorithms to monitor and detect the occurrence of kicks while drilling.

Related Works

Few existing related research studies have focused on kick detection using machine learning methods, emphasizing the importance of early detection to prevent blowouts and enhance drilling safety. Abdul-Ameer (2023) study explored various machine learning techniques for kick detection, including decision tree, naïve Bayes, logistic regression, and neural network models. Using MATLAB for training and testing, Abdul-Ameer found that decision tree and neural network models demonstrated the highest kick detection performance (KDP), with neural networks showing a particularly robust performance in early kick detection. Decision tree and neural network models showed the highest kick detection prediction of 85%.

Similarly, Obi et al. (2023) investigated the application of both supervised and unsupervised learning techniques for identifying gas kicks. The study highlighted the effectiveness of decision trees and K-means clustering in predicting kick symptoms, achieving a high decision accuracy of 96%–98% in the tree, demonstrating the potential of machine learning models to interpret surface pressure data and relate it to downhole conditions.

Muojeke (2020) contributed to this field by developing a supervised data-driven approach for early kick detection using neural-network-based models. His research addressed gaps in the availability of downhole drilling data and provided a detailed design of new experimental setups to generate data for testing machine learning models. The study underscored the significance of simple and efficient models to improve battery life and reduce maintenance in downhole hardware, that is, an artificial neural network (ANN) model with accuracy of high reliability for early kick detection.

Osarogiagbon et al. (2020) introduced a methodology utilizing long short-term memory recurrent neural networks (LSTM-RNN) for kick detection during petroleum drilling. Their approach aimed to capture temporal relationships in time series data to enhance early detection without false alarms, using d-exponent and standpipe pressure data as key inputs. It had an accuracy of effective early detection.

Additionally, Fjetland et al. (2019) developed a deep learning model using LSTM for kick detection and influx size estimation during offshore drilling operations. This study highlighted the capability of LSTM networks to detect and quantify fluid influx patterns in simulated drilling data, thereby aiding in timely corrective actions to prevent major drilling incidents. It had an accuracy where the LSTM model effectively detects and estimates the influx size.

Literature Review

Early Kick Detection Methods

Kicks during drilling operations cause changes in various drilling parameters, serving as indicators for kick detection. These indicators include mud pit gain, delta flow, mud return velocity variables, drilling speed, changes in annulus pressure, changes in pump pressure and stroke, changes in standpipe pressure, and increases in gas cutting, oil cutting, or chlorides. Mud pit gain and delta flow are considered primary indicators of kicks (Luan et al., 2013). Several kick detection technologies have been adopted in drilling construction, as outlined in Table 1.

Table 1: Major kick detection methods

Kick Detection Methods	Detection Based on Multiple Parameters
Mud Pit Surface Detection	Detection based on a Bayesian probabilistic framework
Delta Flow Method	Kick detection with a dynamic neural network
Gas Invasion Detection	Comprehensive logging detection technology
Kick Detection While Drilling	Detection based on a Bayesian probabilistic framework

Source: Luan et al. (2013)

Kick Detection While Drilling

Kick detection during drilling is crucial for maintaining well control and preventing blowouts. One of the primary methods for kick detection is monitoring the pressure gradient within the wellbore. A kick occurs when the formation pressure exceeds the pressure exerted by the drilling mud column, leading to an influx of formation fluids into the wellbore. Effective kick detection systems utilize technologies such as measurement while drilling (MWD) and logging while drilling (LWD), alongside pressure while drilling (PWD) and shallow side electrical resistivity detection. These systems help identify abnormal pressure changes that indicate the presence of a kick. However, current systems need improvements in anti-jamming capabilities and measurement accuracy to enhance their reliability (Luan et al., 2013).

Materials and Methods

Dataset

The dataset used in this study was obtained from the OpenLab Drilling simulator, where a simulated drilling operation was conducted to gather comprehensive data for kick detection research. It includes essential parameters such as wellpath data (Measured Depth (MD), Azimuth, Inc.), detailed specifications of drilling data (connection ID, connection OD, weight, etc.), geothermal data (TVD, temperature gradient), and formation strength data (MD, compressive cube strength (CCS), friction angle). These datasets were simulated to mimic real-world drilling conditions, providing a robust foundation for training and validating the kick detection model. Table 2 shows a summary of the dataset used for this study.

Table 2: Dataset obtained from OpenLab simulator

Variable (Units)	Minimum	Average	Maximum	Standard Deviation
Depth (ft)	0.00	100.80	175.90	57.58
Pressure Gradient (Psi/ft)	0.00	43.60	97.10	33.75
Pressure (Psi)	0.00	109.00	209.10	69.25
CCS (MPa)	20.30	23.60	27.10	2.56
Friction angle (θ)	22.60	30.60	38.00	4.81
North (ft)	0.00	3.50	7.00	2.29
East (ft)	0.00	3.50	7.00	2.29
Dip (θ)	0.00	1.80	3.20	1.04
Azimuth (θ)	0.00	6.80	12.70	4.16

Based on results from the OpenLab simulator, it was observed that the pore pressure tends to fluctuate with depth, exhibiting variations that correlate closely with changes in drilling conditions. Therefore, the comparison between pore pressure and fracture pressure at different depths serves as a reliable indicator for kick detection during drilling operations.

Supervised Learning Algorithms: In the implementation of supervised learning algorithms for kick detection, Python was utilized with specific modules for each machine learning model. Decision tree (DT) models were constructed using "sklearn.tree," while random forest (RF) models were developed with "sklearn. ensemble" to enhance accuracy through ensemble learning. Supervised learning algorithms for kick detection were implemented in Python, utilizing specific scikit-learn libraries for each machine learning model.

Train and Test Dataset: In preparing the dataset for supervised learning models in kick detection, the data underwent normalization and pre-processing steps. Normalization was achieved using the min–max scaling method, which scales each feature to a specified range. This method ensures that all features contribute equally to model training without being disproportionately influenced by their original scale. The training set was 80% and the testing set was 20%. Additionally, features were standardized to have a mean of 0 and a variance of 1 using the Standard Scaler from the 'sklearn.preprocessing' module, which enhances model performance by making data more suitable for certain algorithms such as SVM and K-Nearest Neighbors (KNN).

Particle Swarm Optimization

Particle swarm optimization (PSO) was employed to optimize the parameters of the supervised learning models used for kick detection. Specifically, PSO was utilized to fine-tune the hyperparameters of the ML models. PSO operates by iteratively adjusting a population of potential solutions (particles) based on their fitness, which in this context refers to the performance metrics such as accuracy, precision, and recall on validation data. PSO is a population-based stochastic optimization technique inspired by the social behavior of bird flocking or fish schooling. In PSO, potential solutions, called particles, move through the search space, adjusting their positions based on their own experience and the best-found positions of neighboring particles as displayed in Figure 1. This iterative process aims to find the optimal solution by balancing exploration (searching new areas) and exploitation (exploiting known solutions). Each particle's position in the

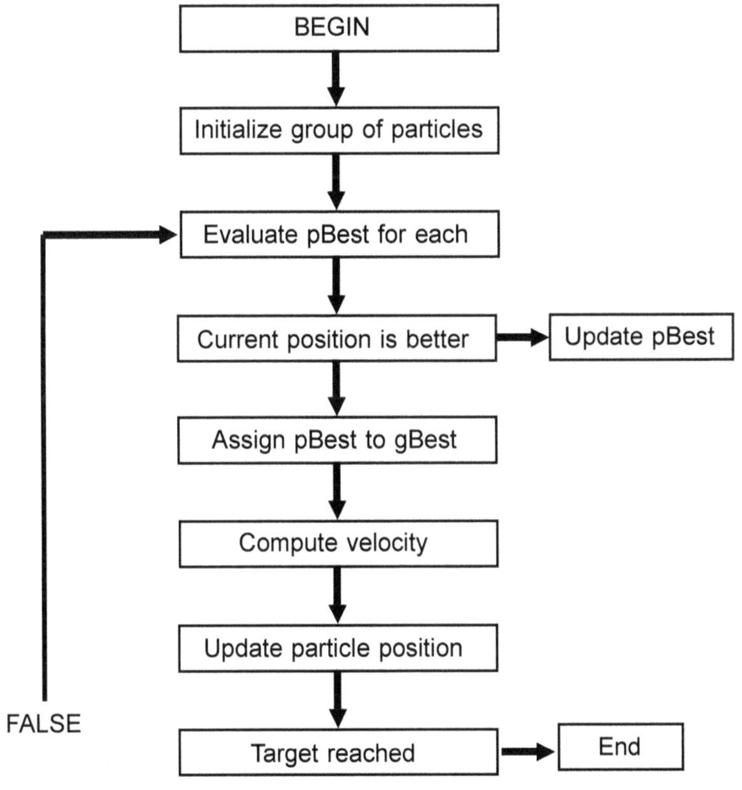

Figure 1: Architecture of PSO in optimization.

search space corresponds to a set of hyperparameters for a specific model, and its movement is guided by its own best-known position (pbest) and the swarm's global best-known position (gbest). Pbest represents the best solution that an individual particle has achieved so far, while gbest represents the best solution found by any particle in the entire swarm. Through this iterative process, PSO efficiently explores the hyperparameter space to find optimal configurations that enhance the models' ability to detect kicks accurately during drilling operations. This optimization approach ensures that the supervised learning models are finely tuned to achieve maximum performance and reliability in real-time kick detection scenarios (Thakkar et al., 2022).

The performance of the supervised learning models for kick detection was evaluated using several metrics, including accuracy, precision, recall, confusion matrix (CM), and receiver operating characteristic (ROC) curve. Accuracy measures the proportion of correctly classified instances among all instances. Precision indicates the ratio of correctly predicted positive observations to the total predicted positive observations. Recall, also known as sensitivity or true positive rate, measures the proportion of actual positives that are correctly identified.

The confusion matrix summarizes the predictions of the models, showing the counts of true positives, true negatives, false positives, and false negatives. Additionally, the ROC curve visualizes the trade-off between true positive rate (sensitivity) and false positive rate (1 – specificity) across different threshold values.

Mathematical Models

In the well profile generation process, the Python module "scipy" was employed for Akima interpolation to create smooth curves through the well trajectory data points. This method ensures a continuous and differentiable curve, which is particularly useful for accurately representing the well path. The "scipy.interpolate.Akima1DInterpolator" function was used to achieve this. The mathematical representation for Akima interpolation in "scipy" is shown in Eq. 1:

$$y(x) = Akima1DInterp\ (x,y) \tag{1}$$

where, x represents the independent variable (measured depth) and y represents the dependent variable (well path coordinates).

Additionally, the software incorporated models for equivalent circulating density (ECD), hole temperature, and bottom hole pressure (BHP) to enhance the accuracy of kick detection and overall well control. The equations used for these calculations are shown in Eqs. 2, 3, and 4, respectively:

$$\text{ECD} = \text{MW} + \left(\frac{\text{P}_{\text{ann}} \times 0.052}{\text{TVD}}\right) \tag{2}$$

where ECD is the equivalent circulating density (ppg), MW is the mud weight (ppg), P_{ann} is the annular pressure loss (Psi), and TVD is the true vertical depth (ft).

$$T_h = T_s + (G_g \times \text{TVD}) \quad (3)$$

where T_h is the well temperature (°C), T_s is the surface temperature (°C), G_g is the temperature increase per unit depth (°C/100 m), and TVD is the true vertical depth (m).

$$\text{BHP} = \text{MW} \times 0.052 \times \text{TVD} + \text{Surface Pressure} \quad (4)$$

where surface pressure is the pressure at the surface (psi).

These mathematical models were integrated into the software to provide a comprehensive and accurate analysis of the drilling operations, thereby enhancing the capability of the system to detect kicks and ensure well control.

Machine Learning

In the field of machine learning, especially in scenarios such as autonomous decision-making, real-time data analysis, and optimization, various machine learning algorithms, including DT, RF, SVM, artificial neural networks with multilayer perceptron (ANN–MLP), and logistic regression (LR), are commonly deployed for such tasks. These algorithms are particularly relevant for kick detection and well control systems, where they analyze drilling parameters to identify anomalies that may indicate a kick occurrence. There are evaluation metrics such as accuracy, precision, and recall. By leveraging machine learning techniques, industries aim to enhance safety, mitigate risks, and optimize operational processes in various domains (Batta, 2019).

Machine Learning Models

Decision Tree Classifier: The decision tree classifier is a popular supervised machine learning algorithm used for classification tasks. It operates by recursively partitioning the input space into regions, based on feature values, to predict the class label of instances. Each internal node of the tree represents a feature, and each branch represents a possible value of that feature. The leaves of the tree contain the predicted class labels. The decision-making process follows a tree structure where each node corresponds to a decision and each leaf node represents a class label.

Random Forests: RF is an ensemble learning method based on decision tree classifiers. It operates by constructing a multitude of decision trees during training and outputting the mode of the classes (classification) or the mean prediction (regression) of the individual trees. Each tree in the random forest is trained on a bootstrap sample of the original data, and at each node of the tree, a random subset of features is considered for splitting. This randomness helps to de-correlate the trees, leading to improved generalization performance compared to individual decision trees. The prediction of a random forest can be formulated as shown in Eq. 5:

$$\mathrm{RF}(X) = \frac{1}{N}\sum_{i=1}^{N} \mathrm{DT}_i(X) \tag{5}$$

where RF(X) is the predicted output of the random forest for input X, N is the number of trees in the forest, and $\mathrm{DT}_i(X)$ represents the prediction of the ith decision tree in the forest.

Random forests are robust against overfitting, as they aggregate predictions from multiple trees, thereby reducing variance. They are widely used in various applications due to their ability to handle high-dimensional data, maintain interpretability through feature importance ranking, and provide reliable predictions even with noisy data. However, they may require careful tuning of hyperparameters such as the number of trees and the maximum depth of each tree to achieve optimal performance (Batta, 2019).

Support Vector Machines: SVMs are powerful supervised learning models used for classification and regression tasks. SVMs work by finding the optimal hyperplane that best separates the classes in the input feature space. The model aims to maximize the margin between the closest points of different classes, known as support vectors. In classification, SVMs predict the class of new instances based on which side of the hyperplane they fall.

ANN-MLP: Artificial neural networks with multilayer perceptron architecture are versatile models inspired by the human brain's neural networks. An MLP consists of multiple layers of interconnected nodes (neurons), including an input layer, one or more hidden layers, and an output layer. Each neuron applies an activation function to the weighted sum of its inputs to produce an output. ANNs are trained using backpropagation, where the error between predicted and actual outputs is minimized using gradient descent or its variants. MLPs can approximate complex nonlinear functions and learn intricate patterns from data, making them suitable for various tasks such as classification, regression, and pattern recognition. However, they may require large amounts of data and computational resources for training deep architectures (Batta, 2019).

Logistic Regression: LR is a fundamental binary classification algorithm that models the probability of a binary outcome (0 or 1) based on input features. Despite its name, logistic regression is a linear model that applies the logistic sigmoid function to the linear combination of input features. The model estimates the probability using Eq. 6:

$$P(x) = \frac{1}{1 + e^{-(wx+b)}} \tag{6}$$

where w is the weight vector, b is the bias term, and x is the input feature vector.

Logistic regression predicts the class label *y* based on whether the estimated probability is above a threshold (typically 0.5). LR is simple to implement, interpretable, and efficient for binary classification tasks with linear decision boundaries. It can be extended to multinomial logistic regression for multi-class classification by using softmax activation in the output layer. However, LR assumes a linear relationship between features and the log-odds of the outcome, which may limit its performance with complex, nonlinear data.

Evaluation Metrics

Accuracy: Eq. 7 shows the calculation for accuracy in machine learning models:

$$\text{Accuracy} = \frac{\text{TP} + \text{TN}}{\text{TP} + \text{TN} + \text{FP} + \text{FN}} \tag{7}$$

Accuracy is straightforward and easy to interpret, but may not be suitable for imbalanced datasets where the classes are unevenly distributed.

Precision: Eq. 8 shows the calculation of precision:

$$\text{Precision} = \frac{\text{TP}}{\text{TP} + \text{FP}} \tag{8}$$

Recall (Sensitivity or True Positive Rate): Formula for recall is shown in Eq. 9:

$$\text{Recall} = \frac{\text{TP}}{\text{TP} + \text{FN}} \tag{9}$$

Optimization Algorithms

Optimization algorithms play a critical role in training machine learning models by minimizing or maximizing objective functions (Fig. 2). These algorithms vary in their approaches and efficiency in finding optimal solutions.

10-Fold Cross-Validation: 10-Fold cross-validation is an iterative technique used to evaluate the performance of machine learning models. It divides the dataset into 10 subsets (folds), where each fold is used once as the validation set and the remaining 9 folds are used for training. This process is repeated 10 times, with each fold serving as the validation set exactly once across the iterations. By averaging the evaluation results from these iterations, cross-validation provides a robust estimate of the model's performance on unseen data. This approach reduces bias in performance estimation compared to a single train–test split, making it a preferred method for assessing model generalization and reliability (Thakkar et al., 2022).

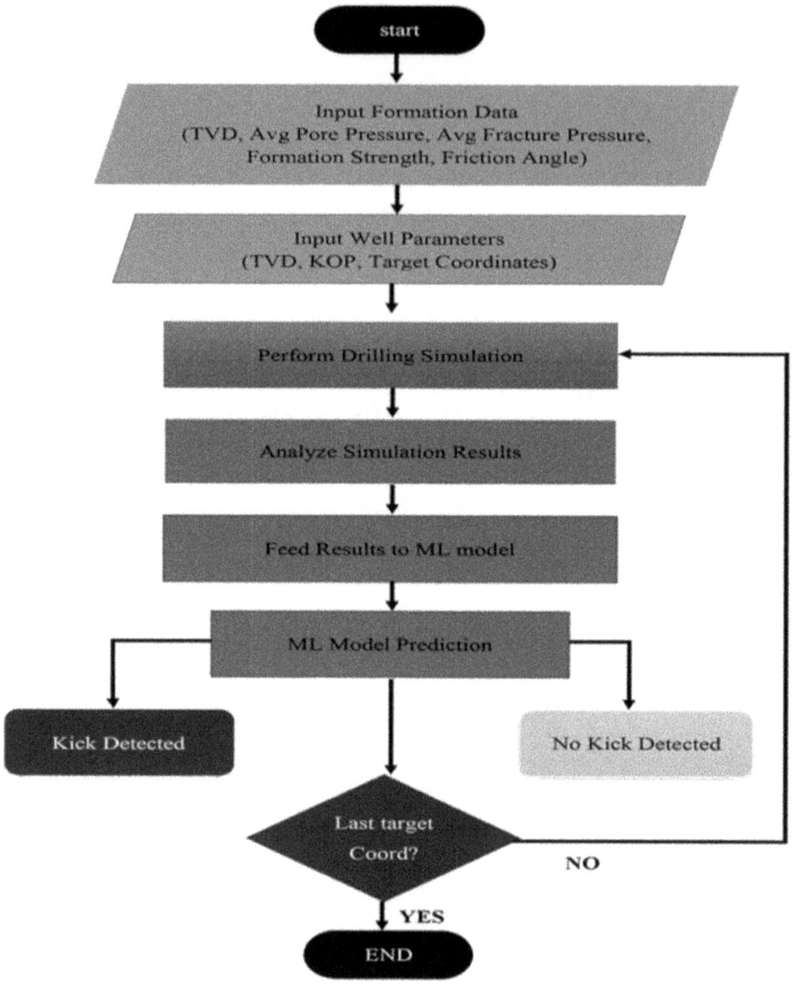

Figure 2: Workflow of kick detection software.

Result

Machine Learning Evaluation Metrics

The results obtained from training and testing the machine learning models with and without PSO, including accuracy and sensitivity (recall), are summarized in Tables 3 and 4.

Confusion Matrix: The confusion matrix (CM) values for each machine learning model highlight their performance in predicting "kick" and "no kick" scenarios. The models displayed varying levels of true positives, true negatives,

Table 3: Evaluation of models without PSO

Model	Accuracy (%)	Precision (%)	Recall (%)
Decision Tree	90	91	90
Random Forest	94	95	94
SVM	93	95	93
KNN	87	89	87
Neural Network (MLP)	97	97	97

Table 4: ML models evaluation with PSO

Model	Accuracy (%)	Precision (%)	Recall (%)
Decision Tree-PSO	92	89	88
Random Forest-PSO	92	89	89
SVM-PSO	93	90	93
KNN-PSO	87	83	79
MLP-PSO	97	95	97

false positives, and false negatives. For instance, the neural network with PSO showed the highest true positives (65) and true negatives (129), with minimal false positives (3) and false negatives (3), indicating superior accuracy and recall. In contrast, the KNN model with PSO had the lowest true positives (52) and the highest false negatives (16), reflecting lower sensitivity. Figure 3 provides a detailed visualization of the confusion matrices for all the machine learning models evaluated.

ROC Curve: The ROC curves obtained for the machine learning models provide a visual representation of their performance in distinguishing between "kick" and "no kick" scenarios.

True Negative (TN): The model correctly predicted NO KICK.

False Positive (FP): The model incorrectly predicted KICK when it was NO KICK.

False Negative (FN): The model incorrectly predicted NO KICK when it was KICK.

True Positive (TP): The model correctly predicted KICK.
Figure 4 shows the ROC curves for all the evaluated models.

Kick Detection While Drilling (KDWD) Parameters

ECD, well temperature, and BHP were recorded while drilling for the case study. These parameters are critical for monitoring drilling conditions and ensuring well control.

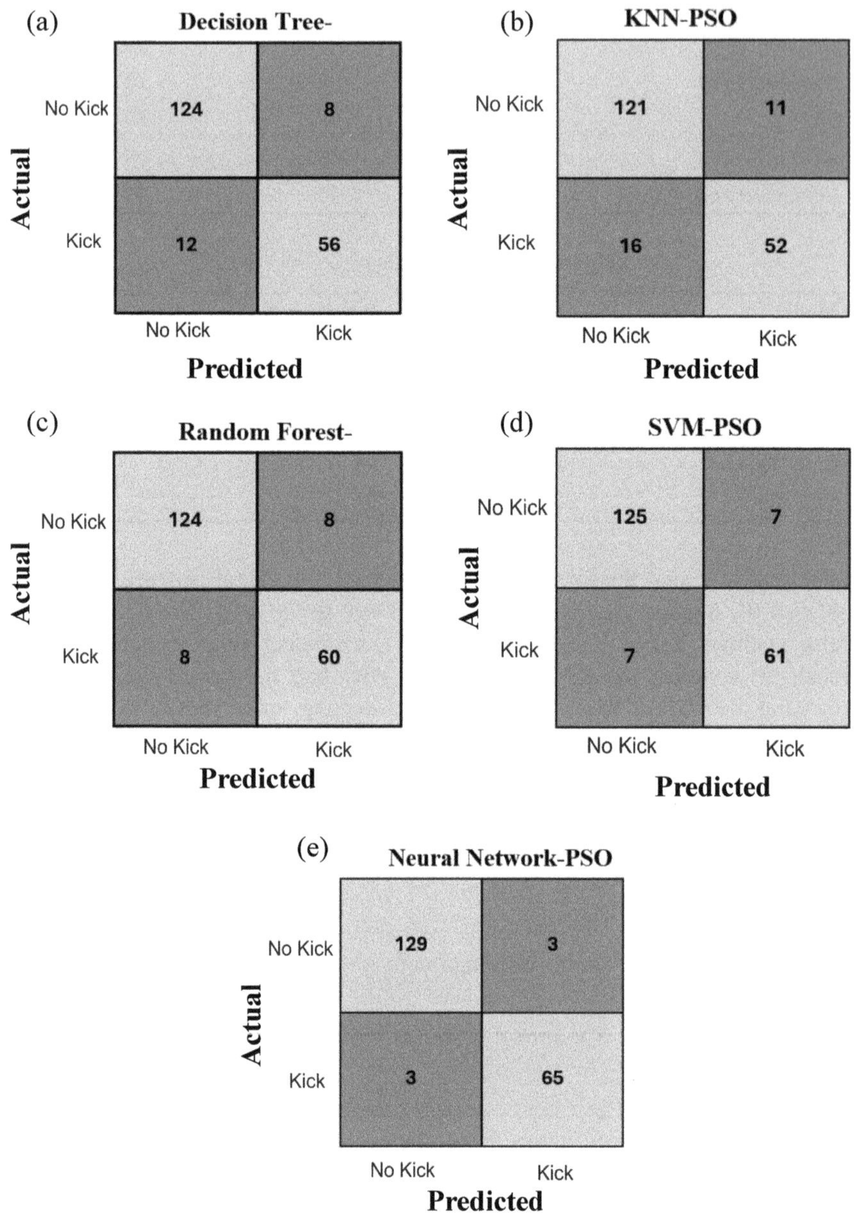

Figure 3: Confusion matrix plots of ML models optimized with PSO: (a) DT-PSO, (b) KNN-PSO, (c) RF-PSO, (d) SVM-PSO, and (e) ANN-MLP-PSO.

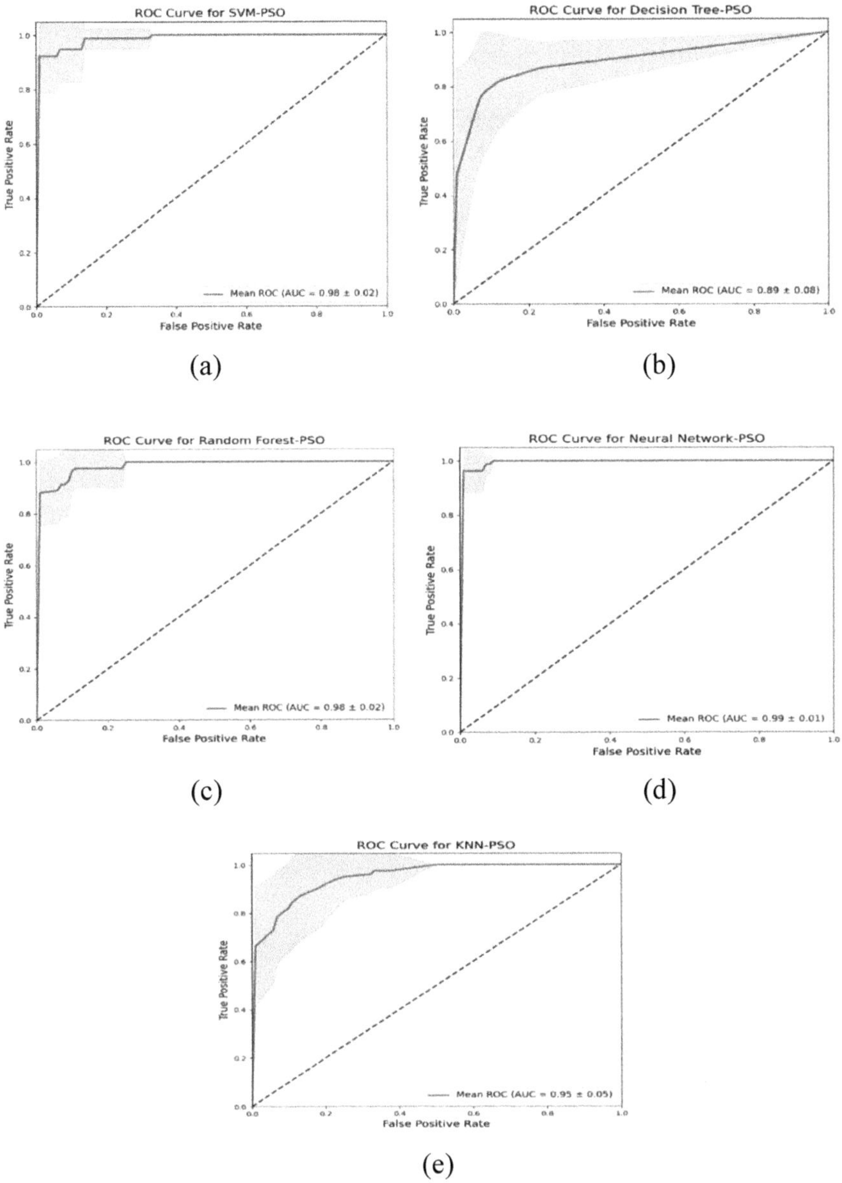

Figure 4: ROC–AUC curves for (a) SVM-PSO, (b) decision tree-PSO, (c) random forest-PSO, (d) neural network-PSO, and (e) KNN-PSO.

The plot shows fluctuating values of ECD as the depth increases, indicating variation in the density of the circulating fluid, which is key in kick detection during drilling. The temperature generally increases with depth, with some fluctuations observed. This is typical behavior in well drilling, as the temperature increases with deeper layers.

The BHP increases with depth, indicating pressure variations at different depths. Maintaining BHP is critical to preventing a blowout. These graphs help visualize the relationship between depth and key drilling parameters, assisting in monitoring well conditions and preventing potential issues such as kicks. It predicts by a criterion of high BHP, high ECD, and high Bottom hole temperatue (BHT) to indicate a kick if otherwise, no kick. The plots of these parameters are shown in Figure 5, providing a visual representation of the variations encountered during the drilling operation.

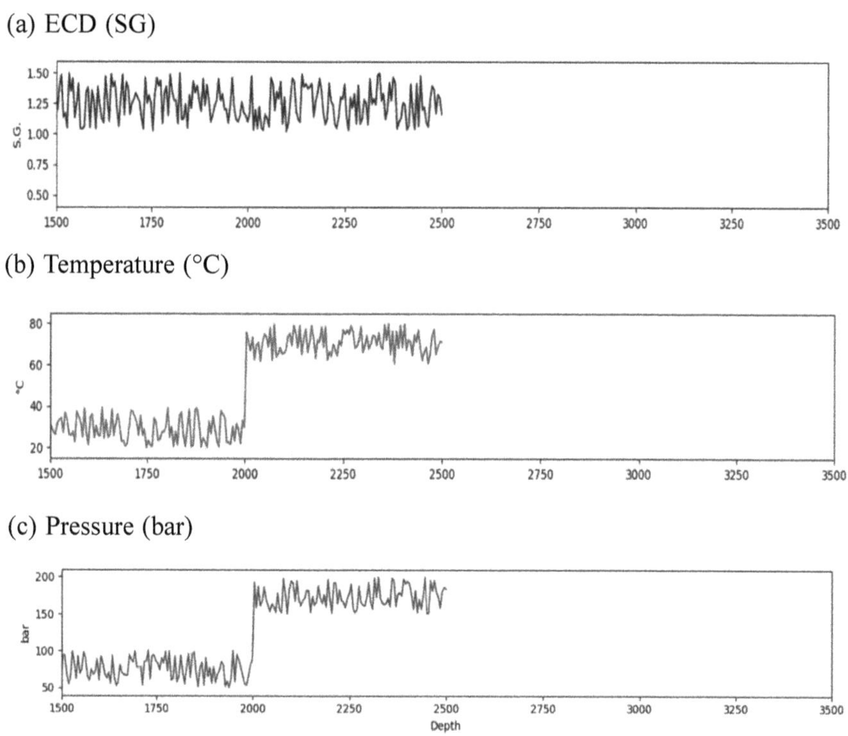

Figure 5: Plots of ECD (SG), well temperature (°C), and BHP (bar) during drilling.

Drilling Simulation: For the drilling simulation, data output from the OpenLab simulator was utilized to train the model. The drilling simulation conducted with the kick detection software accurately predicted the well path, demonstrating software validation. Figure 6 illustrates this validation process.

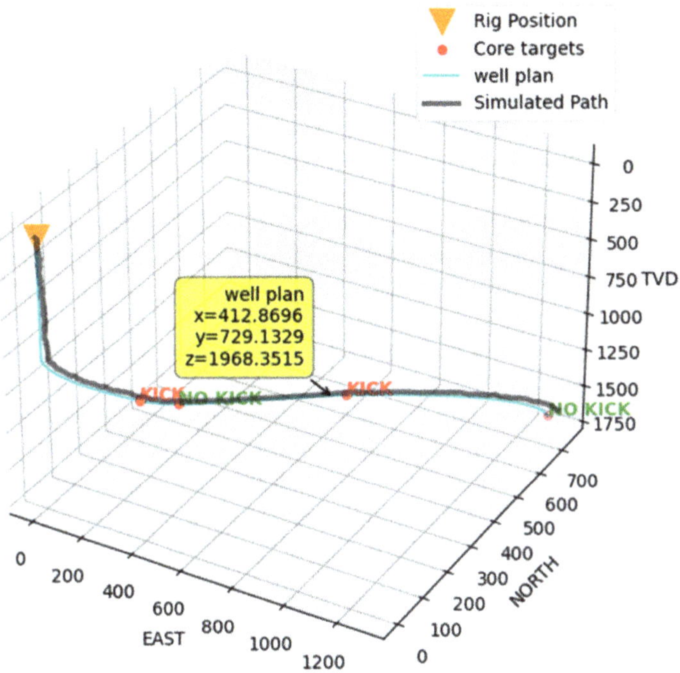

Figure 6: Drilling simulation of model.

The detection of kicks was obtained for each target coordinate. The results obtained for the predictions are shown in Table 5.

Table 5: Kick detection predictions

Target Coordinate (X,Y,Z)	Kick Prediction
238,137,1164	KICK
337,194,1223	NO KICK
759,438,1327	KICK
1284,741,1677	NO KICK
121,549,1933	KICK
209,697,2047	NO KICK
232,701,2039	NO KICK
269,708,2025	KICK
369,723,1985	NO KICK
450,733,1953	KICK
109,587,1931	NO KICK

Discussion

The results of this study highlight the effectiveness of machine learning (ML) models in kick detection and their application in drilling operations. This section discusses the performance of the best-performing ML model and its integration into the kick detection software, along with detailed insights into the obtained results.

Among the machine learning models evaluated, the neural network with PSO optimization emerged as the top performer, achieving an impressive accuracy of 97%. This model demonstrated balanced precision and recall scores, making it well-suited for real-time kick detection applications. The neural network model, identified as the best-performing ML model, was integrated into the kick detection software developed in this study.

The case study involving offshore rig simulation provided foundational data for validating both the ML models and the kick detection software. Figure 6 illustrates the simulated well path, demonstrating the software's capability to navigate complex subsurface targets accurately. The kick detection software interface (Fig. 7) exemplifies the practical application of ML algorithms in real-time operations.

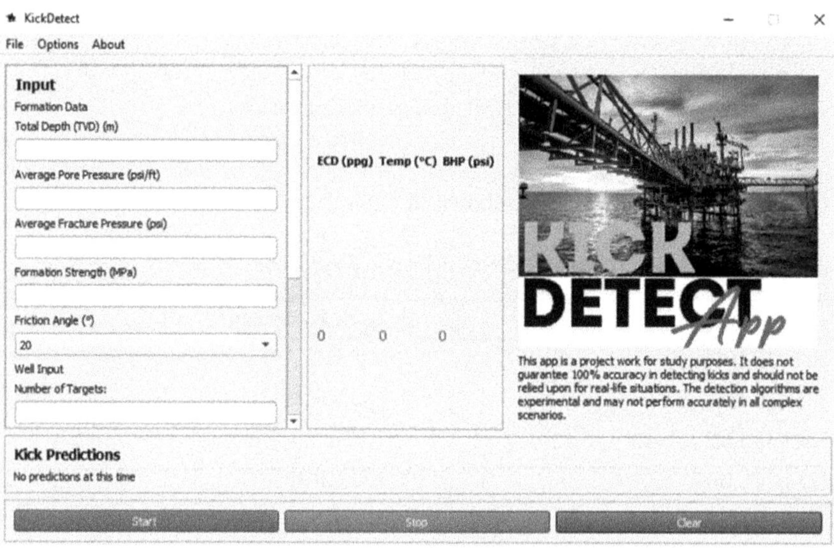

Figure 7: Kick detect software interface.

Conclusions

This study aimed to enhance well control practices in the petroleum industry by developing a Python-based kick detection model, designed to simulate and analyze well profiles, thereby enabling more effective management of wellbore stability and reducing the risk of blowouts and other well control incidents. A thorough examination of the results reveals that by integrating advanced data analytics and machine learning, specifically the MLP-PSO model, the app achieved high performance in accurately predicting kicks during drilling operations, with an accuracy of 97%. The results demonstrate the models' effectiveness in providing real-time monitoring of critical parameters such as pressure, temperature, and ECD. This capability not only reduces the risk of blowouts but also supports safer and more efficient drilling operations.

The findings emphasize the importance of combining cutting-edge technologies like machine learning with established drilling practices to address industry challenges. The developed software serves as a reliable tool for early kick detection, ensuring better decision-making and operational safety. Future work could expand the scope of kick detection by monitoring additional parameters and exploring advanced machine learning techniques to further improve accuracy and adaptability in complex drilling environments.

Acknowledgments

We would like to thank anonymous reviewers for their contributions to this manuscript. Also, we want to thank the University of Mines and Technology, Tarkwa, Ghana, GNPC School of Petroleum Studies, Petroleum and Natural Gas Engineering Department for their immense support.

References

Abdul-Ameer, H. (2023). Machine Learning Methods for Kick Detection. Unpublished PhD Thesis. University of Calgary, Calgary, Canada, pp. 305.

Batta, M. (2019). Machine learning algorithms – A review. *Technical Report*, 10(21275), 1–5.

Fjetland, A. K., Zhou, J., Abeyrathna, D. & Gravdal, J. E. (2019, June). Kick detection and influx size estimation during offshore drilling operations using deep learning. *In: 2019 14th IEEE Conference on Industrial Electronics and Applications* (*ICIEA*) (pp. 2321–2326). IEEE.

Luan, S., Sun, X., Wang, K. & Geng, A. (2013). Summary in research on kick detection technology. *Advanced Materials Research*, 821(1), 1422–1425.

Muojeke, S., Venkatesan, R. and Khan, F. (2020). Supervised data-driven approach to early kick detection during drilling operation. *Journal of Petroleum Science and Engineering*, 192, 107324.

Obi, C. E., Falola, Y., Manikonda, K., Hasan, A. R., Hassan, I. G. & Rahman, M. A. (2023). A machine learning approach for gas kick identification. *SPE Drilling & Completion*, 38(04), 663–681.

Osarogiagbon, A., Muojeke, S., Venkatesan, R., Khan, F. and Gillard, P. (2020). A new methodology for kick detection during petroleum drilling using long short-term memory recurrent neural network. *Process Safety and Environmental Protection*, 142, 126–137.

Skalle, P. (2009). *Pressure Control During Oil Well Drilling*. 6th edition. Bookboon Publications. Norway, pp. 153.

Thakkar, H. K., Shukla, H. & Sahoo, P. K. (2022). Metaheuristics in classification, clustering, and frequent pattern mining. *In: Chapter 2 of Cognitive Data Science in Sustainable Computing: Cognitive Big Data Intelligence with a Metaheuristic Approach*, 1st ed. Academic Press. Cambridge, Massachusetts, pp. 21–70.

CHAPTER

9

Nanopore Sensing for Single-Entity Analysis: Challenges, Enhancements, and Future with AI and Computational Methods

Eugene Gyasi Agyemang

Materials Sci. & Eng. and Chemistry & Biochemistry, University of Arkansas, Fayetteville, USA

Introduction

In early September 2016, a Faraday Discussion held on the campus of the University of York, UK, officially introduced the concept of "single-entity electrochemistry" (SEE), a unifying theme that connects the diverse materials and techniques used to study individual entities—such as single particles, molecules, cells, and reactions—distinct from the traditional field of ensemble electrochemistry (Crooks, 2016; Actis et al., 2016). Since the development of scanned probe microscopies, such as Scanning Ion Conductance Microscopy (SICM), Scanning Electrochemical Microscopy (SECM), Scanning Electrochemical Cell Microscopy (SECCM), and Atomic Force Microscopy (AFM), single-entity electrochemistry has grown into a clear and well-defined approach for electrochemical measurements (Baker, 2018; Lu et al., 2022).

The surge in nanotechnology and the increasing need for quantitative biochemical measurements at the single-cell and single-molecule levels have fostered the growth of SEE (Baker, 2018; Nguyen, 2020; Ying, 2020). The Coulter counter, patch-clamp technique, single-cell electrochemistry, and single-molecule redox cycling represent four pivotal experiments that have validated the feasibility of SEE, offering both conceptual inspiration and proof of practical applicability (Baker, 2018). This chapter delves into resistive pulse sensing, widely recognized

*E-mail: egyasiag@uark.edu

DOI: 10.1201/9781003506348-9

as nanopore sensing, a groundbreaking technique rooted in the Coulter counter principle for characterizing individual particles. The fusion of artificial intelligence and nanopore technology highlights exciting advancements in combining AI-driven algorithms with nanopore experiments, revolutionizing data analysis and streamlining experimental protocols.

The term "single" in SEE can broadly refer to any individual "entity," ranging from a single molecule to a single tissue or nanoparticle cluster (Crooks, 2016; Baker, 2018). Advancing single-entity electrochemical measurements is essential for improving our ability to study and understand (bio)chemical systems (Wang et al., 2023). These techniques are vital for uncovering the molecular mechanisms behind a range of issues, from developing efficient catalysts for batteries and fuel cells to understanding cell function and pathology (Xu et al., 2019; Zhou et al., 2019). Examining individual (bio)molecules within bulk solutions can assist in detecting rare subpopulations, concealed intermediates, dynamic interactions, and diverse reaction pathways (Muthukumar et al., 2015; Li et al., 2020). Studying a single cell provides a unique opportunity to investigate the complex biochemical processes occurring both inside the cell and between neighboring cells. This approach allows for a deeper understanding of the distinct molecular and functional differences that differentiate individual cells, revealing important insights into their behavior, interactions, and roles within tissues or systems (Tang et al., 2022; Goines & Dick, 2019; Goines et al., 2018; Xia et al., 2005). The understanding of molecular mechanisms that regulate cell development, function, and pathology, in this way, could pave the way for breakthroughs in disease prevention and treatment strategies.

Examining nanomaterials at the single-particle level unveils their unique structural and functional variations, offering a depth of insight that conventional ensemble techniques, such as macroscale cyclic voltammetry and coulometry, fail to provide. These traditional methods merely capture averaged activity across vast interfaces, obscuring the diverse behaviors of individual particles, underscoring the value and necessity of single-entity approaches (Zhou et al., 2019; Luo et al., 2014; German et al., 2015; Joudeh & Linke, 2022).

Resistive Pulse Sensing of Single Entities

In 1949, Wallace H. Coulter patented the Coulter counter, a device designed to count particulates suspended in a fluid (Marshall, 1968). It operated by detecting ion current flowing between two electrolyte-filled chambers, connected by a narrow central orifice. Early applications used fluid height differences between the chambers to create hydrostatic pressure, pushing fluid through the channel. As individual particles passed through, they displaced electrolyte momentarily, altering the current and enabling particle counting. Initially, the Coulter counter was used to examine the size and shape of biological cells and, later, viruses. By analyzing numerous particles, it could generate statistical data. More recently, miniaturizing the orifice has allowed the extension of the method to characterize single nanoparticles and biological macromolecules through a technique that is widely known as resistive pulse sensing or nanopore sensing (Marshall, 1968; Du, 2019).

Resistive pulse sensing serves as a versatile methodology for the detection and characterization of individual particles. This technique enables the identification and analysis of singular entities through their modulation of a baseline current as they pass through a nanoscale aperture situated between two electrolyte-filled reservoirs, which are electrically connected via two reference electrodes (Pan et al., 2019; Henriquez et al., 2004). In many applications of this technique, a nanoscale opening is created at the tip of a capillary tube, which is then filled with an electrolyte solution and submerged in an electrolyte-filled bath, as illustrated in Figure 1 (Zhou et al., 2019; Yi et al., 2022; Chang et al., 2020).

On applying a potential bias across the nanopore, the transport of ions through the nanopore, driven by the resulting electric field, generates a current that can be captured using a current amplifier and an electronic data acquisition system. This ionic current establishes a baseline against which measurable deviations occur, as an analyte species temporarily displaces a portion of the electrolyte volume while translocating through the nanopore. The baseline current is contingent on the concentration of the electrolyte species, the magnitude of the applied potential, and the geometry of the nanopore (Pan et al., 2019). The magnitude of the deviation from the baseline current is labeled "Δ*i*" in Figure 1, and the frequency of detectable events, the shape of the current modulation waveform (see

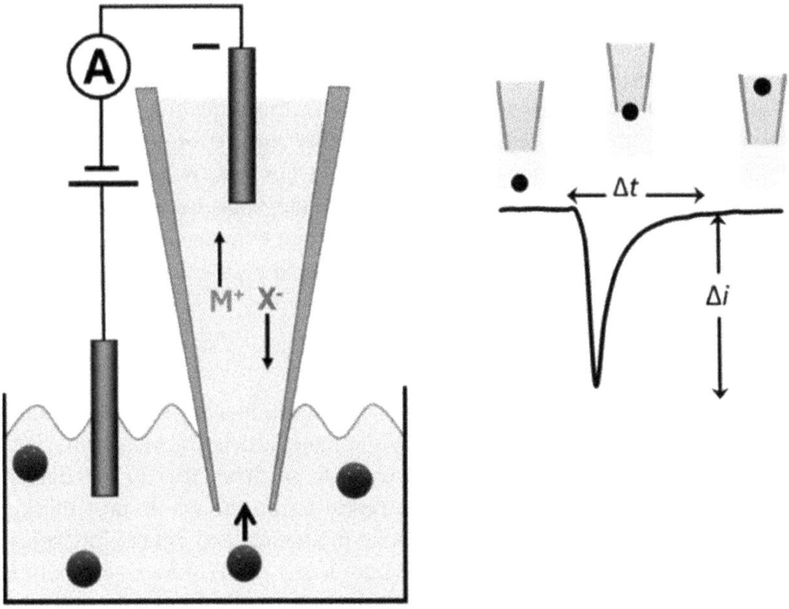

Figure 1: Illustration of a nanopore sensor experimental setup, using a nanopipette as an example.

Figure 1), and the dwell time (labeled Δt in Figure 1) have been shown to reflect the size, concentration, shape, and surface charge of the translocating species, respectively (Luo et al., 2014; Muthukumar et al., 2015; Pan et al., 2019; Luo et al., 2014; Kozak et al., 2011). The sensitivity of a nanopore is intrinsically linked to the relative dimensions of the analyte and the pore; specifically, measurements are more sensitive when the ratio of analyte diameter to nanopore diameter approaches unity.

The apparatus establishes an electric circuit by applying a potential bias across the nanopore at the tip of the electrolyte-filled nanopipette, with one electrode inside the pipette and the other in the external electrolyte bath. Molecules are driven through the nanopore toward the oppositely charged electrode by the applied electric field or a pressure gradient.

The sensitivity of a nanopore is intrinsically linked to the relative dimensions of the analyte and the pore; specifically, measurements are more sensitive when the ratio of analyte diameter to nanopore diameter approaches unity (Kozak et al., 2011; Pan et al., 2019).

Ensemble analytical techniques, such as dynamic light scattering (DLS), are commonly utilized to investigate systems comprising nanometer-sized entities dissolved or suspended in a solvent; however, they are limited in their ability to provide comprehensive insights into the dynamics, energetics, and kinetics at the single-entity level, as they primarily report on average properties (Luo et al., 2014). Small, localized inhomogeneities in particulate compositions have been demonstrated to induce colloidal instability. Furthermore, it is well established within the field of chemistry that even minor variations in size, shape, and surface charge can lead to significant alterations in properties at the nanoscale (Luo et al., 2014). Additionally, subtle modifications in the sequence and number of amino acids can drastically affect the functionality and conformation of proteins, including enzymes, in solution (Chau et al., 2021). For these reasons, single-object readout techniques are highly beneficial and advantageous. Recent advancements in nanofabrication have facilitated the study of single molecules and individual cells with unprecedented resolution, thereby expanding the field of single-entity electrochemistry (Baker, 2018; Tita, 2012).

Advantages of Nanopore Sensing

Single-entity detection methods measure one at a time allowing minor populations to be identified against a background. While nanoparticle tracking analysis can be used to track single entities, it can only do so down to ~30 nm diameter (Filipe et al., 2010), precluding it from reporting on the whole host of single molecules and nanoparticles such as protein monomers and dimers for typically sized proteins. Resistive pulse sensing (Luo et al., 2014) shown schematically in Figure 1 is a label-free method of characterizing single nanoscale entities. As shown in Figure 1, the entities, depicted as black spherical nanoparticles, are suspended in an electrolyte solution and driven one by one through a constriction between two chambers, compelled by either an electric field (for analytes with a net surface charge) or a pressure gradient (for neutral analytes).

This approach offers a simple, label free, versatile, highly sensitive, and real-time analysis of single entities. Also, nanopore sensing can be seamlessly integrated with artificial intelligence (AI), amplifying its capabilities. AI enhances data analysis, automates experimental workflows, and significantly improves classification accuracy, making it a robust tool for precision diagnostics, biomolecule identification, and nanomaterial characterization.

Nanopore sensing has successfully been applied to report the sequence of single DNA molecules, each nucleotide having a different size leading to a different current (Wang et al., 2021). The protein channel carefully matched to the diameter of the DNA molecule was required to deliver a satisfactory signal to noise.

Resistive pulse sensing has also been used to study protein aggregation in the formation of amyloid-like fibrils (Martyushenko et al., 2015) and large biomolecular assemblies such as viruses (Harms et al., 2011; Zhou, K. et al., 2011). Pore sizes can be tailored to one size of entity (in these cases being large enough to allow the whole fibril/virus to pass).

Challenges and Opportunities

Challenges in nanopore detection of single entities often arise because certain analytes are difficult to detect due to their rapid translocation through the nanopore and their relatively small size, which is often smaller than the smallest stable glass nanopores that can be fabricated (German et al., 2013; Mardoum et al., 2018). Dekker and colleagues demonstrated that most proteins translocate at time scales that exceed the detection limits of patch-clamp amplifiers, which typically operate within a time range of tens of microseconds. Consequently, these proteins often remain undetectable as they pass through the nanopore without generating a sufficiently strong signal which does not stand out from background noise (Muthukumar et al., 2015). These issues necessitate exceptionally high sensitivity and temporal resolution from the sensing apparatus. A common solution to enhance sensitivity involves adjusting the ratio of analyte diameter to pore diameter; however, very small nanopores typically exhibit low physical and chemical stability over extended periods. Additionally, the fabrication of such diminutive nanopores poses significant technical challenges regarding the feasibility and reproducibility of the manufacturing process (Kozak et al., 2011; Zhou et al., 2019; Pan et al., 2019).

State-of-the-art measurements of the smallest samples rely on advanced current amplifiers, chemical modifications of the nanopore to slow translocation rates, and enhancements to the analyte's signal strength (Kozak et al., 2011; Luo et al., 2014). While these approaches can be effective in certain scenarios, they also present their own challenges. Chemical agents and markers may interfere with direct measurements of the analyte and often require complex execution, while high-end electronics can be prohibitively expensive or not yet commercially available. Moreover, it has been demonstrated that even with the most sophisticated current amplifiers and data acquisition systems, shot noise remains an inherent physical limitation (Luo et al., 2014; Drung & Krause, 2015; Liang, 2022).

Edwards et al. introduced a modified Coulter counter that utilizes programmable data acquisition hardware to improve its functionality (Edwards et al., 2015). After detecting the resistive pulse of a nanoparticle passing through the nanopipette orifice, they reversed the voltage, altering the direction of the driving force on the particle. This detection-reversal cycle enabled a single nanoparticle to pass through the orifice thousands of times, allowing for statistical capture of the distribution of resistive pulses and precise determination of the mean pulse amplitude. This reported monitoring sub-angstrom resolution through careful measurement of the mean blocking current.

Also, emerging evidence suggests that incorporating a polymer into the electrolyte within one chamber can enhance the signal-to-noise ratio in nanopore measurements (Chau et al., 2020; Chau et al., 2021; Marcuccio et al., 2023). This technique was pioneered by Actis et al. when they studied the poly(ethylene glycol) (PEG)-induced signal enhancement for DNA molecules translocating a glass nanopore (Chau et al., 2020; Marcuccio et al., 2023). The effect appears to

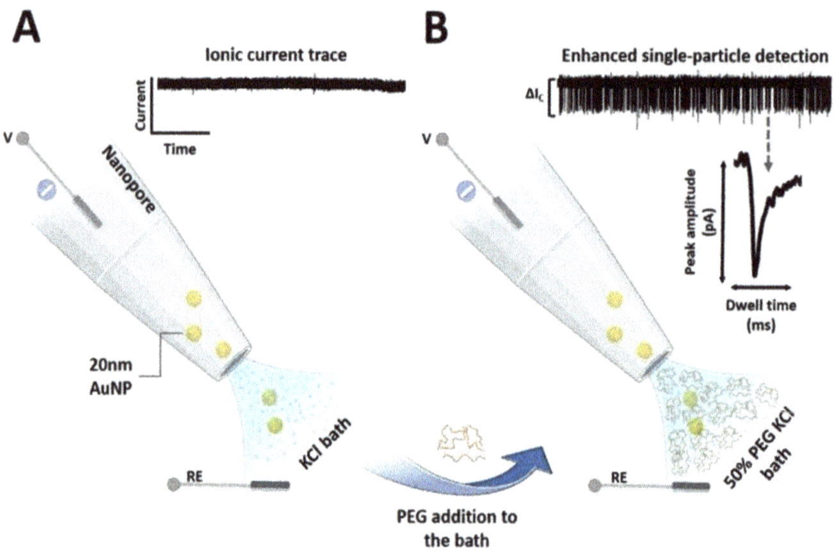

Figure 2: Enhanced nanoparticle detection in a polymer electrolyte bath. (A) Diagram showing 20 nm AuNPs translocating through the nanopore in a 50 mM KCl bath at –500 mV, with no events detected. (B) Setup using 50% (w/v) PEG in 50 mM KCl, showing translocation events for 20 nm AuNPs at –500 mV, with a representative peak and dwell time (Reproduced from Chau et al., 2024).

generalize to measuring entities beyond nucleic acids, including proteins (Chau et al., 2020) and nanoparticles (Confederat et al., 2023; Gyasi Agyemang et al., 2024). Many researchers have integrated hydrogels into nanopores to enhance the signal fidelity of the measured pulses, aiming to improve the accuracy and reliability of the readings (Acharya et al., 2020; Al Sulaiman et al., 2018; Al Sulaiman et al., 2021; Sun et al., 2021). The incorporation of a polymer electrolyte significantly enhances nanoparticle translocation signals, allowing previously undetectable events to be resolved (Figure 2).

Types of Nanopores

Solid-State Nanopores

Solid-state nanopores can be fabricated by creating nanometer-sized apertures in solid materials, such as synthetic polymer substrates or silicon nitride (SiN), by chemical or physical approaches or through laser pulling techniques to form nanochannels in quartz or glass (Li et al., 2001; Yuan et al., 2018; Yi et al., 2022). Various other inorganic materials have been cited in the literature for use in fabricating solid-state nanopores including graphene (DNA Translocation through Graphene Nanopores) and h-BN (Zhou et al., 2013). These nanopores can be produced in cylindrical (Petrossian et al., 2007) or conical (Lan et al., 2011) geometries and are advantageous due to their ease, speed, and cost-effectiveness in manufacturing (Kozak et al., 2011). Conical nanopores enhance signal fidelity by concentrating the point of highest resistance at the tip of the conical channel and hence are widely used in many applications in the literature (Kozak et al., 2011; Luo et al., 2014; Zhou et al., 2019). Direct drilling of nanopores can be accomplished using a focused ion beam with various ions (Li et al., 2001). Alternative approaches to creating solid-state nanopores in synthetic polymer membranes utilize existing nanofabrication techniques, specifically electron beam lithography (Storm et al., 2003) and reactive ion etching (Goodwin et al., 2023) or controlled (electro)chemical breakdown of the membrane (Kwok et al., 2014). In reactive ion etching, reactive gases such as oxygen, argon, or chlorine are ionized creating reactive ions, which then bombard the material surface, etching away exposed areas in a controlled manner (Goodwin et al., 2023). Controlled breakdown involves applying a strong electric field across a membrane to generate a small leakage current, with a sudden spike signaling the formation of a nanopore, after which the voltage is promptly turned off to prevent membrane damage (Kwok et al., 2014).

Laser-assisted pipette pulling is a fabrication technique that creates nanopores in glass or quartz capillaries by applying heat while pulling from both ends of the capillary (Brown et al., 2008).

Due to their versatility and straightforward fabrication, nanopipettes have been employed to detect a range of analytes, including nucleic acids, proteins, and nucleic acid–protein complexes (Yi et al., 2022). Figure 3 illustrates various nanopore fabrication techniques.

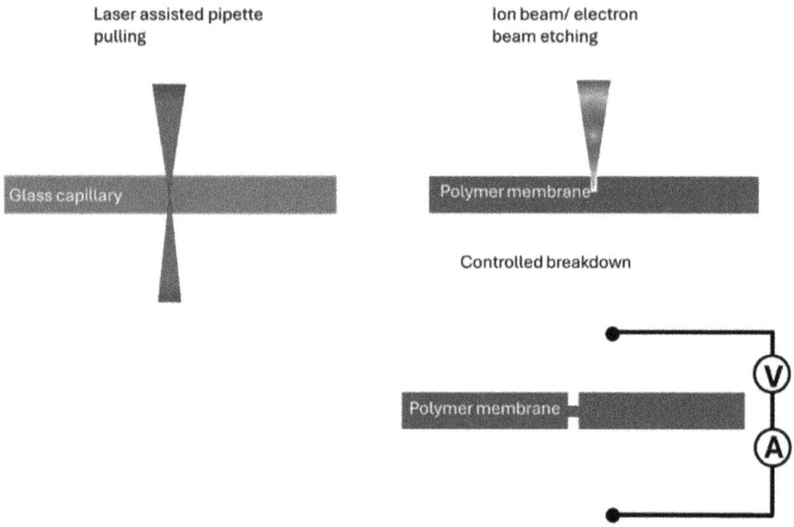

Figure 3: Various nanopore fabrication techniques utilized to create nanopores with precise dimensions and properties.

Biological Nanopores

Biomolecules that assemble into nanometer-sized pores establish a highly controlled environment, permitting only single molecules to occupy each pore (Zhang et al., 2022). Numerous such channels and pores embedded in the cell membrane regulate the transport of molecules into and out of the cell, a process crucial for cellular biochemistry and essential life-sustaining functions (Lewis et al., 1990). Biological nanopores naturally form and embed themselves into lipid bilayers, providing a reliable chemical and structural makeup, which presents a considerable benefit compared to solid-state nanopores that depend on the accuracy of their fabrication methods (Houghtaling et al., 2018). These nanopores, ranging in size from angstroms to tens of nanometers, can be customized to meet specific sensing needs, and their electrostatic and steric properties can be fine-tuned by selecting native pores with distinct characteristics for highly targeted applications (Zhang et al., 2022). By leveraging the natural properties of these biological systems, we can utilize these naturally occurring nanoscale proteinaceous pores to investigate single molecules. Patch clamp is a laboratory technique used to measure the electrical activity of individual cells or small parts of a cell by monitoring the response of single-ion channels (Kornreich, 2007). It involves using a fine glass pipette to isolate a tiny patch of the cell membrane. This method is commonly used to study how cells respond to various signals or drugs, and it helps researchers understand cell function in areas such as neuroscience, cardiology, and physiology.

Experiments utilizing biological nanopores typically demonstrate low variability and can support continuous data collection for several hours (McCabe & Forstner, 2013). In contrast, solid-state nanopores show significantly higher variability and are more prone to clogging (Awasthi et al., 2020). However, the lipid bilayer supporting biological pores is relatively delicate, which restricts the experimental conditions that can be applied, such as voltage, chemical additives, pH, temperature, and external pressure (Zhang et al., 2022). Protein-based nanopores embedded within lipid bilayers provide a highly controlled environment for selective molecular transport and sensing (Figure 4).

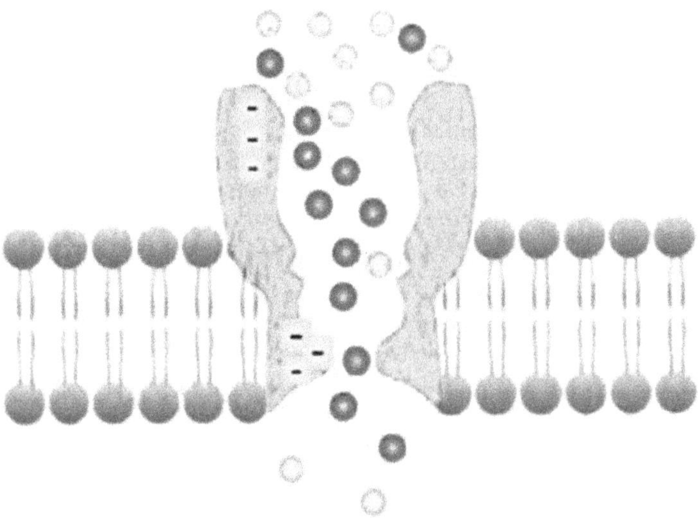

Figure 4: A schematic illustration depicting a protein-based nanopore embedded within a lipid bilayer support for selective transport or sensing applications (Modified from Bhatti et al., 2021).

Computational Simulations for Augmented Data Interpretation

Computational models and simulations are crucial for understanding single-entity processes that occur at fast time scales beyond the capabilities of current instrumentation. They also allow for the rapid study of how experimental systems respond to parameters and conditions that would be difficult or time-consuming to alter directly in physical experiments. Computational modeling approaches have become essential tools in advancing nanopore-based technology (Edwards et al., 2018). This section showcases instances where researchers have used simulations alongside experiments to enhance their understanding of complex physical systems, illustrating the valuable role simulations play in revealing insights that might otherwise remain elusive.

The use of simulations to tease out the finer details of experimental phenomena is not new to the field of nanopore science. White and Bund studied ion current rectification in conical nanopores using finite-element simulations of KCl solutions (White & Bund, 2008). By solving the Nernst–Planck, Poisson, and Navier–Stokes equations, they found that rectification results from voltage-dependent conductivity near the pore entrance. They were also able to conclude from the results of the simulations that the effect is most pronounced at intermediate ion concentrations, where a balance between surface charge screening at high concentrations and ion availability at low concentrations impacts conductivity, causing voltage dependence to disappear at either extremum.

Lan et al. conducted a foundational study on resistive pulse signals by combining dynamic electric force calculations for nanoparticle trajectories with finite-element simulations, modeling ion fluxes and tracking the position- and time-dependent velocities of nanoparticles moving through conical nanopores (Lan et al., 2011). These simulations were then applied to predict the resistive pulse current–time responses, allowing for a comparison of experimental and theoretical pulse heights and translocation times. Their findings demonstrated that nanoparticle size could be effectively differentiated by pulse height, with a smaller influence from translocation time, when utilizing a conical glass nanopore for sensing nanoparticles.

Marcuccio et al. used finite-element modeling alongside experimental data to show that adding PEG to the external solution creates a significant imbalance in the transport properties of cations and anions, drastically altering the current response of the nanopore (Marcuccio et al., 2023). In their experiments, a glass nanopipette filled with KCl solution was immersed in a bath containing "50% w/v PEG 35K + KCl." Through finite-element simulations, they uncovered the mechanism behind the polymer-enhanced signals. The model revealed that the asymmetric current response in the PEG–electrolyte system was driven by a voltage-dependent ion distribution and transport at the nanopipette tip, resulting in ion depletion or accumulation in the nanopore region. A key finding from their model was that the increased translocation signals were primarily due to the decreased diffusion coefficients of cations in the PEG phase, along with the interactions between translocating molecules and the PEG/no-PEG interface.

Lan et al. investigated the role of voltage-dependent ion distributions within nanopores in generating unique nanoscale flow behaviors, which are valuable for studying ion transport in porous materials (Lan et al., 2016). By using finite-element simulations, they examined how changes in pressure affect position-dependent ionic conductivities. Their findings revealed that combining pressure with asymmetric electroosmotic flow could produce new electrical phenomena, such as negative differential resistance. In this case, the current through a conical nanopore decreases as the applied voltage increases. This effect is the result of a positive feedback loop between ion distributions and electroosmotic flow (EOF), which creates bistable behavior in both fluid flow and electrical current at a critical voltage. Modelling and simulations are routinely used to explain and predict experimental behavior in the field of nanopore sensing (Baker, 2018). Various studies have employed quantum mechanical (QM), molecular mechanical (MM), and hybrid QM/MM modeling methods to uncover the intricate dynamics of molecules navigating nanopores, shedding light on their complex behaviors and interactions (Trick et al., 2013; Lynch et al., 2020; Choudhary et al., 2022; Chau et al., 2024).

Applications of Nanopore Sensing for Nanoparticles Biomolecules

Single Nanoparticle Sensing

The development of nanopore-based sensors using resistive pulse sensing revolutionized the detection and characterization of nanoscale species (Jodeyri & Taghipoor, 2024). Strong correlations between resistive pulse characteristics and particle properties have been established, enabling researchers to analyze particles in commercial devices and research setups (Lan et al., 2011).

Lan et al. conducted a foundational study involving the preparation, characterization, and analytical application of silica nanochannels in the size range of 5–100 nm (Lan et al., 2011). They utilized the nanochannels to sense single 40-nm polystyrene nanoparticles. A linear response was observed between the detection rate and the concentration of nanoparticles in the range of 0–25 nM, establishing the utility of the nanopore-based methods in measuring concentrations.

Confederat et al. reported the development of nanopore sensors enhanced by a polymer electrolyte system, which allows for the analysis of heterogeneous nanoparticle mixtures at low ionic strength (Confederat et al., 2023). They demonstrated the versatility of this method by characterizing a range of nanoparticle types, including metallic nanospheres of different sizes, plasmonic nanostars with varying levels of branching, and protein-based spherical nucleic acids with different oligonucleotide loadings. Gyasi et al. in a follow-up work on the PEG-enhanced nanopore report that incorporating PEG–electrolyte inside the nanopore improves nanoparticle detection at low ionic strength (Gyasi Agyemang et al., 2024). As a proof of concept, they demonstrated multimodal analysis of a nanoparticle sample by combining the polymer electrolyte nanopore sensor with nano-impact electrochemistry. They propose that this combined approach allows for multiparametric analysis of nanoparticle systems, providing electrochemical reactivity data from nano-impact electrochemistry alongside additional information on nanoparticle size, shape, and surface charge through nanopore measurements.

Yazbeck et al. present a novel electrokinetic tweezer for single nanoparticle manipulation and characterization, based on electrokinetic trapping near a low-aspect-ratio nanopore (Yazbeck et al., 2019). The study demonstrates that variations in nanoparticle size and zeta potential significantly influence ionic current and trap stiffness. Furthermore, they demonstrate that both trap stiffness and nanoparticle position can be easily controlled by adjusting the applied voltage and buffer concentration, offering a straightforward method for particle manipulation. The authors suggest that further development of this electrokinetic tweezer could lead to new advancements in nanophotonics, drug delivery, and biosensing. Nanopore sensing has proven to be a powerful tool for detecting and characterizing nanoparticles, offering precise insights into their size, shape, and other key properties. As highlighted in this chapter, nanopore-based resistive pulse sensing has been successfully employed in various studies, ranging from analytical applications to proof-of-concept experiments. The technique is suitable for probing nanoparticles in diverse environments, providing valuable data for advancing material characterization.

Single Nucleic Acids, Proteins, and Peptides

Understanding how cellular function relates to molecular composition is crucial for developing next-generation therapeutics, but it requires accurately measuring all types of molecules within cells. Nanometer-sized tips allow for highly accurate targeting with minimal cellular disruption, making them perfect for delivering materials, sampling cellular content, and conducting real-time monitoring and measurements from individual cells (Chau et al., 2024).

Several large corporations, most notably Oxford Nanopore, have developed commercial DNA sequencing tools based on nanopore technology, and these products have demonstrated significant potential (Maitra et al., 2012; Steinbock & Radenovic, 2015). Beyond DNA sequencing, nanopore-based single-molecule tools could also be applied to analyze small molecules, peptides, light-driven reactions, chemical reactions, protein folding and unfolding, enzymatic activity, and binding processes (Ying et al., 2020; Ying et al., 2022).

Choi et al. present the integration of patch-clamp electrophysiology with subcellular shotgun proteomics through high-resolution mass spectrometry (HRMS) (Choi et al., 2022). By recording electrical activity, they identified dopaminergic neurons in the brain's substantia nigra pars compacta. Through multiple experiments, they quantified around 275 distinct proteins from digests representing Somal aspirates in cultured neurons and 157 unique proteins in single neurons. This quantification highlights the ability of biological nanopores to effectively discriminate proteomic differences between Somal and neuronal protein samples, offering valuable insights into their distinct characteristics.

Salehirozveh et al. developed a compact, portable, and leakproof flow cell made from poly(methyl methacrylate) (PMMA) specifically for nanopore experiments (Salehirozveh et al., 2023). They introduced an innovative in situ poly-L-lysine (PLL)-g-PEG functionalization method, creating nonadhesive nanopores suitable for measuring disease-related biomarkers. Their functionalized nanopores showed several enhanced properties, including extended nanopore lifespan and improved protein translocation events, increased current blockade amplitudes, and higher translocation frequencies. Using this system, they accurately measured the volume of a single alpha-1 antitrypsin protein (253 nm^3), closely matching known literature values.

Early detection of neurodegenerative diseases such as Alzheimer's, multiple sclerosis (MS), and amyotrophic lateral sclerosis (ALS) has been shown to improve the effectiveness of therapeutic treatments (Pedersen & Heegaard, 2013). Protein aggregates or specific genetic markers—clear indicators of these diseases (Ross & Poirier, 2004; Pedersen & Heegaard, 2013)—could be rapidly and cost-effectively identified using nanopore technology, enabling diagnosis at much earlier stages, enabling timely medical intervention.

Chau et al. describe a solid-state nanopore system enhanced with a polymer–electrolyte solution to detect and analyze RNA fragments of the chikungunya virus (CHIKV) under biologically relevant conditions. The researchers found that the polymer–electrolyte environment controls how RNA moves through the nanopore and linked this behavior to the specific salt properties. They identified cesium bromide (CsBr) as the best salt to pair with PEG for strong signal amplification. Using this setup, they successfully identified CHIKV RNA fragments from approximately 300 to 2000 nucleotides and distinguished structural differences between RNA that was transcriptionally folded and that which was refolded.

Seibertz et al. utilized an automated patch-clamp technique to investigate a comprehensive range of electrophysiological parameters in isolated swine atrial and ventricular cardiomyocytes (Seibertz et al., 2022). Key parameters such as action potential, L-type calcium current, and basal inward rectifier current were reliably measured. Pharmacological tests demonstrated that this method is suitable for drug screening applications using native cardiomyocyte material. They recorded multiple parameters from individual cells in a high-throughput setup, improving data quantity and quality. These protocols serve as a good starting point for enhanced mechanistic and phenotypical studies in human cardiac electrophysiology.

Ryan et al. examined ricin, a well-known biotoxin, using Oxford Nanopore Technologies (ONT) direct RNA sequencing (Ryan et al., 2022). This study provides the first clear evidence that specific base modifications in ricin poisoning is linked to ribosome-inactivating protein activity, confirmed through the use of a ricin-neutralizing antibody. The method is transferrable for detecting and quantifying depurination events induced by ricin-like toxins. The authors suggest that this approach could be valuable in both assessing novel therapeutic interventions and diagnosing ricin exposure.

Leveraging the PEG-enhanced nanopore technique, Chau et al. successfully monitored endonuclease activity in real time under physiological salt conditions (Chau et al., 2024). This technique enabled them to observe the cleavage action of the restriction enzyme SwaI on dsDNA across various digestion buffers and to assess off-target activity of CRISPR–Cas9 ribonucleoprotein (RNP) endonuclease when single base pair mismatches were present. Their work opens new possibilities for detailed analysis of enzymatic behavior and specificity. These examples highlight the breadth of the applications of nanopore-based approaches for biomolecule detection and characterization, across varied disciplines including toxicology, pharmacology, electrophysiology, virology, and neurobiology.

Readers seeking a more comprehensive literature review on the advancements and applications of biological nanopores are encouraged to consult references (Ross & Poirier, 2004; Pedersen & Heegaard, 2013; Guo et al., 2015; Wang et al., 2018; Bhatti et al., 2021; Seibertz et al., 2022; Salehirozveh et al., 2023; Wei et al., 2023).

Fusion of Artificial Intelligence and Nanopore-Based Technology

Low throughput and demand of significant experimenter expertise limit traditional patch-clamp techniques and other nanopore-based methods. However, advancements in machine learning and artificial intelligence are set to transform electrochemical analysis, enabling high-throughput, experimenter-independent approaches (Seibertz et al., 2022). Artificial intelligence enhances its predictive and anomaly-detection capabilities through continuous data learning, efficiently managing vast datasets via pattern recognition and self-optimization to advance diagnostic accuracy without manual intervention (Deng, 2018; Dutt et al., 2024). AI methods automate signal analysis and reveal hidden patterns within the noisy ionic currents of patch-clamp and nanopore-based techniques, uncovering insights previously overlooked by researchers (Richter-Laskowska et al., 2024). Machine learning and statistical testing tools have been reported to offer powerful capabilities for detecting DNA modifications by analyzing ion current signals in nanopore-based direct DNA sequencing (Wan et al., 2022). Supervised machine learning utilizes classifiers to extract key features from the nanopore pulse signals, such as peak height and width and assigns them to specific analytes (Arima et al., 2021). Common classification models, such as naïve Bayes, support vector machine (SVM), and k-nearest neighbors, along with random forest and rotation forest, have shown strong performance (Arima et al., 2021). In their recent review, Dutt et al. highlight AI's transformative potential for rapid biomolecule identification and biomarker quantification (Dutt et al., 2024). They propose that combining a portable nanopore reader with AI algorithms to analyze purified bodily fluids enables autonomous detection and measurement of biomarkers in complex solutions, such as blood, opening new frontiers in medical diagnostics.

Taniguchi et al. introduce a highly sensitive nanopore-based method combined with artificial intelligence for detecting novel coronaviruses, providing a simpler alternative to traditional reverse transcription-polymerase chain reaction (RT-PCR), which requires RNA extraction from clinical samples (Taniguchi et al., 2021). Their "artificially intelligent nanopore" system integrates machine learning software, a portable high-speed current measuring device, and scalable semiconducting nanopore modules. The method accurately identified four coronaviruses of comparable size—HCoV-229E, SARS-CoV, MERS-CoV, and SARS-CoV-2—achieving "90% sensitivity and 96% specificity in 5 minutes." The machine learning algorithm they developed resolves overlapping signals, enabling precise nanoparticle identification with a single waveform. The authors emphasize that high-throughput, high-accuracy detection tools, such as the one they developed, could be pivotal in managing disease outbreaks, such as the global COVID-19 pandemic of 2020.

Dutt and colleagues examined the translocation of four proteins of similar size through nanopores in thin silicon nitride membranes paired with high-bandwidth amplifiers (Dutt et al., 2023). Signal classification into various clusters based on specific attributes, combined with a machine learning algorithm, improved protein identification accuracy, achieving an F-value and specificity of 96.4% for the four proteins. Their work suggests that leveraging high-bandwidth technology, advanced clustering, and machine learning can enable precise, label-free protein identification.

Arima et al. developed a machine-learning-based diagnostic method for influenza using nanopore-resistive pulse data to distinguish individual virions based on their physical characteristics (Arima et al., 2018). They employed selective resistive-pulse sensing to detect negatively charged flu viruses in physiological media, leveraging electroosmotic flow to filter contaminants at the Si_3N_4 pore. They achieved 68% accuracy in identifying allotypes at the single-virus level, with classification accuracy exceeding 95% when pulse data from over 20 virions were combined. They suggest that this straightforward and adaptable method holds promise for rapid, on-site testing of different flu strains.

Jena and Pathak introduced an AI nanopore method that eliminates the need to extract signature transmission functions for each nucleotide in a DNA strand (Jena & Pathak, 2023). By integrating supervised machine learning with transverse quantum transport and graphene nanopores, their optimized model predicts the transmission function of any nucleotide after training on various orientations, achieving a root-mean-square error as low as 0.062. They report 96.01% classification accuracy for unlabeled nucleotides based on transmission readouts. Their AI nanopore approach shows potential in reducing experimental burdens while enabling rapid and high-precision DNA sequencing.

Ma et al. introduce a cell detection method that uses differential interference contrast (DIC) images, integrating clarity evaluation and machine learning for cell identification in brain tissue slices (Ma et al., 2021). They utilize principal component analysis to assess image clarity, followed by the k-means algorithm to classify images into clear or blurred categories, achieving 99.1% precision. Clear images are then selected for further analysis, with neuron detection performed using a sliding window approach and a pre-trained GoogLeNet classifier, effectively identifying viable cells and offering a valuable tool for automating patch clamping.

Koos et al. developed an automated tool for electrophysiological recordings in label-free tissue slices (Koos et al., 2021). Their system independently automated the detection of cells, micropipette movement and positioning, and data acquisition. Using deep learning, the model is trained on a new database of unlabeled brain tissue neurons. The tool provides high-quality measurements from both human and rodent neurons and supports further molecular and anatomical analysis. Additionally, their system integrates seamlessly into existing setups, enabling parallelization and accelerating the discovery process.

The integration of artificial intelligence with existing nanopore technology is steadily growing with the potential for revolutionizing scientific research becoming increasingly evident. The synergy between AI and nanofabrication is opening new avenues for more efficient and precise experimental setups, enabling faster and more accurate data collection and analysis. This evolution is not only enhancing the capabilities of nanopore technology but also paving the way for automation in complex processes, reducing human intervention and increasing throughput. As these advancements progress, we can expect transformative changes in fields ranging from medical diagnostics to environmental monitoring and epidemiology, marking a new era in scientific exploration and innovation.

Conclusion

Since its official recognition as a unique field in 2016, SEE has matured into a cohesive and transformative framework within electrochemical measurement. A landmark experiment that epitomizes the practical utility of SEE is the Coulter counter, which has inspired advancements in techniques for analyzing individual entities. Its miniaturized counterpart, often referred to as resistive pulse sensing or nanopore sensing, represents a significant leap forward by enabling the detection and characterization of single entities one at a time. This capability is particularly valuable for identifying rare or minor populations within complex samples, which might otherwise be masked by dominant components in ensemble measurements.

In contrast, conventional ensemble techniques, such as dynamic light scattering, are inherently limited in their scope. These methods report averaged properties, offering little insight into the nuanced dynamics, energetics, or kinetics that govern individual entities. By bridging this gap, SEE not only enhances our understanding of these microscopic processes but also opens avenues for advancements in diverse fields such as materials science, diagnostics, and molecular biology. Nanopores can be fabricated using either solid-state or biological approaches, each offering unique advantages and applications. Solid-state nanopores are typically created in materials such as polymer or glass membranes, while biological nanopores are derived from proteinaceous ion channels that occur naturally in cell membranes or are engineered from self-assembling biomolecules. These nanopores have found extensive applications in detecting nanoparticles and biomolecules such as nucleic acids and proteins, demonstrating their versatility and utility across diverse scientific fields.

Despite their remarkable potential, nanopore technologies face certain physical limitations that researchers are actively working to overcome. For instance, the high-speed multipass Coulter counter developed by Edward and coworkers addresses throughput constraints, while the Actis group's PEG-enhanced nanopore sensing introduces innovative signal fidelity enhancements to improve sensitivity and reliability. Such advancements not only mitigate existing challenges but also expand the applicability of nanopore-based systems, solidifying their role as a cornerstone in modern analytical and diagnostic techniques.

The advent of artificial intelligence and related technologies has opened new avenues for automating various aspects of nanopore-based technology, ranging from experimental methodologies to real-time data analysis. These advancements significantly amplify the utility of nanopore sensing, cementing its importance within the toolkit of single-entity electrochemistry. By bridging the gap between traditional ensemble techniques and cutting-edge single-entity methodologies, nanopore sensing has established itself as an indispensable tool for probing the fundamental properties of individual particles and molecules.

Moreover, the integration of computational tools and AI-driven approaches holds immense promise for expanding the scope and precision of nanopore applications. This synergy is not only fostering innovation but also enabling high-throughput, experimenter-independent approaches, transforming how electrochemical analysis is conducted. As numerous research groups work toward the fusion of nanopore technologies with artificial learning methods, advancements in machine learning and AI are poised to revolutionize the field. These developments promise to unlock new opportunities for discovery and application, driving the evolution of electrochemical analysis into an era defined by unparalleled efficiency and accuracy.

References

Acharya, S., Jiang, A., Kuo, C., Nazarian, R., Li, K., Ma, A. et al. (2020). Improved measurement of proteins using a solid-state nanopore coupled with a hydrogel. *ACS Sens*, 5(2), 370–376. https://doi.org/10.1021/acssensors.9b01928.

Actis, P., Bentley, C. L., Edwards, M. A. & Jacobse, L. (2016). Highlights from the Faraday discussion on single entity electrochemistry, York, UK. *Chem. Commun.*, 52(97), 13934–13940. https://doi.org/10.1039/C6CC90523D.

Al Sulaiman, D., Cadinu, P., Ivanov, A. P., Edel, J. B. & Ladame, S. (2018). Chemically modified hydrogel-filled nanopores: A tunable platform for single-molecule sensing. *Nano Lett.*, 18(9), 6084–6093. https://doi.org/10.1021/acs.nanolett.8b03111.

Al Sulaiman, D., Gatehouse, A., Ivanov, A. P., Edel, J. B. & Ladame, S. (2021). Length-dependent, single-molecule analysis of short double-stranded DNA fragments through hydrogel-filled nanopores: A potential tool for size profiling cell-free DNA. *ACS Appl. Mater. Interfaces*, 13(23), 26673–26681. https://doi.org/10.1021/acsami.1c01145.

Arima, A., Tsutsui, M., Harlisa, I. H., Yoshida, T., Tanaka, M., Yokota, K. et al. (2018). Selective detections of single-viruses using solid-state nanopores. *Sci Rep.*, 8(1), 16305. https://doi.org/10.1038/s41598-018-34665-4.

Arima, A., Tsutsui, M., Washio, T., Baba, Y. & Kawai, T. (2021). Solid-state nanopore platform integrated with machine learning for digital diagnosis of virus infection. *Anal. Chem.*, 93(1), 215–227. https://doi.org/10.1021/acs.analchem.0c04353.

Awasthi, S., Sriboonpeng, P., Ying, C., Houghtaling, J., Shorubalko, I., Marion, S. et al. (2020). Polymer coatings to minimize protein adsorption in solid-state nanopores. *Small Methods*, 4(11), 2000177. https://doi.org/10.1002/smtd.202000177.

Baker, L. A. (2018). Perspective and prospectus on single-entity electrochemistry. *J. Am. Chem. Soc.*, 140(46), 15549–15559. https://doi.org/10.1021/jacs.8b09747.

Bhatti, H., Jawed, R., Ali, I., Iqbal, K., Han, Y., Lu, Z. et al. (2021). Recent advances in biological nanopores for nanopore sequencing, sensing and comparison of functional variations in MspA mutants. *RSC Adv.*, 11(46), 28996–29014. https://doi.org/10.1039/D1RA02364K.

Brown, A. L., Johnson, B. E. & Goodman, M. B. (2008). Making patch-pipettes and sharp electrodes with a programmable puller. *Journal of Visualized Experiments* (*JoVE*), 20, e939. https://doi.org/10.3791/939.

Chang, M., Morgan, G., Bedier, F., Chieng, A., Gomez, P., Raminani, S. & Wang, Y. (2020). Review—Recent advances in nanosensors built with pre-pulled glass nanopipettes and their applications in chemical and biological sensing. *J Electrochem Soc.*, 167(3), 037533. https://doi.org/10.1149/1945-7111/ab64be.

Chau, C. C., Hewitt, E. W. & Actis, P. (2021). The role of macromolecular crowding in single-entity electrochemistry: Friend or foe? *Current Opinion in Electrochemistry*, 25, 100654. https://doi.org/10.1016/j.coelec.2020.100654.

Chau, C. C., Maffeo, C. M., Aksimentiev, A., Radford, S. E., Hewitt, E. W. & Actis, P. (2024). Single molecule delivery into living cells. *Nat Commun*, 15(1), 4403. https://doi.org/10.1038/s41467-024-48608-3.

Chau, C. C., Weckman, N. E., Thomson, E. E. & Actis, P. (2024). Solid-state nanopore real time assay for monitoring Cas9 endonuclease reactivity. bioRxiv, 2024.09.20.612695. https://doi.org/10.1101/2024.09.20.612695.

Chau, C. C., Radford, S. E., Hewitt, E. W. & Actis, P. (2020). Macromolecular crowding enhances the detection of DNA and proteins by a solid-state nanopore. *Nano Lett.*, 20(7), 5553–5561. https://doi.org/10.1021/acs.nanolett.0c02246.

Choi, S. B., Polter, A. M. & Nemes, P. (2022). Patch-clamp proteomics of single neurons in tissue using electrophysiology and subcellular capillary electrophoresis mass spectrometry. *Anal. Chem.*, 94(3), 1637–1644. https://doi.org/10.1021/acs.analchem.1c03826.

Choudhary, A., Maffeo, C. & Aksimentiev, A. (2022). Multi-resolution simulation of DNA transport through large synthetic nanostructures. *Phys. Chem. Chem. Phys.*, 24(5), 2706–2716. https://doi.org/10.1039/D1CP04589J.

Confederat, S., Lee, S., Vang, D., Soulias, D., Marcuccio, F., Peace, T. I. et al. (2023). Next-generation nanopore sensors for enhanced detection of nanoparticles (preprint). *Biophysics*. https://doi.org/10.1101/2023.03.31.534385.

Confederat, S., Lee, S., Vang, D., Soulias, D., Marcuccio, F., Peace, T. I. et al. (2023). Next-generation nanopore sensors based on conductive pulse sensing for enhanced detection of nanoparticles. *Small n/a* (n/a), 2305186. https://doi.org/10.1002/smll.202305186.

Crooks, R. M. (2016). Concluding remarks: Single entity electrochemistry one step at a time. *Faraday Discuss*, 193(0), 533–547. https://doi.org/10.1039/C6FD00203J.

Deng, L. (2018). Artificial intelligence in the rising wave of deep learning: The historical path and future outlook [Perspectives]. *IEEE Signal Processing Magazine*, 35(1), 180–177. https://doi.org/10.1109/MSP.2017.2762725.

DNA Translocation through Graphene Nanopores | Nano Letters. https://pubs.acs.org/doi/10.1021/nl102069z?src=getftr&utm_source=iopp&getft_integrator=iopp (accessed 2024-11-16).

Drung, D. & Krause, C. (2015). Excess current noise in amplifiers with switched input. *IEEE Trans. Instrum. Meas.*, 64(6), 1455–1459. https://doi.org/10.1109/TIM.2015.2398958.

Du, X., Liu, X., Blackar, M. J., Zhou, Y., Wang, H., Liu, F. & He, Z. (2019). Coulter Method for the Detection of Physiological Parameters. protocols.io https://dx.doi.org/10.17504/protocols.io.6wehfbe.

Dutt, S., Karawdeniya, B., Bandara, Y. M. & Kluth, P. (2024). Nanopore sensing and machine learning: Future of biomarker analysis and disease detection. *Future Science OA*, 10(1), 2340882. https://doi.org/10.2144/fsoa-2023-0226.

Dutt, S., Shao, H., Karawdeniya, B., Bandara, Y. M. N. D. Y., Daskalaki, E., Suominen, H. et al. (2023). High accuracy protein identification: Fusion of solid-state nanopore sensing and machine learning. *Small Methods*, 7(11), 2300676. https://doi.org/10.1002/smtd.202300676.

Edwards, M. A., German, S. R., Dick, J. E., Bard, A. J. & White, H. S. (2015). High-speed multipass Coulter counter with ultrahigh resolution. *ACS Nano*, 9(12), 12274–12282. https://doi.org/10.1021/acsnano.5b05554.

Edwards, M. A., Robinson, D. A., Ren, H., Cheyne, C. G., Tan, C. S. & White, H. S. (2018). Nanoscale electrochemical kinetics & dynamics: The challenges and opportunities of single-entity measurements. *Faraday Discuss*, 210(0), 9–28. https://doi.org/10.1039/C8FD00134K.

Filipe, V., Hawe, A. & Jiskoot, W. (2010). Critical evaluation of nanoparticle tracking analysis (NTA) by nanosight for the measurement of nanoparticles and protein aggregates. *Pharm Res*, 27(5), 796–810. https://doi.org/10.1007/s11095-010-0073-2.

German, S. R., Hurd, T. S., White, H. S. & Mega, T. L. (2015). Sizing individual Au nanoparticles in solution with sub-nanometer resolution. *ACS Nano*, 9(7), 7186–7194. https://doi.org/10.1021/acsnano.5b01963.

German, S. R., Luo, L., White, H. S. & Mega, T. L. (2013). Controlling nanoparticle dynamics in conical nanopores. *J. Phys. Chem. C*, 117(1), 703–711. https://doi.org/10.1021/jp310513v.

Goines, S. & Dick, J. E. (2019). Review—Electrochemistry's potential to reach the ultimate sensitivity in measurement science. *J. Electrochem. Soc.*, 167(3), 037505. https://doi.org/10.1149/2.0052003JES.

Goodwin, M. J., Harteveld, C. A. M., Boer, M. J. & de Vos, W. L. (2023). Deep reactive ion etching of cylindrical nanopores in silicon for photonic crystals. *Nanotechnology*, 34(22), 225301. https://doi.org/10.1088/1361-6528/acc034.

Guo, B.-Y., Zeng, T. & Wu, H.-C. (2015). Recent advances of DNA sequencing via nanopore-based technologies. *Science Bulletin*, 60(3), 287–295. https://doi.org/10.1007/s11434-014-0707-6.

Gyasi Agyemang, E., Confederat, S., Mohanan, G., Sani, M. A., Chau, C., Charnock, D. et al. (2024). Multimodal nanoparticle analysis enabled by a polymer electrolyte nanopore combined with nanoimpact electrochemistry. *Faraday Discuss*. https://doi.org/10.1039/D4FD00143E.

Harms, Z. D., Mogensen, K. B., Nunes, P. S., Zhou, K., Hildenbrand, B. W., Mitra, I. et al. (2011). Nanofluidic devices with two pores in series for resistive-pulse sensing of single virus capsids. *Analytical Chemistry*, 83(24), 9573–9578. https://doi.org/10.1021/ac202358t.

Henriquez, R. R., Ito, T., Sun, L. & Crooks, R. M. (2004). The resurgence of Coulter counting for analyzing nanoscale objects. *Analyst*, 129(6), 478–482. https://doi.org/10.1039/B404251B.

Houghtaling, J., List, J. & Mayer, M. (2018). Nanopore-based, rapid characterization of individual amyloid particles in solution: Concepts, challenges, and prospects. *Small*, 14(46), e1802412. https://doi.org/10.1002/smll.201802412.

Jena, M. K. & Pathak, B. (2023). Development of an artificially intelligent nanopore for high-throughput DNA sequencing with a machine-learning-aided quantum-tunneling approach. *Nano Lett.*, 23(7), 2511–2521. https://doi.org/10.1021/acs.nanolett.2c04062.

Jodeyri, Z. & Taghipoor, M. (2024). Multivariate analysis of nanoparticle translocation through a nanopore to improve the accuracy of resistive pulse sensing. *Phys. Chem. Chem. Phys.*, 26(6), 5097–5105. https://doi.org/10.1039/D3CP05565E.

Joudeh, N. & Linke, D. (2022). Nanoparticle classification, physicochemical properties, characterization, and applications: A comprehensive review for biologists. *Journal of Nanobiotechnology*, 20(1), 262. https://doi.org/10.1186/s12951-022-01477-8.

Koos, K., Oláh, G., Balassa, T., Mihut, N., Rózsa, M., Ozsvár, A. et al. (2021). Automatic deep learning-driven label-free image-guided patch clamp system. *Nature Communications*, 12, 936. https://doi.org/10.1038/s41467-021-21291-4.

Kornreich, B. G. (2007). The patch clamp technique: Principles and technical considerations. *Journal of Veterinary Cardiology*, 9(1), 25–37. https://doi.org/10.1016/j.jvc.2007.02.001.

Kozak, D., Anderson, W., Vogel, R. & Trau, M. (2011). Advances in resistive pulse sensors: Devices bridging the void between molecular and microscopic detection. *Nano Today*, 6(5), 531–545. https://doi.org/10.1016/j.nantod.2011.08.012.

Kwok, H., Briggs, K. & Tabard-Cossa, V. (2014). Nanopore fabrication by controlled dielectric breakdown. *PLOS ONE*, 9(3), e92880. https://doi.org/10.1371/journal.pone.0092880.

Lan, W.-J., Edwards, M. A., Luo, L., Perera, R. T., Wu, X., Martin, C. R. et al. (2016). Voltage-rectified current and fluid flow in conical nanopores. *Acc. Chem. Res.*, 49(11), 2605–2613. https://doi.org/10.1021/acs.accounts.6b00395.

Lan, W.-J., Holden, D. A., Zhang, B. & White, H. S. (2011). Nanoparticle transport in conical-shaped nanopores. *Anal. Chem.*, 83(10), 3840–3847. https://doi.org/10.1021/ac200312n.

Lewis, D. L., Lechleiter, J. D., Kim, D., Nanavati, C. & Clapham, D. E. (1990). Intracellular regulation of ion channels in cell membranes. *Mayo Clinic Proceedings*, 65(8), 1127–1143. https://doi.org/10.1016/S0025-6196(12)62726-8.

Li, J., Stein, D., McMullan, C., Branton, D., Aziz, M. J. & Golovchenko, J. A. (2001). Ion-beam sculpting at nanometre length scales. *Nature*, 412(6843), 166–169. https://doi.org/10.1038/35084037.

Li, Y., Yang, C. & Guo, X. (2020). Single-molecule electrical detection: A promising route toward the fundamental limits of chemistry and life science. *Acc. Chem. Res.*, 53(1), 159–169. https://doi.org/10.1021/acs.accounts.9b00347.

Liang, Y.-X. (2022). Low-noise large-bandwidth transimpedance amplifier for measuring scanning tunneling shot noise spectra in cryogenic STM and its applications. *Ultramicroscopy*, 234, 113466. https://doi.org/10.1016/j.ultramic.2022.113466.

Lu, S.-M., Li, M.-Y. & Long, Y.-T. (2022). Dynamic chemistry interactions: Controlled single-entity electrochemistry. *J. Phys. Chem. Lett.*, 13(21), 4653–4659. https://doi.org/10.1021/acs.jpclett.2c00960.

Luo, L., German, S. R., Lan, W.-J., Holden, D. A., Mega, T. L. & White, H. S. (2014). Resistive-pulse analysis of nanoparticles. *Annu Rev Anal Chem (Palo Alto Calif)*, 7, 513–535. https://doi.org/10.1146/annurev-anchem-071213-020107.

Lynch, C. I., Rao, S. & Sansom, M. S. P. (2020). Water in nanopores and biological channels: A molecular simulation perspective. *Chem. Rev.*, 120(18), 10298–10335. https://doi.org/10.1021/acs.chemrev.9b00830.

Ma, Y., Cai, Y., Wang, Z., Sun, M. & Zhao, X. (2021). Visual detection of cells in brain tissue slice for patch clamp system. *In: 2021 IEEE 11th Annual International Conference on CYBER Technology in Automation, Control, and Intelligent Systems* (*CYBER*), pp. 521–526. https://doi.org/10.1109/CYBER53097.2021.9588141.

Maitra, R. D., Kim, J. & Dunbar, W. B. (2012). Recent advances in nanopore sequencing. *Electrophoresis*, 33(23), 3418–3428. https://doi.org/10.1002/elps.201200272.

Marcuccio, F., Soulias, D., Chau, C. C. C., Radford, S. E., Hewitt, E., Actis, P. et al. (2023). Mechanistic study of the conductance and enhanced single-molecule detection in a polymer–electrolyte nanopore. *ACS Nanosci. Au*, 3(2), 172–181. https://doi.org/10.1021/acsnanoscienceau.2c00050.

Mardoum, W. M., Gorczyca, S. M., Regan, K. E., Wu, T.-C. & Robertson-Anderson, R. M. (2018). Crowding induces entropically-driven changes to DNA dynamics that depend on crowder structure and ionic conditions. *Frontiers in Physics*, 6.

Marshall, K., Edmundson, I. C., Kellie, J. L. F. & Wood, W. M. (1968). The Coulter counter. *Proc. Soc. Anal. Chem.*, 5(12), 239–242. https://doi.org/10.1039/SA9680500239.

Martyushenko, N., Bell, N. A. W., Lamboll, R. D. & Keyser, U. F. (2015). Nanopore analysis of amyloid fibrils formed by lysozyme aggregation. *Analyst*, 140(14), 4882–4886. https://doi.org/10.1039/C5AN00530B.

McCabe, I. P. & Forstner, M. B. (2013). Polymer supported lipid bilayers. *OJBIPHY*, 03(01), 59–69. https://doi.org/10.4236/ojbiphy.2013.31A008.

Muthukumar, M., Plesa, C. & Dekker, C. (2015). Single-molecule sensing with nanopores. *Physics Today*, 68(8), 40–46. https://doi.org/10.1063/PT.3.2881.

Nguyen, T. H. T., Lee, J., Kim, H.-Y., Nam, K. M. & Kim, B.-K. (2020). Current research on single-entity electrochemistry for soft nanoparticle detection: Introduction to detection methods and applications. *Biosensors and Bioelectronics*, 151, 111999. https://doi.org/10.1016/j.bios.2019.111999.

Pan, R., Hu, K., Jiang, D., Samuni, U. & Mirkin, M. V. (2019). Electrochemical resistive-pulse sensing. *J Am Chem Soc*, 141(50), 19555–19559. https://doi.org/10.1021/jacs.9b10329.

Pedersen, J. T. & Heegaard, N. H. H. (2013). Analysis of protein aggregation in neurodegenerative disease. *Anal. Chem.*, 85(9), 4215–4227. https://doi.org/10.1021/ac400023c.

Petrossian, L., Wilk, S. J., Joshi, P., Hihath, S., Goodnick, S. M. & Thornton, T. J. (2007). Fabrication of cylindrical nanopores and nanopore arrays in silicon-on-insulator substrates. *Journal of Microelectromechanical Systems*, 16(6), 1419–1428. https://doi.org/10.1109/JMEMS.2007.908435.

Richter-Laskowska, M., Wawrzkiewicz-Jalowiecka, A., Bies, A. & Trybek, P. (2024). Machine learning methods for the analysis of the patch-clamp signals. *In:* Furini, S. (ed.), *Potassium Channels: Methods and Protocols*, Springer US: New York, NY, pp. 249–270. https://doi.org/10.1007/978-1-0716-3818-7_15.

Ross, C. A. & Poirier, M. A. (2004). Protein aggregation and neurodegenerative disease. *Nat Med*, 10(7), S10–S17. https://doi.org/10.1038/nm1066.

Ryan, Y., Harrison, A., Trivett, H., Hartley, C., David, J., Clark, G. C. et al. (2022). RIpore: A novel host-derived method for the identification of ricin intoxication through Oxford nanopore direct RNA sequencing. *Toxins*, 14(7), 470. https://doi.org/10.3390/toxins14070470.

Salehirozveh, M., Kure Larsen, A.-K., Stojmenovic, M., Thei, F. & Dong, M. (2023). In-situ PLL-g-PEG functionalized nanopore for enhancing protein characterization. *Chemistry – An Asian Journal*, 18(17), e202300515. https://doi.org/10.1002/asia.202300515.

Seibertz, F., Rapedius, M., Fakuade, F. E., Tomsits, P., Liutkute, A., Cyganek, L. et al. (2022). A modern automated patch-clamp approach for high throughput electrophysiology recordings in native cardiomyocytes. *Commun Biol*, 5(1), 1–10. https://doi.org/10.1038/s42003-022-03871-2.

Steinbock, L. J. & Radenovic, A. (2015). The emergence of nanopores in next-generation sequencing. *Nanotechnology*, 26(7), 074003. https://doi.org/10.1088/0957-4484/26/7/074003.

Storm, A. J., Chen, J. H., Ling, X. S., Zandbergen, H. W. & Dekker, C. (2003). Fabrication of solid-state nanopores with single-nanometre precision. *Nature Mater*, 2(8), 537–540. https://doi.org/10.1038/nmat941.

Sun, X., Agate, S., Salem, K. S., Lucia, L. & Pal, L. (2021). Hydrogel-based sensor networks: Compositions, properties, and applications—A review. *ACS Appl. Bio Mater.*, 4(1), 140–162. https://doi.org/10.1021/acsabm.0c01011.

Tang, H., Wang, H., Zhao, D., Cao, M., Zhu, Y. & Li, Y. (2022). Nanopore-based single-entity electrochemistry for the label-free monitoring of single-molecule glycoprotein–boronate affinity interaction and its sensing application. *Anal. Chem.*, 94(14), 5715–5722. https://doi.org/10.1021/acs.analchem.2c00860.

Taniguchi, M., Minami, S., Ono, C., Hamajima, R., Morimura, A., Hamaguchi, S. et al (2021). Combining machine learning and nanopore construction creates an artificial intelligence nanopore for coronavirus detection. *Nat Commun*, 12(1), 3726. https://doi.org/10.1038/s41467-021-24001-2.

Tita, D. F. (2012). Fabrication and Characterization of Thinner Solid-State Nanopores.

Trick, J. L., Wallace, E. J., Bayley, H. & Sansom, M. S. P. (2013). Simulations and modelling of biomimetic nanopores. *Biophysical Journal*, 104(2, Supplement 1), 527a. https://doi.org/10.1016/j.bpj.2012.11.2916.

Wan, Y. K., Hendra, C., Pratanwanich, P. N. & Göke, J. (2022). Beyond sequencing: Machine learning algorithms extract biology hidden in nanopore signal data. *Trends in Genetics*, 38(3), 246–257. https://doi.org/10.1016/j.tig.2021.09.001.

Wang, Q., Lin, J., Li, S., Tian, H., Zhang, D. & Xin, Q. (2023). Label-free detection of single living bacteria: Single-entity electrochemistry targeting metabolic products. *Anal. Chem.*, 95(35), 13082–13090. https://doi.org/10.1021/acs.analchem.3c01517.

Wang, S., Zhao, Z., Haque, F. & Guo, P. (2018). Engineering of protein nanopores for sequencing, chemical or protein sensing and disease diagnosis. *Current Opinion in Biotechnology*, 51, 80–89. https://doi.org/10.1016/j.copbio.2017.11.006.

Wang, Y., Zhao, Y., Bollas, A., Wang, Y. & Au, K. F. (2021). Nanopore sequencing technology, bioinformatics and applications. *Nat Biotechnol*, 39(11), 1348–1365. https://doi.org/10.1038/s41587-021-01108-x.

Wei, X., Penkauskas, T., Reiner, J. E., Kennard, C., Uline, M. J., Wang, Q. et al. (2023). Engineering biological nanopore approaches toward protein sequencing. *ACS Nano*, 17(17), 16369–16395. https://doi.org/10.1021/acsnano.3c05628.

White, H. S. & Bund, A. (2008). Ion current rectification at nanopores in glass membranes. *Langmuir*, 24(5), 2212–2218. https://doi.org/10.1021/la702955k.

Xia, F., Jin, W., Yin, X. & Fang, Z. (2005). Single-cell analysis by electrochemical detection with a microfluidic device. *Journal of Chromatography A*, 1063(1), 227–233. https://doi.org/10.1016/j.chroma.2004.12.022.

Xu, W., Zou, G., Hou, H. & Ji, X. (2019). Single particle electrochemistry of collision. *Small*, 15(32), 1804908. https://doi.org/10.1002/smll.201804908.

Yazbeck, R., Alibakhshi, M. A., Schoppe, J. V., Ekinci, K. L. & Duan, C. (2019). Characterization and manipulation of single nanoparticles using a nanopore-based electrokinetic tweezer. *Nanoscale*, 11(47), 22924–22931. https://doi.org/10.1039/C9NR08476B.

Yi, W., Li, X., He, X., Yue, F. & Wang, T. (2022). Glass nanopipette sensing of single entities. *Journal of Electroanalytical Chemistry*, 909, 116106. https://doi.org/10.1016/j.jelechem.2022.116106.

Ying, Y.-L., Hu, Z.-L., Zhang, S., Qing, Y., Fragasso, A., Maglia, G. et al. (2022). Nanopore-based technologies beyond DNA sequencing. *Nat Nanotechnol.*, 17, 1136–1146 https://doi.org/10.1038/s41565-022-01193-2.

Ying, Y.-L., Wang, J., Leach, A. R., Jiang, Y., Gao, R., Xu, C. et al. (2020). Single-entity electrochemistry at confined sensing interfaces. *Sci. China Chem.*, 63(5), 589–618. https://doi.org/10.1007/s11426-020-9716-2.

Yuan, Z., Wang, C., Yi, X., Ni, Z., Chen, Y. & Li, T. (2018). Solid-state nanopore. *Nanoscale Res Lett*, 13(1), 56. https://doi.org/10.1186/s11671-018-2463-z.

Zhang, J., Zhou, J., Pan, R., Jiang, D., Burgess, J. D. & Chen, H.-Y. (2018). New frontiers and challenges for single-cell electrochemical analysis. *ACS Sens.*, 3(2), 242–250. https://doi.org/10.1021/acssensors.7b00711.

Zhang, M., Chen, C., Zhang, Y. & Geng, J. (2022). Biological nanopores for sensing applications. *Proteins: Structure, Function, and Bioinformatics*, 90(10), 1786–1799. https://doi.org/10.1002/prot.26308.

Zhou, K., Li, L., Tan, Z., Zlotnick, A. & Jacobson, S. C. (2011). Characterization of Hepatitis B virus capsids by resistive-pulse sensing. *J. Am. Chem. Soc.*, 133(6), 1618–1621. https://doi.org/10.1021/ja108228x.

Zhou, Y., Wang, D., Li, C., Hu, P. & Jin, Y. (2019). Resistive-pulse sensing and surface charge analysis of a single nanoparticle collision at a conical glass nanopore. *Anal Chem*, 91(12), 7648–7653. https://doi.org/10.1021/acs.analchem.9b00553.

Zhou, Z., Hu, Y., Wang, H., Xu, Z., Wang, W., Bai, X. et al. (2013). DNA translocation through hydrophilic nanopore in hexagonal boron nitride. *Sci Rep*, 3(1), 3287. https://doi.org/10.1038/srep03287.

CHAPTER
10

Identification and Evaluation of Shallow Water Flow-Prone Facies Using Advanced Seismic Interpretation and Machine Learning: A Case Study from the Malay Basin, Malaysia

Dejen Teklu Asfha[1*], Ismailalwali Babikir[1*], Abdul Halim Abdul Latiff[1], Haylay Tsegab Gebretsadik[2], Alidu Rashid[1] and Daniel Asante Otchere[3]

[1] Centre for Subsurface Imaging (CSI), Geoscience Department, Universiti Teknologi PETRONAS, Seri Iskandar 32610, Malaysia

[2] Southeast Asia Clastic and Carbonate Research Laboratory (SEACARL), Geoscience Department, Universiti Teknologi PETRONAS, Seri Iskandar 32610, Malaysia

[3] Institute for Computational and Data Sciences, The Pennsylvania State University, University Park, PA 16802, USA

Introduction

Shallow water flow (SWF) is a geological phenomenon characterized by the unexpected influx of water and sediment into a wellbore during drilling operations. In deep-water basins, SWF is a damaging geohazard that can lead to serious issues when drilling occurs (Purnomo and Ghosh, 2018). This influx typically originates from shallow, overpressured zones located beneath the seabed. There have been reports of SWF in various deep-water basins located 100–1000 meters below the seabed in continental margin environments (Razmi et al., 2023). These zones can develop due to rapid sedimentation, which can trap fluids and increase pore pressure within the sediments (Li et al., 2019). For overpressure to develop, the rate of overburden deposition must be greater than the rate of sediment dewatering (Asfha et al., 2019). The increase in the formation pressure over the normal hydrostatic pressure generally shows a

*Corresponding authors: dejen_22010091@utp.edu.my; ismailalwali@utp.edu.my

DOI: 10.1201/9781003506348-10

tendency to correlate with the rate of sediment deposition in the area (Asfha et al., 2024). Hence, the presence of water can weaken the formation and cause sand grains to detach and be produced along with fluid (Asfha et al., 2024; Safaei et al., 2023). When drilling operations penetrate these overpressured zones, the pressure differential can cause a sudden and forceful flow of water and sediment up the wellbore (Ren et al., 2018). The complexities associated with SWF arise primarily from the pressure differentials between overpressured sediments and the wellbore, which can result in excessive water inflow into the wellbore, potentially leading to well control issues and even the loss of the well itself (Guo et al., 2020). Figure 1 illustrates how a shallow water blowout might happen in deep-water drilling.

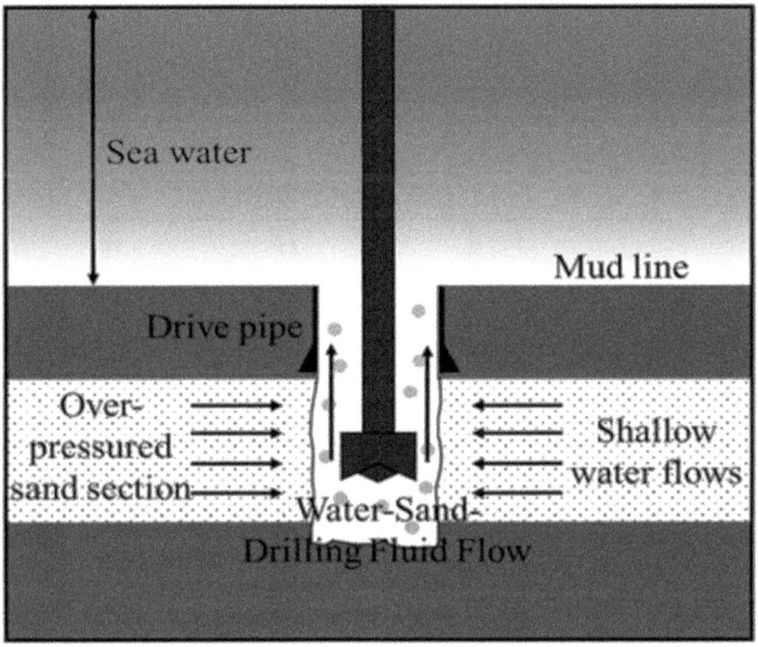

Figure 1: Shallow water flow blowout in deep-water drilling (Modified from Ren et al. (2018).

As illustrated in Figure 2 in the following text, a critical drilling hazard known as SWF can arise in offshore drilling operations when pore pressure, fracture pressure, and overburden pressure converge at shallow depths beneath the mudline. The figure depicts this convergence, where the hydrostatic pressure, pore pressure, and lithostatic pressure gradients approach each other near the seafloor. Consequently, the drilling margin, which represents the safe operating window between pore pressure and fracture gradient, significantly narrows.

The highlighted SWF zone in Figure 2 indicates the interval where the formation is highly susceptible to uncontrolled fluid influx into the wellbore.

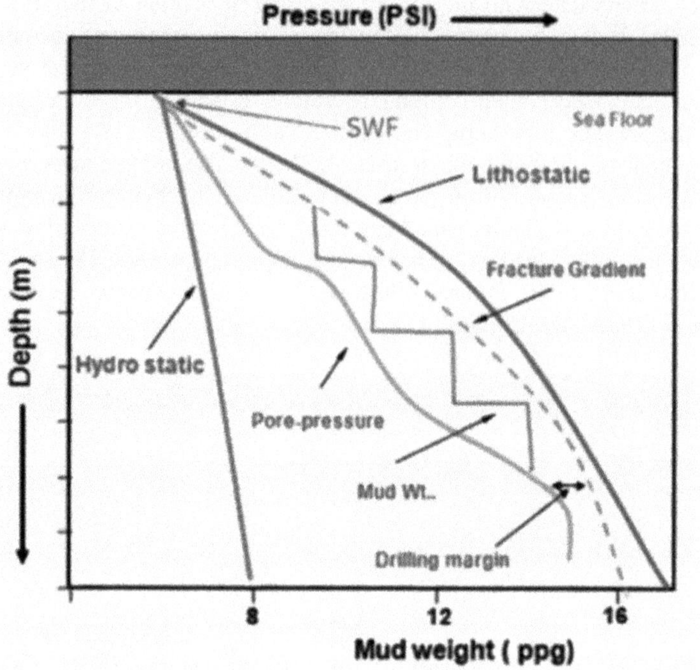

Figure 2: Typical pressure versus depth profile in a deep-water basin. At shallow depths, pores, fractures, and overburden pressures are close to each other, which is a considerably hazardous shallow water flow (Satti, 2016).

This occurs because the reduced effective stress, resulting from the converging pressure gradients, weakens the formation's ability to contain formation fluids. When the mud weight falls below the elevated pore pressure in this shallow zone, a shallow water flow event can be triggered, posing a significant risk to drilling safety and well integrity.

The early detection of SWF-prone zones is critical for ensuring safe and efficient drilling operations. Identifying these hazards before drilling allows operators to design appropriate well plans, including casing programs and mud weights, to mitigate potential risks (Flemings et al., 2002). Detecting SWF-prone zones in advance is critical because failure to do so can trigger costly and dangerous consequences. For instance, the uncontrolled influx of water and sediment can severely compromise wellbore stability, potentially leading to collapse or a loss of the well's structural integrity. Furthermore, these SWF events can exert immense pressure on casing strings, risking deformation or even complete failure. Beyond operational concerns, uncontrolled flows pose significant environmental risks by potentially releasing sediments and fluids into the marine ecosystem, causing detrimental ecological impacts. This underscores the urgency of your work in protecting the environment while conducting drilling operations.

Finally, the necessity for SWF incident remediation often demands substantial time and resources, inevitably resulting in project delays and escalating overall costs.

The ability to predict and map SWF-prone zones early in the exploration phase is therefore essential for minimizing risks and ensuring the success of drilling campaigns. Advanced seismic interpretation techniques, such as seismic attribute analysis, seismic inversion, and amplitude-versus-offset (AVO) analysis, play a pivotal and powerful role in the early detection of these hazards, providing a sophisticated and accurate means of risk assessment (Chopra and Marfurt, 2007; Dutta, 2002). Traditionally, the evaluation of SWF hazards relies on integrating well logs, seismic data, and geotechnical information. Well logs, such as resistivity, sonic, and density logs, provide direct measurements of lithology, porosity, and pore pressure, which are essential for identifying overpressured zones (Flemings et al., 2002). However, in many frontier exploration areas, well logs may be unavailable or limited, leaving 3D seismic data as the primary tool for subsurface evaluation.

3D seismic data offers a powerful means of characterizing SWF-prone facies, even without well logs. Seismic attributes, such as amplitude, coherence, and seismic-inversion-derived properties, can be used to identify key geological features associated with SWF hazards, including high-porosity sands, overpressured zones, and fluid accumulations (Chopra and Marfurt, 2007; Rutherford and Williamst, 1989). Advances in seismic interpretation techniques, such as AVO analysis and geobody extraction, further enhance the ability to delineate SWF-prone zones and assess their spatial distribution (Dutta, 2002).

Seismic attribute analysis has become a cornerstone of modern geophysical interpretation, widely employed for delineating sedimentary features such as channels and sand bodies. Attributes such as amplitude, coherence, and spectral decomposition provide crucial insights into lithological and fluid variations. However, the selection of the most predictive attributes often relies on interpreter experience and can be subjective, leading to potential inconsistencies. Furthermore, in frontier or data-constrained environments, the absence of well logs makes it challenging to validate these interpretations and quantitatively link attributes to specific drilling hazards such as SWF.

This is where the research gap becomes apparent. While seismic attributes are widely used for channel and sand body identification, few studies have systematically integrated attribute ranking with machine learning (ML) classification to address SWF hazards in data-constrained settings. Many existing approaches remain qualitative or depend on well-based calibration, which is not always available.

This study focuses on the evaluation of SWF-prone facies using 3D seismic data in frontier exploration areas where well logs are unavailable. By leveraging advanced seismic interpretation techniques and ML, we aim to identify and map potential SWF hazards, providing critical insights for safe and efficient drilling operations. The results of this not only demonstrate the value of 3D seismic data as a standalone tool for SWF hazard assessment but also underline its potential as a robust workflow that can be applied to other regions with similar challenges.

Geological and Stratigraphic Setting

The Malay Basin is a Tertiary extensional rift basin located offshore Peninsular Malaysia (Figure 3). It was formed during the Late Eocene to Early Oligocene period, and then underwent thermal subsidence and sedimentation during the Early Miocene. Overpressure is generally encountered at shallower depths in the basin center than in the basin flanks because of the convex-upward top-of-overpressure surface (Satti et al., 2016). The onset of overpressure in the basin is structurally controlled, occurring at shallowest depths in the basin center and deepening toward the flanks as shown in Figure 4. This widespread overpressure, when combined with the presence of unconsolidated, highly porous sand bodies at shallow depths, creates ideal conditions for SWF. The sand-rich elements of the meandering fluvial system such as point bars, channel fills, and crevasse splays are particularly prone to this hazard due to their high porosity and permeability, which can facilitate the rapid flow of overpressured water and sand during drilling for hydrocarbons.

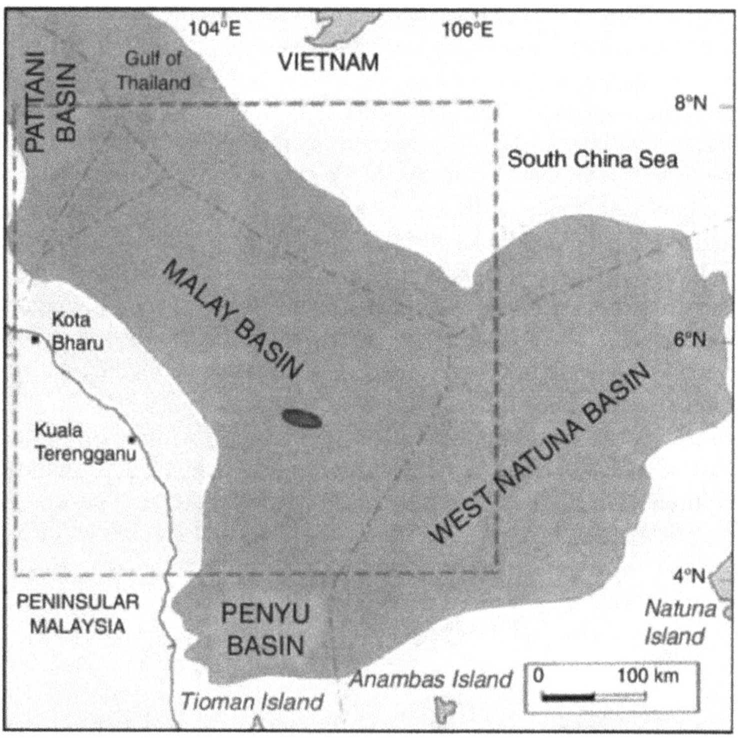

Figure 3: Location map of the Malay Basin (Nazri & Latiff, 2016).

The primary mechanism for overpressure in the basin center is believed to be disequilibrium compaction. High subsidence and sedimentation rates during basin extension cause overpressure development (Madon, 2007). The Malay Basin extends ~500 km northwest (NW) and ~200 km wide (Madon et al., 2004; Tjia, 1998). Its northern part shifts northward, aligning with structures in the Gulf of Thailand. The basin is filled with ~14 km of sediments from the pre-upper Oligocene onward, concealing any surface depression (Tjia, 1998). It formed in the Early Tertiary due to crustal extension from strike-slip movement along an NW-trending shear zone (Madon, 1997). A basement saddle from Bintang to Bergading separates it from a smaller sub-basin in the Malaysia–Thailand Joint Development Area (JDA) (Madon et al., 2004). The western and southeastern basement consists mainly of Cretaceous granitoids, Jurassic-Cretaceous sediments, and older metasediments linked to Peninsular Malaysia. Sedimentation began in continental settings, with a brief marine incursion in the late Early Miocene, followed by dominant marine conditions since the Pliocene (Tjia, 1998). Tectonic evolution includes a pre-Late Oligocene transtensional phase, quiet subsidence until the Middle Miocene, and subsequent compression and folding due to regional transpression until the Late Miocene. Later, epeirogenic subsidence led to cross faults in anticlines (Ramli, 1988; San, 1987). The initial transpressive phase was linked to the India–Asia collision (Tapponnier et al., 1982), while left-lateral movement along basement faults formed en echelon grabens (Tjia, 1994).

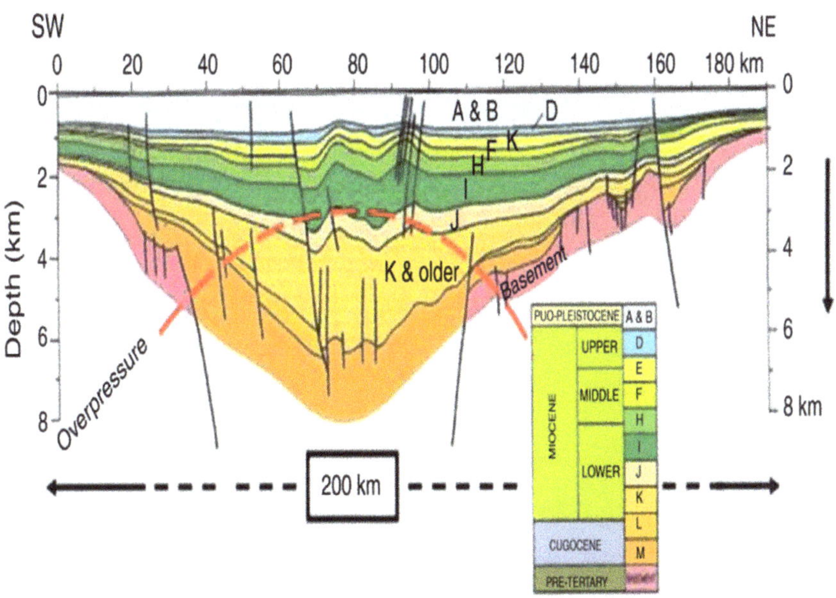

Figure 4: The onset of overpressure in the Malay Basin. The depth to the start of overpressure is shallowest in the basin center and deepens toward the basin flanks (Satti et al., 2016).

Middle to Late Miocene tectonic inversion was driven by changes in regional plate movements, including halted South China Basin expansion and Australian Plate advancement. This inversion caused right-lateral displacement of axial basement faults, forming anticlines and uplifting southern basin rocks (Tjia, 1998). The basin comprises Oligocene-to-Recent period continental to shallow marine sediments, stratigraphically classified from older Group M to younger Group A as shown in Figure 5 in the following text. It hosts two petroleum systems: Oligocene–Miocene lacustrine shales and Middle Miocene coal and carbonaceous shales (Madon et al., 1999).

The Malay Basin's stratigraphy is rich in palynomorphs, while foraminifera and nannofossils appear mainly during marine transgressions. Nine peak abundance periods ("TR" zones) mark these intervals. Two key unconformities

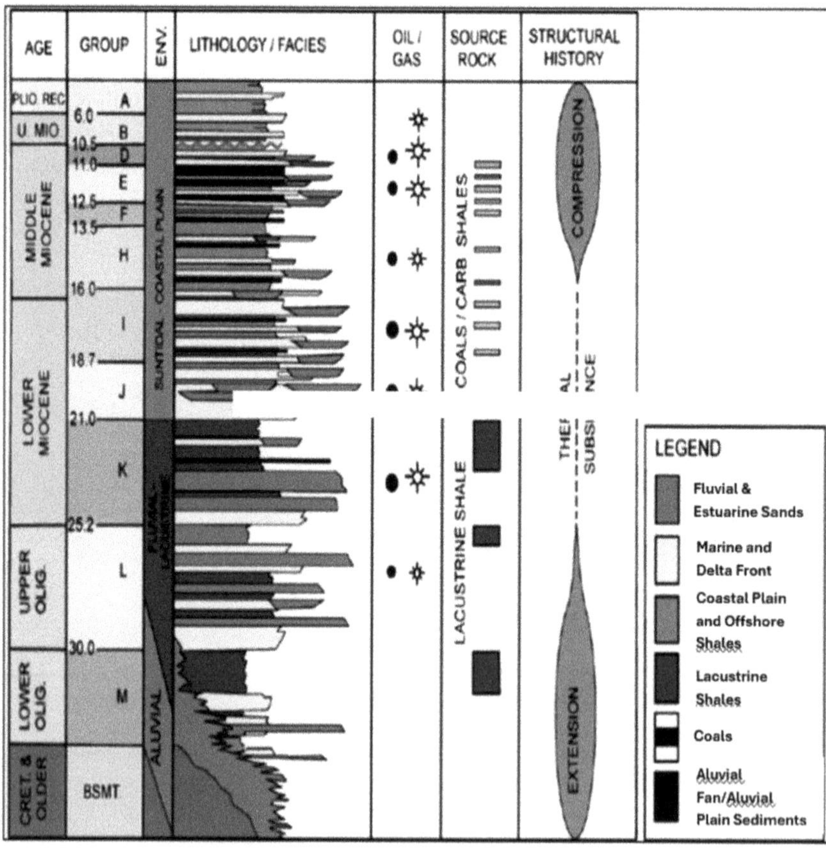

Figure 5: Stratigraphy of Malay Basin from Group A to M showing the dominant lithofacies in fluvial, estuary sand and marine with delta front sands. The right-most column is displaying the major tectonic events that have affected this basin (Modified after Jamaludin et al., 2019).

exist: the Middle-Miocene unconformity (10.0–10.5 Ma), linked to sea-level drop, and a 7.8 Ma unconformity, associated with tectonic inversion (Yakzan et al., 1996). According to Bishop (2002), the offshore Malay Basin is a Tertiary oil and gas province with half-grabens filled by lacustrine shales and continental clastics. South China Sea transgressions shaped its evolution, forming the Gulf of Thailand. Hydrocarbons from Oligocene–Miocene shales and Miocene coals are mainly trapped in late Miocene anticlines from half-graben inversion. The Northern Malay Basin has a central gas-rich sector, flanked to the sides and south by mixed oil and gas zones. Non-associated gas in the central zone is mainly in Groups D and E reservoirs within anticlinal traps formed by late Miocene basin inversion (Madon et al., 2004). Total Organic Carbon (TOC) in lacustrine shales of the southern Malay Basin ranges from 0.35 to 2.00 wt%, with kerogen types I–III. Group L shales are the most oil-prone, with TOC of 0.45–1.95 wt% and Hydrogen Index (HI) of 300–400, indicating Type II kerogens. Variations in source rock quality among Groups K, L, and M are likely due to organic input and preservation factors (Muhamad and Jamil, 2010). The Malay Basin's sedimentary layers, dating from the Eocene to Lower Oligocene, comprise fluvial–lacustrine deposits with some marine influence. Predominantly siltstone, the sequence includes fine-grained sand and volcanic tuff. Correlation and seismic data suggest Eocene sediments extend into unexplored areas, indicating basin formation by the mid-Eocene (Kessler et al., 2020). Inter-distributary bay/offshore, tidal flat, subaqueous delta platform/prodelta, tide-influenced channel/bar, mouth bar, and mangrove/overbank deposits associations were found in the Malay basin. Evidence of tidal action includes rhythmites, thick heterolithic successions, mud drapes, bidirectional current indicators, and an impoverished marine trace fossil assemblage. Three-dimensional seismic data show tidal influence through flaring channel mouths, cuspate meanders, high sinuosity, and dendritic to parallel channel networks with dead-end and through-flowing tributaries (Manshor et al., 2022). Five depositional environments based on log, biostratigraphy, and core data were identified in the basin, and these include in-distributary channel, delta plain (flood plain shale), crevasse splay, delta front shale (prodelta), and mouth bar (Kamilah et al., 2022). Five lithofacies were identified in the Group-B shallow clastic reservoir of the Malay Basin: laminated mudstone (LM), bioturbated mudstone (MB), heterolithic mud-dominated (MH), heterolithic sand-dominated (SH), and bioturbated sandstone (SB). LM and SB exhibit moderate to good reservoir quality, while SH and MH range from high to low. Reservoir quality declines with increased clay-rich sandstone, thin bedding, and heterolithic intervals. Bioturbation worsens quality by redistributing clay and reducing pore space. Trace fossils indicate a fully marine depositional environment (Bashiron et al., 2023).

Study Area

The SP field is approximately 200 km offshore from Kuala Terengganu, off the coast of Peninsular Malaysia. It was first discovered in 1970 through the SP-1 well (Mohamad and Mohamed, 2003). The field was later appraised in 1998 by drilling the SP-2 well, which was about 100–200 meters downdip of SP-1. SP-2 reached a total depth of 1,781 meters, confirming the gas extension in the shallower reservoirs, while detecting water in the deeper sand layers (Mohamad and Mohamed, 2003). This highly productive gas field is located on an east–west trending anticline with a four-way dip closure. It was formed due to transgressional movements of northwest-trending dextral shear along basement faults in the Malay Basin (Jamaludin et al., 2019). Exploration efforts in this field primarily target the Lower to Middle Miocene units, specifically Group J, I, and H strata (Jamaludin et al., 2019).

The Shallow Fluvial System of the Malay Basin

A basin-wide, near-surface fluvial system was dominant in the Malay Basin during the late Pleistocene time. High-resolution seismic data covering extensive areas of the basin reveal several fluvial depositional elements, including straight channels, meandering channels, point bars, crevasse splays, and levees (Al-Masgari et al., 2021; Alqahtani et al., 2015; Babikir et al., 2022; Miall, 2002). Seismic attributes have been widely utilized to analyze and predict sand distribution and associated facies within these channels, with high-amplitude responses indicating sand-rich facies and low-amplitude responses corresponding to muddy facies (Babikir et al., 2022; Babikir et al., 2019b; Ghosh et al., 2014)

Two types of channel configurations exist in the Malay Basin: low sinusoidal channels filled with clay or silt, and high sinusoidal channels filled with sand in the form of point bars, which have potential as hydrocarbon reservoirs. Several channels have been identified and characterized (Ghosh et al., 2014).

Data and Method

Predicting the potential for shallow water flow requires a solid understanding of the depositional environments that are conducive to sand deposits. Once these depositional environments are clearly defined, the next step is to estimate the presence of overpressure. In other words, predicting overpressured sand involves two distinct steps: forecasting the shallow sands and assessing the pressure that may exist within those sand bodies.

The presence of sand can be most effectively predicted by utilizing seismic attributes and geomorphological concepts, assuming no offset wells nearby. Various techniques are available for identifying sand deposits. The best practice is to employ as many additional techniques as possible for the proposed location.

We implemented the following workflow (Figure 6) to identify sand-prone geomorphologies and facies based on seismic attributes and our knowledge of their representation in the study area and relevant interval. Furthermore, we incorporated machine learning to improve the prediction of sand within the interval.

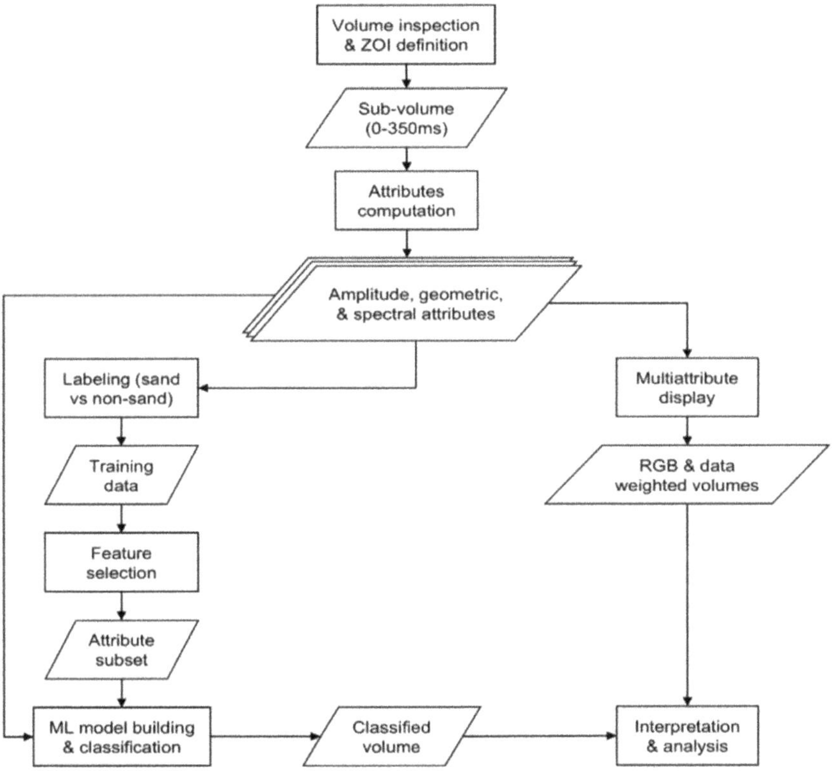

Figure 6: Workflow of the study.

Volume Inspection and Zone of Interest Definition

By applying dense time slicing, we systematically explored and analyzed the seismic volume to identify the near-surface interval dominated by the fluvial system and channelized features. We then cropped the volume to a Two-way traveltime (TWT) interval of interest ranging from 0 to 350 ms (Figure 7).

Seismic Attributes

Seismic attributes serve as a powerful tool for enhancing subsurface interpretation, identifying geological features, predicting lithology and facies distribution, and improving reservoir characterization. In this study, we computed 30 seismic attributes that are particularly sensitive to lithologic variations and valuable for identifying channels and sand deposits (Table 1). Volume-based seismic attributes provide insights into the bulk distribution of sands within a channelized unit (McConnell, 2000), while time slices of 3D seismic data reveal patterns indicative of channels and channel complexes.

Table 1: Applied attributes names, abbreviations, and types

#	Attribute Name	Abbreviation	Type
1	Chaos Attribute	chaos	Geometric
2	Instantaneous Frequency	inst_freq	Spectral
3	Instantaneous Phase	inst_phase	Geometric
4	Original Amplitude	origin_ampl	Amplitude
5	Relative Acoustic Impedance	rai	Impedance
6	Reflection Intensity	refl_intens	Amplitude
7	Root Mean Square Amplitude	rms_ampl	Amplitude
8	Spectral Magnitude Component at 5 Hz	smc_5hz	Spectral
9	Spectral Magnitude Component at 10 Hz	smc_10hz	Spectral
10	Spectral Magnitude Component at 15 Hz	smc_15hz	Spectral
11	Spectral Magnitude Component at 20 Hz	smc_20hz	Spectral
12	Spectral Magnitude Component at 25 Hz	smc_25hz	Spectral
13	Spectral Magnitude Component at 30 Hz	smc_30hz	Spectral
14	Spectral Magnitude Component at 35 Hz	smc_35hz	Spectral
15	Spectral Magnitude Component at 40 Hz	smc_40hz	Spectral
16	Spectral Magnitude Component at 45 Hz	smc_45hz	Spectral
17	Spectral Magnitude Component at 50 Hz	smc_50hz	Spectral
18	Spectral Magnitude Component at 55 Hz	smc_55hz	Spectral
19	Spectral Magnitude Component at 60 Hz	smc_60hz	Spectral
20	Spectral Magnitude Component at 65 Hz	smc_65hz	Spectral
21	Spectral Magnitude Component at 70 Hz	smc_70hz	Spectral
22	Spectral Magnitude Component at 75 Hz	smc_75hz	Spectral
23	Spectral Magnitude Component at 80 Hz	smc_80hz	Spectral
24	Spectral Magnitude Component at 85 Hz	smc_85hz	Spectral
25	Spectral Magnitude Component at 90 Hz	smc_90hz	Spectral
26	Spectral Magnitude Component at 95 Hz	smc_95hz	Spectral
27	Spectral Magnitude Component at 100 Hz	smc_100hz	Spectral
28	Structural Smoothing	struct_smooth	Geometric
29	Sweetness Attribute	sweet	Amplitude
30	Variance Attribute	var	Geometric

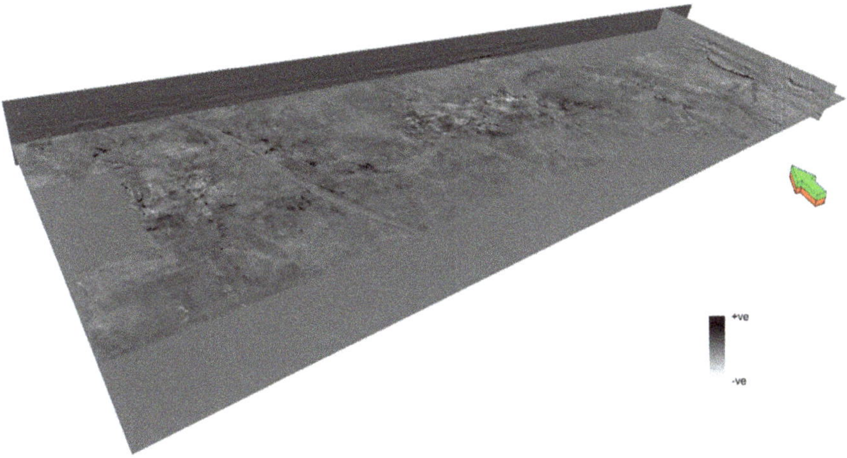

Figure 7: Volume cropped to the shallow section (0–350 ms).

Amplitude Attributes: Amplitude attributes, such as RMS amplitude and sweetness, are commonly used to identify lithologic variations and hydrocarbon accumulations (Koson et al., 2013). In young sedimentary basins such as the Malay Basin, sand-prone depositional features within fluvial systems, such as point bars, are typically characterized by high-amplitude anomalies on amplitude attribute maps.

Geometric Attributes: Geometric attributes such as variance and curvature are essential for delineating faults and channel edges (Pigott et al., 2013). For example, the variance attribute measures the similarity between waveforms or traces in 2D or 3D seismic volumes. It is particularly useful for highlighting discontinuities, such as faults, and stratigraphic boundaries when viewed in map view or time slices.

Spectral Attributes: Spectral seismic attributes are derived from the frequency domain of seismic data and provide insights into subsurface features by analyzing the distribution of energy across different frequency bands. These attributes enhance the detection of subtle geological features such as fractures, faults, and lithological boundaries, which might not be easily visible in the time domain. By highlighting variations in frequency content, spectral attributes can aid in characterizing reservoir properties, improving fault interpretation, and identifying hydrocarbon-bearing zones.

Multi-Attribute Display

Multi-attribute display techniques, such as co-rendering, enhance visualization by blending multiple attribute volumes, adding dimensionality to the interpretation, and revealing geological patterns that may not be evident from a single attribute (Babikir et al., 2019a; Marfurt, 2015). In this study, we apply alpha-weighted and red–green–blue (RGB) blending to the extracted attributes to improve data interpretability.

Spectral decomposition in combination with variance can be used to identify and analyze channels accurately (Li et al., 2015). Variance illuminates the channel edges, while spectral decomposition represents the channel thickness. Moreover, these attributes provide information about channel body continuity and fill variability (Koson et al., 2013).

To develop a depositional model for the interval of interest at the proposed location, seismic data was used. Once a depositional model has been established, analogs can be used to identify the most likely formations for sand deposits. The identification of SWF potential from seismic data is mainly based on both the evaluation of seismic reflector character (amplitude, architecture, other seismic attributes, etc.) and the recognition of depositional facies (channels, slumps, debris flows, etc.) (Dutta et al., 2010; Lu et al., 2005; Ostermeier et al., 2000; Wu et al., 2018; Zhang et al., 2018).

Machine Learning

ML classification is a valuable tool in seismic interpretation and reservoir characterization, enabling the prediction of facies, lithology, and other rock properties with improved efficiency and accuracy. By leveraging data-driven algorithms, ML models can identify patterns and relationships within seismic and well data that may not be easily captured through traditional methods. In this study, we developed and trained multiple ML models to predict and classify sand deposits within the targeted interval, aiming to enhance subsurface characterization and improve evaluation. A binary classification was performed to classify sand from non-sand using selected attributes as input.

Labeling: Supervised ML relies on training and testing models using a subset of the data. To ensure good model accuracy, manual facies classification or picking is performed on a representative number of samples to generate training and testing datasets. Based on our geomorphological interpretation and study objectives, we define two classes: sand-prone and non-sand-prone.

Data Labeling and Training Set Creation: The supervised machine learning workflow required a robustly labeled dataset for model training and validation. Because no well log control exists at the target shallow interval within the study area, the labeling strategy relied on established seismic geomorphological principles, supported by analogs from neighboring fields and by the characteristic seismic responses of fluvial depositional systems.

- **Geological Rationale for Labels:** In fluvial settings, sand-rich, porous facies such as channel fills and point bars typically display strong reflection amplitudes owing to their high acoustic impedance contrast with surrounding fine-grained floodplain deposits. Conversely, mud-dominated facies (e.g., overbank and floodplain deposits) are characterized by weak reflections or seismically transparent zones due to their lower impedance contrast. This amplitude–facies relationship has been consistently observed in analogous shallow stratigraphic intervals and forms the geological foundation for the present labeling approach.

- **Labeling Implementation:** Following this rationale, a single representative time slice at 330 ms was selected because it clearly depicts the geomorphology of a meandering channel system. Two primary facies classes were defined based on their seismic expression (Figures 8 and 9):
  - Class 0—Sand-Prone Facies: High-amplitude, channelized, and laterally continuous reflections interpreted as sand-filled channels and point-bar deposits.
  - Class 1—Non-Sand (Mud-Prone) Facies: Low-amplitude or seismically transparent areas representing floodplain or overbank deposits.
- **Dataset Construction and Size:** A total of 22,000 sample points were systematically selected: 11,000 points from Class 0 and 11,000 points from Class 1. This ensured a perfectly balanced training dataset, mitigating potential model bias toward a majority class.

Figure 8: RMS amplitude time slice at 330 ms showing the geomorphic expression of meandering channels and associated floodplain areas within the study interval.

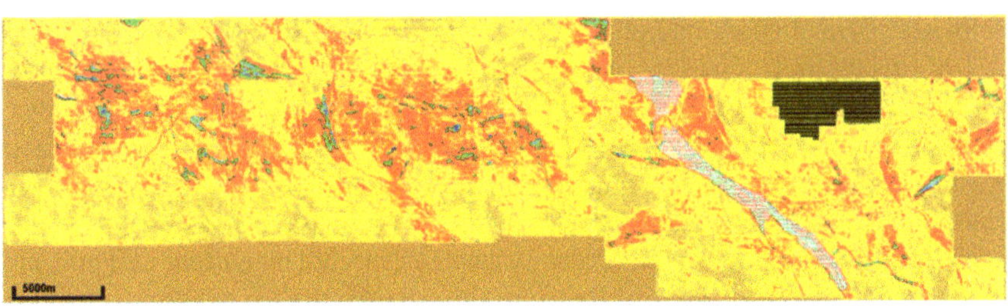

Figure 9: Corresponding labeled time slice at 330 ms. White-hatched regions denote sand-prone facies (Class 0); black-hatched regions denote non-sand facies (Class 1). A total of 22,000 sample points (11,000 per class) were extracted from these masks for model training and validation.

- **Ensuring Labeling Consistency (Inter-Annotator Consistency):** The entire labeling process was performed by a single, experienced interpreter. This approach of using a single annotator is a recognized methodology to eliminate inter-annotator variability entirely, thereby ensuring absolute consistency in the labeling criteria across all 22,000 data points. The labels were derived directly from the clear and unambiguous geomorphological features on the seismic slice, as described in the earlier text, forming a consistent and reliable ground truth dataset for model development.

Feature Selection: Attribute selection, also referred to as variable or feature selection, involves identifying the most relevant subset of attributes for machine learning predictions. This process aims to optimize model performance by reducing computational time, minimizing overfitting, and enhancing overall accuracy (Babikir et al., 2024; Cai et al., 2018; Jiliang and Salem, 2014). We applied the feature importance analysis of random forest to rank attributes based on their contribution to the classification.

ML Model Building and Application: In this study, five commonly used machine learning algorithms, including support vector machine, decision tree, random forest, neural network, and logistic regression, were implemented for lithology classification. These algorithms were selected due to their established effectiveness and ability to handle the complexities of seismic data. Their performance was assessed using selected subsets of attributes.

Results and Interpretation

Figure 10 presents single-attribute displays highlighting the varying expressions of channels and fluvial features within the interval. In general, sand-rich fluvial features, such as sand-filled channels, point bars, scroll bars, and crevasse splays, exhibit strong reflections on amplitude attributes. Figure 13 presents the sand-prone fluvial facies in the zone of interest of the study area. In contrast, muddy facies, including floodplains and mud-filled channels, produce weak amplitude responses. Additionally, geometric attributes such as variance effectively delineate channel edges and other stratigraphic boundaries.

Multi-attribute displays using data-weighted and RGB color blending techniques are used to enhance visualization and improve the interpretability of the features of interest. Figure 11 combines RMS amplitude and variance attributes through data-weighting, aiming to integrate channel edges and other feature boundaries with amplitude expressions in a single display. On the other hand, the RGB blend in Figure 12 combines three spectral magnitude components at 20, 35, and 50 Hz. Spectral decomposition interpretation is often enhanced using an RGB color blend, where high-, medium-, and low-frequency spectral magnitude components are co-visualized.

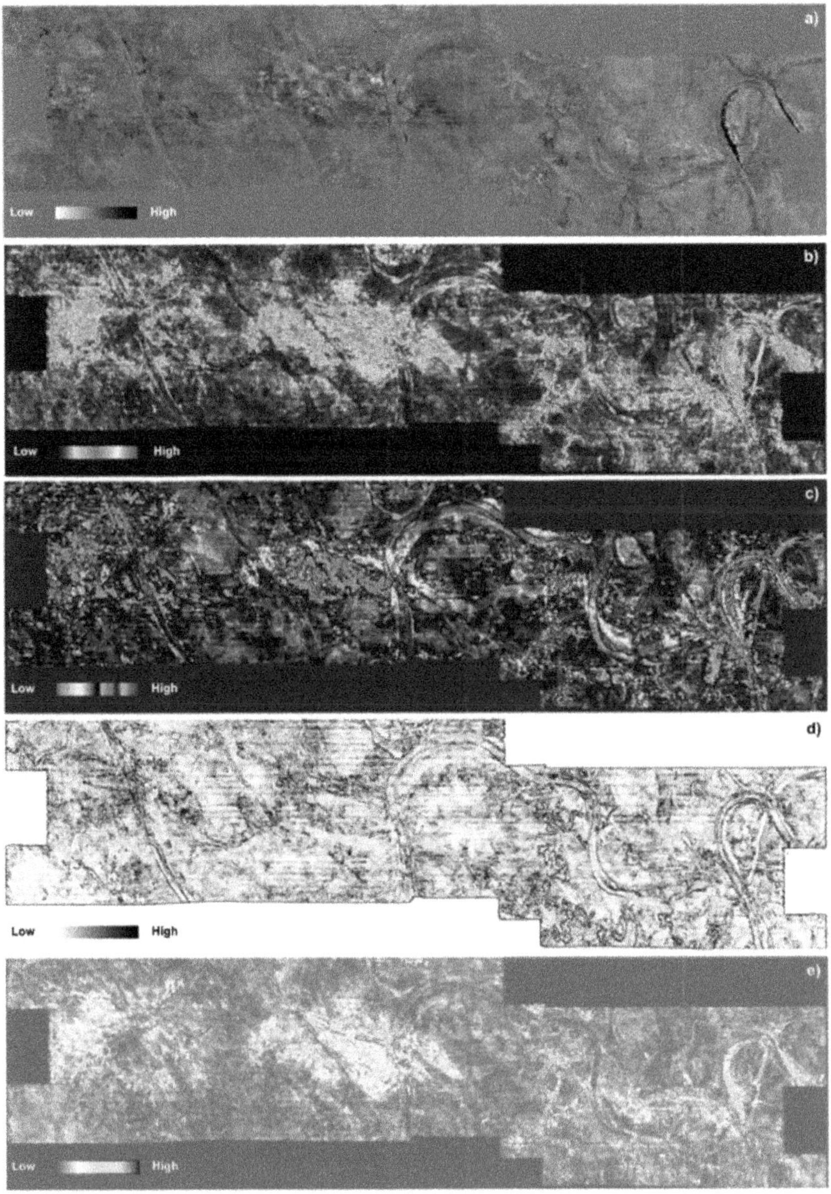

Figure 10: Single attribute display at 231 ms of (a) original seismic, (b) RMS amplitude, (c) relative acoustic impedance, (d) variance, and (e) 35 Hz spectral component (time slice = 231 ms).

Figure 11: The RMS amplitude is co-rendered with the variance attribute to enhance visualization and improve the interpretability of the channel system (time slice = 231 ms).

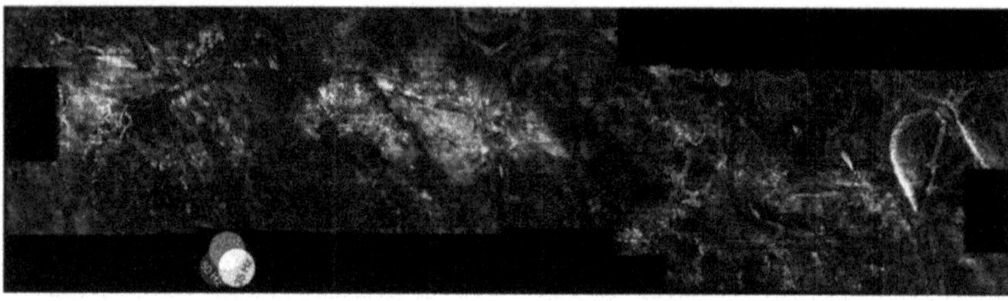

Figure 12: RGB color blend of spectral magnitude components with R = 20 Hz, G = 35 Hz, and B = 50 Hz (time slice = 231 ms).

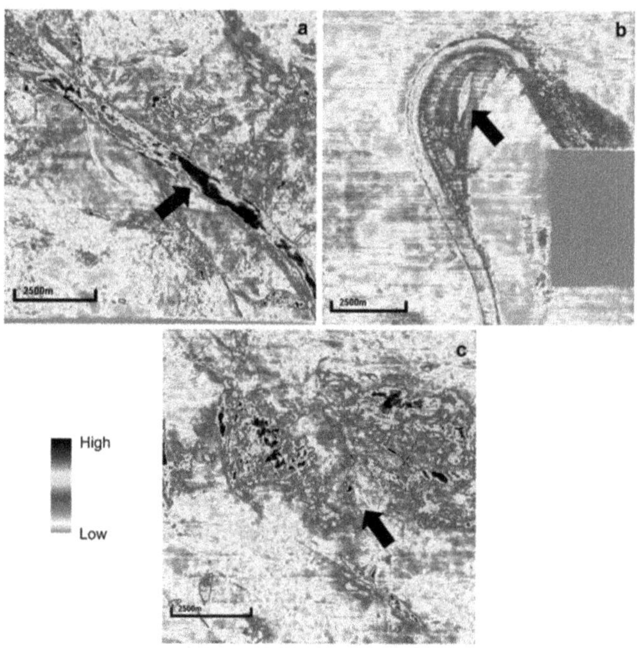

Figure 13: Sand-prone fluvial facies. (a) sand-filled channel, (b) point bar, and (c) crevasse splay.

The selection of the most important seismic attributes from a set of 30 was performed using the feature importance ranking of random forest. Figure 14 presents the attribute ranking, highlighting RMS amplitude, reflection strength, and several spectral magnitude components as the top-ranked features. This is expected, as the differentiation between sand and mud is reflected in their amplitude responses—high for sand and low for mud.

A reliable prediction of sand deposits was achieved using machine learning and validated through attribute analysis and seismic geomorphology. As this shallow interval lacks logging data, the classification results align well with our understanding of attribute expressions and the area's seismic geomorphology. Figure 15 illustrates the classification results, where channel sands, point bars, and crevasse splays are correctly identified as sand-prone features.

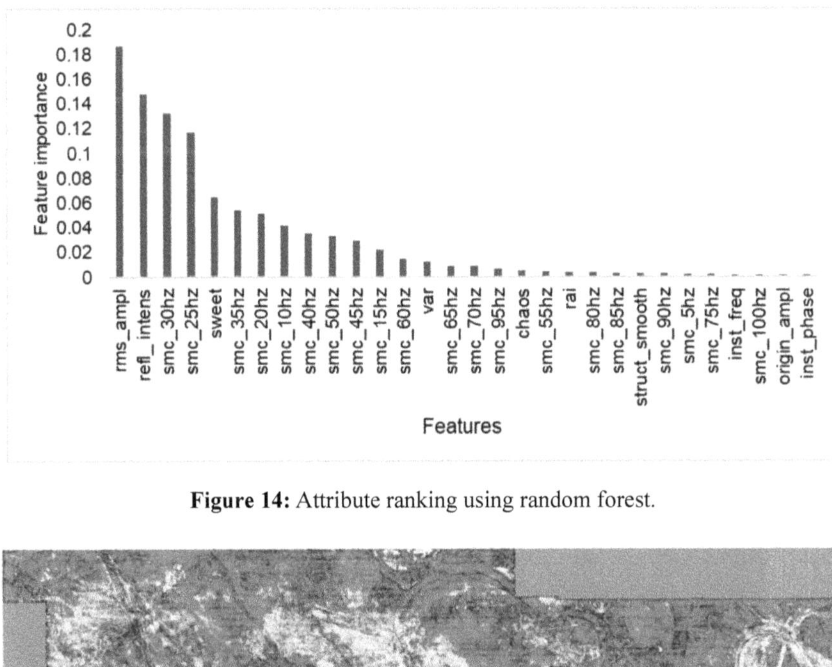

Figure 14: Attribute ranking using random forest.

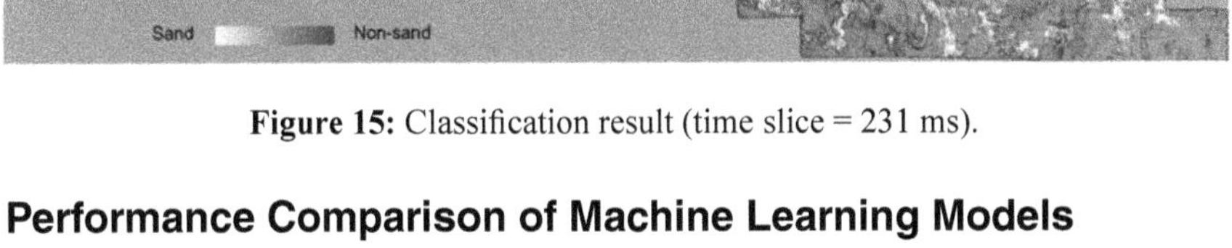

Figure 15: Classification result (time slice = 231 ms).

Performance Comparison of Machine Learning Models

To comprehensively evaluate the most effective algorithm for differentiating between sand and non-sand facies, nine supervised machine learning models were trained and tested. A balanced dataset was constructed to minimize class imbalance, comprising 11,000 training samples for each class (sand and non-sand), resulting in a total of 22,000 labeled samples for model development. The predictive performance of each model was assessed using standard evaluation metrics, including accuracy, precision, recall, F1-score, and computational wall time, as summarized in Table 2. All models demonstrated high classification accuracy (≥98%), reflecting the strong separability of sand and non-sand facies in the seismic attribute space. However, the random forest model exhibited the best overall performance, achieving accuracy and F1-score of 99.65%, followed closely by the neural network and AdaBoost classifiers. The ensemble structure of random forest effectively captured nonlinear attribute interactions and provided stable generalization while minimizing overfitting. Figure 16 graphically compares the performance of all models, where random forest's superior accuracy and robustness are clearly visible despite its higher computational cost. This justifies its selection as the optimal model for sand-prone facies classification and subsequent shallow water flow hazard prediction.

Overall Performance and Top Contenders

The results demonstrate that all models achieved exceptionally high performance, with accuracy exceeding 98%, thereby affirming the robustness of the selected

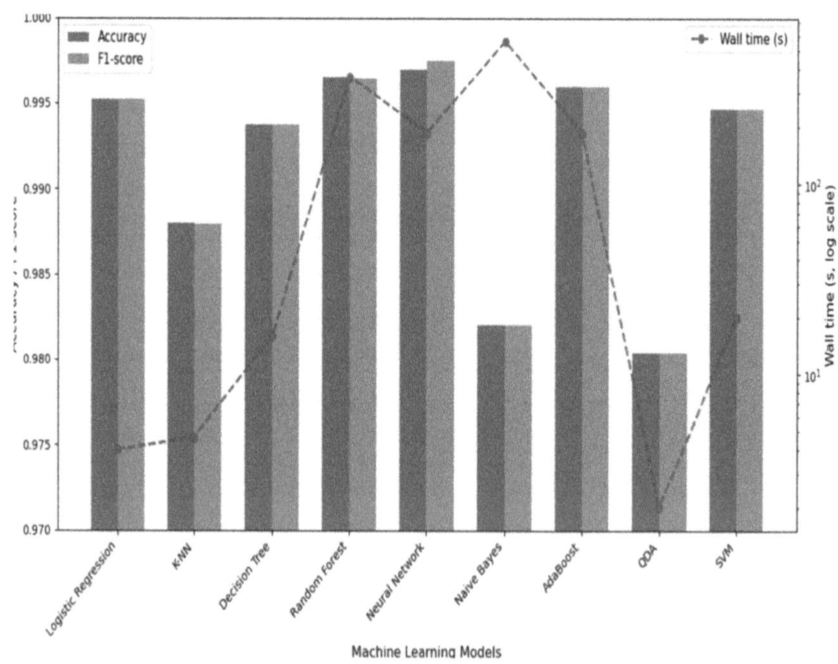

Figure 16: Comparison of machine learning model performance metrics.

Table 2: The performance of nine models/classifiers we trained for the work

Model	Metrics	
Logistic Regression	Accuracy	0.995
	Precision	0.995
	Recall	0.995
	F1-score	0.995
	Wall time (s)	4
K-Nearest Neighbors (K-NN)	Accuracy	0.988
	Precision	0.988
	Recall	0.988
	F1-score	0.988
	Wall time (s)	4.64
Decision Tree	Accuracy	0.994
	Precision	0.994
	Recall	0.994
	F1-score	0.994
	Wall time (s)	15.8
Random Forest	Accuracy	0.997
	Precision	0.996
	Recall	0.997
	F1-score	0.996
	Wall time (s)	363
Neural Network	Accuracy	0.997
	Precision	0.998
	Recall	0.997
	F1-score	0.997
	Wall time (s)	184
Naïve Bayes	Accuracy	0.982
	Precision	0.982
	Recall	0.982
	F1-score	0.982
	CPU time (s)	469
	Wall time (s)	559
AdaBoost	Accuracy	0.996
	Precision	0.996
	Recall	0.996

(Contd.)

	F1-score	0.996
	Wall time (s)	184
Quadratic Discriminant Analysis	Accuracy	0.980
	Precision	0.981
	Recall	0.980
	F1-score	0.980
	Wall time (s)	2
Support Vector Machine	Accuracy	0.995
	Precision	0.995
	Recall	0.995
	F1-score	0.995
	Wall time (s)	20

seismic attributes for classification and validating the efficacy of employing a balanced dataset. The neural network exhibited superior predictive accuracy, attaining the highest scores across all metrics: accuracy (0.997), precision (0.998), recall (0.997), and F1-score (0.997). This superior performance can be attributed to the model's inherent capacity to discern complex, nonlinear decision boundaries within the multi-attribute seismic data. The random forest model closely followed, notably achieving the highest recall (0.997) among all models. This indicates an exceptional capability to minimize false negatives, which is a critical consideration in hazard assessment where the accurate identification of all potential SWF-prone sand facies is paramount. AdaBoost also demonstrated robust and balanced performance, securing third place with consistently high metrics across all evaluated parameters.

Analysis of Other High-Performing Models

Models such as logistic regression and support vector machine delivered excellent results with high accuracy and F1-scores (>0.994). Their strong performance, despite being relatively simpler linear and kernel-based models, suggests that the underlying classification problem, while complex, can be approximated effectively by these well-established algorithms. The decision tree model performed commendably but was consistently outperformed by its ensemble counterpart, random forest, which leverages multiple trees to reduce overfitting and variance, demonstrating the power of ensemble methods.

Consideration of Computational Efficiency

A critical practical consideration pertains to computational cost, specifically quantified as wall time. Logistic regression and quadratic discriminant analysis (QDA) demonstrated superior efficiency, completing both training and prediction phases within a few seconds (4 s and 2 s, respectively). This renders them highly advantageous for rapid prototyping or in scenarios where computational resources are constrained. Conversely, the models exhibiting the highest accuracy, namely, random forest and neural network, necessitated substantially greater processing times (363 s and 184 s, respectively). This inherent trade-off between optimal performance and computational expenditure represents a pivotal factor in the selection of models for operational deployment.

Performance of Baseline Models

The K-NN and naïve Bayes classifiers served as solid baseline models, achieving accuracies of 0.988 and 0.982, respectively. While their performance was slightly lower than the top-tier models, it confirms the general separability of the classes within the feature space. QDA showed similar, competent baseline performance. It is noteworthy that naïve Bayes was unexpectedly slow, which may be due to the implementation and the high dimensionality of the input features.

Discussion and Model Selection Rationale

The choice of a final model is a trade-off between predictive power, computational efficiency, and interpretability, guided by the project's specific objectives.

- **Maximum Predictive Accuracy:** If the primary goal is to create the most reliable SWF hazard map with the highest possible confidence, the neural network is the optimal choice.
- **Balance of Performance and Insight:** If a balance between top-tier performance, computational efficiency, and model interpretability is desired, random forest or AdaBoost are excellent alternatives. Random forest, in particular, provides feature importance scores that can offer valuable geological insights into which seismic attributes contribute most to the classification, a key advantage for geoscientific interpretation.
- **For Operational Efficiency and Speed:** For scenarios requiring rapid results with minimal computational footprint while still maintaining very high accuracy (e.g., for initial screening or when computational resources are constrained), logistic regression presents an outstanding option. Its performance is remarkably competitive, and its speed is unparalleled among the top performers.

In conclusion, while the neural network achieved the highest metric scores, the marginal difference between the top three models (NN, RF, AdaBoost) suggests that the ensemble methods provide a very compelling and potentially more efficient solution for the task of SWF-prone facies identification in the Malay Basin. The successful application of all models, facilitated by the balanced dataset, confirms the efficacy of integrating advanced seismic interpretation with machine learning for robust facies classification.

Conclusions and Recommendations

The characterization of a meandering fluvial system in the Malay Basin, indicative of a gentle slope and slow depositional energy, reveals sand-rich elements such as point bars and crevasse splays as potential sources of SWF. This study successfully established a quantitative, data-driven workflow for identifying these SWF-prone facies in a data-constrained environment by integrating advanced seismic interpretation with ML. The key findings, supported by robust validation metrics, are as follows:

1. **Effective Seismic Attribute Delineation:** Single-attribute displays (e.g., amplitude, variance) and multi-attribute visualization techniques (e.g., RGB blending) effectively highlighted the geomorphological expression of fluvial features, forming the qualitative foundation for ML-based classification.
2. **Quantitative Feature Importance:** The random forest model provided a quantitative ranking of seismic attributes, which was critical for refining the classification process. The analysis revealed that RMS amplitude (~19%) and reflection strength (~15%) were the most dominant predictors, consistent with their strong sensitivity to acoustic impedance contrasts associated with sand-prone facies. Additionally, spectral magnitude attributes in the 20–40 Hz range contributed cumulatively ~45% of the model importance, underscoring the significance of frequency-dependent seismic responses for differentiating lithologies. Other attributes, such as instantaneous frequency, variance, and structural smoothness, exhibited minor but supportive influence (<10%), collectively enhancing model robustness. These results reaffirm that amplitude-driven and mid-frequency spectral attributes are the primary seismic indicators of sand-prone facies in the Malay Basin.
3. **Validated High-Accuracy Classification:** The ML models achieved exceptional performance in binary sand versus non-sand classification. The random forest classifier demonstrated 99.65% accuracy and a 99.65% F1-score, with the neural network and AdaBoost models also exceeding 99.6% accuracy. This high level of performance was validated through seismic geomorphology, confirming the models' ability to accurately map SWF-prone features such as channel sands and point bars without reliance on well logs.
4. **Practical Implications for Risk Mitigation:** The resulting classification maps provide a critical pre-drill risk assessment tool. By precisely mapping sand-prone facies, operators can proactively design safe well paths, optimize casing setting depths, and implement controlled drilling strategies to mitigate the risks posed by overpressured, unconsolidated sands.

To transition this research from a successful case study to a scalable operational tool, future work should focus on:

- **Real-Time Application:** The workflow should be adapted for real-time or near-real-time use during drilling operations. By processing newly acquired seismic volumes or look-ahead seismic data, the model could update hazard maps ahead of the bit, enabling dynamic decision-making.

- **Scalability and Generalization:** The methodology must be tested and calibrated in other basins with different geological settings (e.g., deltaic, turbidite systems) to assess its scalability. Transfer learning techniques could be employed to adapt the pre-trained models to new regions with minimal data, enhancing global applicability.
- **Model Enhancement with Additional Data:** Whenever available, incorporating well logs, core data, and pressure measurements will serve to further validate and refine the models, potentially increasing their resolution and predictive certainty for pore pressure prediction.
- **Advanced Algorithm Exploration:** While tree-based models excelled here, exploring deep learning approaches such as convolutional neural networks (CNNs) could automatically extract more complex spatial features from seismic data, potentially unlocking even higher accuracy.

In conclusion, this investigation moves beyond a qualitative assessment by delivering a quantitatively validated framework with over 99.6% classification accuracy. It underscores the transformative potential of integrating ML with seismic interpretation to de-risk offshore drilling campaigns, ultimately contributing to enhanced safety, reduced costs, and greater operational efficiency in exploration targets worldwide.

Author Contributions

Dejen Teklu Asfha and Ismailalwali Babikir: Writing original draft, review & editing, visualization, validation, resources, project administration, methodology, investigation, formal analysis, data curation, and conceptualization.
Abdul Halim Abdul Latiff: Supervision, reviewing, and editing.
Haylay Tsegab Gebretsadik: Reviewing and editing.
Alidu Rashid: Writing original draft, reviewing, and editing.
Daniel Asante Otchere: Reviewing and editing.

Data Availability

The data used in this research is confidential and cannot be made publicly available.

Competing Interests

The authors declare no competing interests.

Funding

The authors would like to express their sincere gratitude to the Centre for Subsurface Seismic Imaging and Hydrocarbon Prediction (CSI) at the Universiti Teknologi PETRONAS.

References

Al-Masgari, A. A. S., Elsaadany, M., Siddiqui, N. A., Abdul Latiff, A. H., Bakar, A. A., Elkurdy, S. et al. (2021). Geomorphological geometries and high-resolution seismic sequence stratigraphy of Malay basin's fluvial succession. *Applied Sciences* (Switzerland), 11(11), 5156. https://doi.org/10.3390/app11115156

Alqahtani, F. A., Johnson, H. D., Jackson, C. A. L. & Som, M. R. B. (2015). Nature, origin and evolution of a Late Pleistocene incised valley-fill, Sunda Shelf, Southeast Asia. *Sedimentology*, 62(4), 1198–1232. https://doi.org/10.1111/sed.12185

Asfha, D. T., Gebretsadik, H. T., Ismail, W. & Yusoff, W. (2019). Pore pressure prediction using well-logging data in the West Baram Delta, offshore Sarawak Basin, Malaysia. *International Journal of Recent Technology and Engineering*, 8(4), 9172–9178. https://doi.org/10.35940/ijrte.d9050.118419

Asfha, D. T., Gebretsadik, H. T., Latiff, A. H. A. & Rahmani, O. (2024). Predictive pore pressure modeling using well-log data in the West Baram Delta, offshore Sarawak Basin, Malaysia. *Geomechanics and Geophysics for Geo-Energy and Geo-Resources*, 10(1), 196. https://doi.org/10.1007/s40948-024-00903-5

Asfha, D. T., Latiff, A. H. A., Otchere, D. A., Tackie-Otoo, B. N., Babikir, I., Rafi, M. et al. (2024). Mechanisms of sand production, prediction – A review and the potential for fiber optic technology and machine learning in monitoring. *Journal of Petroleum Exploration and Production Technology*, 14(10). https://doi.org/10.1007/s13202-024-01860-1

Babikir, I. A. M., Salim, A. M. A. & Ghosh, D. P. (2019a). Lithogeomorphological facies analysis of Upper Miocene coal-prone fluviodeltaic reservoirs, Northern Malay Basin. *Interpretation*, 7(3), T565–T579. https://doi.org/10.1190/INT-2018-0103.1

Babikir, I. A. M., Salim, A. M. A. & Ghosh, D. P. (2019b). Seismic geomorphology analysis of coal-bearing reservoirs using waveform classification: A case study from the northern Malay basin. *Bulletin of the Geological Society of Malaysia*, 68, 91–97. https://doi.org/10.7186/bgsm68201908

Babikir, I., Abdul Latiff, A. H., Elsaadany, M., Pratama, H., Sajid, M., Mad Sahad, S. et al. (2024). Enhancing machine learning-based seismic facies classification through attribute selection: Application to 3D seismic data from the Malay and Sabah Basins, offshore Malaysia. *Geomechanics and Geophysics for Geo-Energy and Geo-Resources*, 10(1), 138. https://doi.org/10.1007/s40948-024-00846-x

Babikir, I., Salim, A. M. A., Hermana, M., Latiff, A. H. A. & Al-Masgari, A. A. S. (2022). Characterizing the subsea Pleistocene fluvial system of the Sunda shelf, offshore Malaysia, using multiattribute co-rendering and self-organizing maps. *Interpretation*, 10(2), SF27–SF42. https://doi.org/10.1190/int-2021-0005.1

Bashiron, N., Johari, S., Sidek, A., Agi, A., Ariffin, M. H., Supee, A. et al. (2023). Sedimentary facies, depositional environment and reservoir quality of group-B shallow clastic reservoir in the Malay Basin, Malaysia. *SPE Nigeria Annual International Conference and Exhibition*, D032S028R010.

Bishop, M. G. (2002). *Petroleum Systems of the Malay Basin Province, Malaysia*. US Geological Survey.

Cai, J., Luo, J., Wang, S. & Yang, S. (2018). Feature selection in machine learning: A new perspective. *Neurocomputing*, 300, 70–79. https://doi.org/10.1016/j.neucom.2017.11.077

Chopra, S. & Marfurt, K. J. (2007). Seismic attributes for prospect identification and reservoir characterization. *Society of Exploration Geophysicists and European Association of Geoscientists and Engineers*, pp. i–xvi. https://doi.org/10.1190/1.9781560801900.fm

Dutta, N. C. (2002). Deepwater geohazard prediction using prestack inversion of large offset P-wave data and rock model. *The Leading Edge*, 21(2), 193–198.

Dutta, N. C., Utech, R. W. & Shelander, D. (2010). Role of 3D seismic for quantitative shallow hazard assessment in deepwater sediments. *The Leading Edge*, 29(8), 930–942. https://doi.org/10.1190/1.3480006

Flemings, P. B., Stump, B. B., Finkbeiner, T. & Zoback, M. (2002). Flow focusing in overpressured sandstones: Theory, observations, and applications. *American Journal of Science*, 302(10), 827–855. https://doi.org/10.2475/ajs.302.10.827

Ghosh, D., Sajid, M., Ibrahim, N. A. & Viratno, B. (2014). Seismic attributes add a new dimension to prospect evaluation and geomorphology offshore Malaysia. *Leading Edge*, 33(5), 536–545. https://doi.org/10.1190/tle33050536.1

Guo, X., Jin, Y. & Lin, B. (2020). A parametric study of coupled hydromechanical behaviors induced by shallow water flow in shallow sediments in deepwater drilling based on numerical modeling. *Geofluids*, Article ID 2592693. https://doi.org/10.1155/2020/2592693

Jamaludin, S. N. F., Rajasansi, T. K. & Latiff, A. H. A. (2019). Characterization of the Pleistocene fluvial system in Sepat field, Malay basin using seismic attributes analysis. *Platform: A Journal of Engineering*, 3(1), 19–32.

Jiliang, T. & Salem, A. H. L. (2014). *Feature Selection for Classification: A Review* (C. C. Aggarwal (ed.), 1st Edition). Chapman and Hall/CRC. https://doi.org/10.1201/b17320

Kamilah, T., Hermana, M., Almanna, L., Mokhtar, S. S. S. & Yunginger, R. (2022). Depositional environment characterization in deltaic system by utilizing core, well log and seismic attribute data, Northern Malay Basin. *IOP Conference Series: Earth and Environmental Science*, 1003(1), 12013.

Kessler, F. L., Jong, J. & Madon, M. (2020). Sedimentary record of Paleogene sequences in the Penyu and Malay Basins, offshore Peninsular Malaysia. *Berita Sedimentologi*, 46(1), 6–20.

Koson, S., Chenrai, P. & Choowong, M. (2013). Seismic attributes and their applications in seismic geomorphology. *Bulletin of Earth Sciences of Thailand*, 6(1), 1–9.

Li, F., Qi, J. & Marfurt, K. (2015). Attribute mapping of variable-thickness incised valley-fill systems. *Leading Edge*, 34(1), 48–52. https://doi.org/10.1190/tle34010048.1

Li, Luo, X., Zhang, L., Wang, B., Guan, X., Luo, H. & Lei, Y. (2019). Overpressure generation mechanisms and its distribution in the Paleocene Shahejie formation in the Linnan sag, Huimin depression, eastern China. *Energies*, 12(16), 3183. https://doi.org/10.3390/en12163183

Lu, S., McMechan, G. A. & Liaw, A. (2005). Identification of shallow-water-flow sands by V p/V s inversion of conventional 3D seismic data. *Geophysics*, 70(5), O29–O37.

Madon, M. (2007). Overpressure development in rift basins: An example from the Malay Basin, offshore Peninsular Malaysia. *Petroleum Geoscience*, 13(2), 169–180. https://doi.org/10.1144/1354-079307-744

Madon, M. B. H. (1997). Analysis of tectonic subsidence and heat flow in the Malay Basin (offshore Peninsular Malaysia). *Bulletin of the Geological Society of Malaysia*, 41, 95–108. https://doi.org/10.7186/bgsm41199709

Madon, M., Abolins, P., Mohammad Jamaal Hoesni, & Ahmad, M. (1999). Malay Basin. *In:* K. M. Leong (Ed.), *The Petroleum Geology and Resources of Malaysia* (pp. 171–217). Kuala Lumpur, Malaysia: PETRONAS.

Madon, M., Yang, J.-S., Abolins, P., Abu Hassan, R., Yakzan, A. M., & Zainal, S. B. (2004). Petroleum systems of the Northern Malay Basin. *Bulletin of the Geological Society of Malaysia*, 49, 125–134. https://doi.org/10.7186/bgsm49200420

Manshor, N. A., Hassan, M. H. A. & Madon, M. (2022). Tidally-influenced fluvial channel systems from the Miocene Malay Basin, Malaysia: Evidence from core facies and seismic geomorphological analyses. *Marine and Petroleum Geology*, 135, 105384.

Marfurt, K. J. (2015). Techniques and best practices in multiattribute display. *Interpretation*, 3(1), B1–B23. https://doi.org/10.1190/INT-2014-0133.1

McConnell, D. R. (2000, May 1). Optimizing deepwater well locations to reduce the risk of shallow-water-flow using high-resolution 2D and 3D seismic data. *Offshore Technology Conference*. https://doi.org/10.4043/11973-MS

Miall, A. D. (2002). Architecture and sequence stratigraphy of Pleistocene fluvial systems in the Malay Basin, based on seismic time-slice analysis. *AAPG Bulletin*, 86(7), 1201–1216.

Mohamad, H. & Mohamed, I. (2003). Oil discovery in the Sepat Field (Paper 40). *In: Geological Society of Malaysia Petroleum Geology Conference & Exhibition 2003, Program & Abstracts* (p. 122). Kuala Lumpur, Malaysia, 17–18 December 2003.

Muhamad, A. J. & Jamil, A. S. A. (2010). Organic facies variation in lacustrine source rocks in the southern Malay Basin. *Bulletin of the Geological Society of Malaysia*, 56, 27–33.

Nazri, M. R. I. M. & Abdul Latiff, A. H. (2016). New insight of fractured basement in Malay Basin using multi-attributes method. *In: International Conference on Innovation and Technology for Sustainable Built Environment* (*ICIPEG 2016*) (extended abstract).

Ostermeier, R. M., Pelletier, J. H., Winker, C. D., Nicholson, J. W., Rambow, F. H. & Cowan, K. M. (2000). Dealing with shallow-water flow in the deepwater Gulf of Mexico. *Offshore Technology Conference*, OTC-11972.

Pigott, J. D., Kang, M.-H. & Han, H.-C. (2013). First order seismic attributes for clastic seismic facies interpretation: Examples from the East China Sea. *Journal of Asian Earth Sciences*, 66, 34–54.

Purnomo, E. W. & Ghosh, D. P. (2018). *OTC-28276-MS Identifying Shallow Water Flow in Offshore Malaysia Using Multicomponent Data and FWI Approach.* http://onepetro.org/OTCASIA/proceedings-pdf/18OTCA/3-18OTCA/D031S022R001/1192472/otc-28276-ms.pdf/1

Ramli, M. N. (1988). Stratigraphy and palaeofacies development of Carigali's operating areas in the Malay Basin, South China Sea. *Bulletin of the Geological Society of Malaysia*, 22, 153–187. https://doi.org/10.7186/bgsm22198808

Razmi, Z. N., Moghaddas, N. H., Sadeghi, H., Harami, S. R. M. & Farajkhah, N. K. (2023). A review of shallow water flows (SWF) and evaluation of this problem by 3D seismic data in the South Caspian Basin, Iran. *Iranian Journal of Geophysics*, 17(3), 45–63. Iranian Geophyisical Society. https://doi.org/10.30499/IJG.2023.366469.1466

Ren, S., Liu, Y., Gong, Z., Yuan, Y., Yu, L., Wang, Y. et al. (2018). Numerical simulation of water and sand blowouts when penetrating through shallow water flow formations in deep water drilling. *Journal of Ocean University of China*, 17(1), 17–24. https://doi.org/10.1007/s11802-018-3454-5

Rutherford, S. R. & Williamst, R. H. (1989). Amplitude-versus-offset variations in gas sands. *Geophysics*, 54(6), 680–688.

Safaei, A., Asefi, M., Ahmadi, M., Pourshamsi, T., Baloochestanzadeh, S., Khalilnezhad, A. et al. (2023). Chemical treatment for sand production control: A review of materials, methods, and field operations. *Petroleum Science*, 20(3), 1640–1658. https://doi.org/10.1016/j.petsci.2023.02.013

San, N. T. (1987). Trap styles of the Tenggol Arch and the southern part of the Malay Basin. *Bulletin of the Geological Society of Malaysia*, 21, 177–193. https://doi.org/10.7186/bgsm21198710

Satti, I. A. (2016). Overpressure mechanism and prediction in the Malay Basin (Master's thesis). Universiti Teknologi PETRONAS, Seri Iskandar, Perak, Malaysia.

Tapponnier, P., Peltzer, G., Le Dain, A. Y., Armijo, R. & Cobbold, P. (1982). Propagating extrusion tectonics in Asia: New insights from simple experiments with plasticine. *Geology*, 10(12), 611–616.

Tjia, H. D. (1994). Inversion tectonics in the Malay Basin: Evidence and timing of events. *Bulletin of the Geological Society of Malaysia*, 36, 119–126. https://doi.org/10.7186/bgsm36199411

Tjia, H. D. (1998). Origin and tectonic development of Malay-Penyu-West Natuna basins. *Bulletin of the Geological Society of Malaysia*, 42, 147–160. https://doi.org/10.7186/bgsm42199813

Wu, S., Wang, D. & Völker, D. (2018). Deep-sea geohazards in the South China Sea. *Journal of Ocean University of China*, 17, 1–7.

Yakzan, A. M., Harun, A., Nasib, B. M. & Morley, R. J. (1996). Integrated biostratigraphic zonation for the Malay Basin. *Bulletin of the Geological Society of Malaysia*, 39, 157–184. https://doi.org/10.7186/bgsm39199615

Zhang, X., Sun, Y., Wu, S. & Dong, D. (2018). Geophysical signature of the shallow water flow in the Deepwater Basin of the northern South China Sea. *Journal of Ocean University of China*, 17, 791–798.

Index

For Product Safety Concerns and Information please contact our EU
representative GPSR@taylorandfrancis.com
Taylor & Francis Verlag GmbH, Kaufingerstraße 24, 80331 München, Germany

www.ingramcontent.com/pod-product-compliance
Ingram Content Group UK Ltd.
Pitfield, Milton Keynes, MK11 3LW, UK
UKHW022210190926
475702UK00003B/284

* 9 7 8 1 0 3 2 8 1 9 0 9 9 *